Lecture Notes in Physics

Springer-Verlag Berlin Heidelberg GmbH

The Editorial Policy for Proceedings

The series Lecture Notes in Physics reports new developments in physical research and teaching – quickly, informally, and at a high level. The proceedings to be considered for publication in this series should be limited to only a few areas of research, and these should be closely related to each other. The contributions should be of a high standard and should avoid lengthy redraftings of papers already published or about to be published elsewhere. As a whole, the proceedings should aim for a balanced presentation of the theme of the conference including a description of the techniques used and enough motivation for a broad readership. It should not be assumed that the published proceedings must reflect the conference in its entirety. (A listing or abstracts of papers presented at the meeting but not included in the proceedings could be added as an appendix.)

When applying for publication in the series Lecture Notes in Physics the volume's editor(s) should submit sufficient material to enable the series editors and their referees to make a fairly accurate evaluation (e.g. a complete list of speakers and titles of papers to be presented and abstracts). If, based on this information, the proceedings are (tentatively) accepted, the volume's editor(s), whose name(s) will appear on the title pages, should select the papers suitable for publication and have them refereed (as for a journal) when appropriate. As a rule discussions will not be accepted. The series editors and Springer-Verlag will normally not interfere with the detailed editing except in fairly obvious cases or on technical matters.

Final acceptance is expressed by the series editor in charge, in consultation with Springer-Verlag only after receiving the complete manuscript. It might help to send a copy of the authors' manuscripts in advance to the editor in charge to discuss possible revisions with him. As a general rule, the series editor will confirm his tentative acceptance if the final manuscript corresponds to the original concept discussed, if the quality of the contribution meets the requirements of the series, and if the final size of the manuscript does not greatly exceed the number of pages originally agreed upon. The manuscript should be forwarded to Springer-Verlag shortly after the meeting. In cases of extreme delay (more than six months after the conference) the series editors will check once more the timeliness of the papers. Therefore, the volume's editor(s) should establish strict deadlines, or collect the articles during the conference and have them revised on the spot. If a delay is unavoidable, one should encourage the authors to update their contributions if appropriate. The editors of proceedings are strongly advised to inform contributors about these points at an early stage.

The final manuscript should contain a table of contents and an informative introduction accessible also to readers not particularly familiar with the topic of the conference. The contributions should be in English. The volume's editor(s) should check the contributions for the correct use of language. At Springer-Verlag only the prefaces will be checked by a copy-editor for language and style. Grave linguistic or technical shortcomings may lead to the rejection of contributions by the series editors. A conference report should not exceed a total of 500 pages. Keeping the size within this bound should be achieved by a stricter selection of articles and not by imposing an upper limit to the length of the individual papers. Editors receive jointly 30 complimentary copies of their book. They are entitled to purchase further copies of their book at a reduced rate. As a rule no reprints of individual contributions can be supplied. No royalty is paid on Lecture Notes in Physics volumes. Commitment to publish is made by letter of interest rather than by signing a formal contract. Springer-Verlag secures the copyright for each volume.

The Production Process

The books are hardbound, and the publisher will select quality paper appropriate to the needs of the author(s). Publication time is about ten weeks. More than twenty years of experience guarantee authors the best possible service. To reach the goal of rapid publication at a low price the technique of photographic reproduction from a camera-ready manuscript was chosen. This process shifts the main responsibility for the technical quality considerably from the publisher to the authors. We therefore urge all authors and editors of proceedings to observe very carefully the essentials for the preparation of camera-ready manuscripts, which we will supply on request. This applies especially to the quality of figures and halftones submitted for publication. In addition, it might be useful to look at some of the volumes already published. As a special service, we offer free of charge L^AT$_E$X and T$_E$X macro packages to format the text according to Springer-Verlag's quality requirements. We strongly recommend that you make use of this offer, since the result will be a book of considerably improved technical quality. To avoid mistakes and time-consuming correspondence during the production period the conference editors should request special instructions from the publisher well before the beginning of the conference. Manuscripts not meeting the technical standard of the series will have to be returned for improvement.

For further information please contact Springer-Verlag, Physics Editorial Department II, Tiergartenstrasse 17, D-69121 Heidelberg, Germany

Germán Sierra
Miguel A. Martín-Delgado (Eds.)

Strongly Correlated Magnetic and Superconducting Systems

Proceedings of the El Escorial Summer School
Held in Madrid, Spain, 15–19 July 1996

 Springer

Editors

Germán Sierra
Instituto de Matemáticas y Física Fundamental
C.S.I.C., Serrano 123
E-28006 Madrid, Spain

Miguel A. Martín-Delgado
Departamento de Física Teórica I
Universidad Complutense
E-28040 Madrid, Spain

Cataloging-in-Publication Data applied for.

Die Deutsche Bibliothek - CIP-Einheitsaufnahme

Strongly correlated magnetic and superconducting systems :
proceedings of the El Escorial summer school held at Madrid, Spain,
15 - 19 July 1996 / Germán Sierra ; Miguel A. Martín-Delgado
(ed.).
(Lecture notes in physics ; Vol. 478)
ISBN 978-3-662-14136-6   ISBN 978-3-540-49734-9 (eBook)
DOI 10.1007/978-3-540-49734-9

ISSN 0075-8450
ISBN 978-3-662-14136-6

The use of general descriptive names, registered names, trademarks, etc. in this publication does not imply, even in the absence of a specific statement, that such names are exempt
from the relevant protective laws and regulations and therefore free for general use.

Typesetting: Camera-ready by the authors
Cover design: *design & production* GmbH, Heidelberg
SPIN: 10550489   55/3144-543210 - Printed on acid-free paper

Preface

The summer school on "Strongly Correlated Magnetic and Superconducting Systems" took place in El Escorial (Madrid) from 15 to 19 July 1996 within the summer school's organization of Fundación General Universidad Complutense de Madrid.

This summer school was devoted to the study of several related topics such as Heisenberg, t-J and Hubbard models in one and two dimensions, real-space renormalization group methods, quantum phase transitions, non-linear sigma model, one-dimensional electron systems, quantum Monte Carlo methods, spin ladders and layers, quantum Hall Effect.

The audience was very broad and special effort was made to convey the essential ideas of the field and to give the audience (students and experts) a comprehensive introduction to these topics.

In the past few years there has been a large increase in the study of strongly correlated systems (spin systems, electron systems, etc.) in several interdisciplinary fields such as condensed matter physics, statistical field theory and quantum field theory. The issue of the strongly correlated systems had been a working area of paramount importance in condensed matter for some years before exploding worldwide after the discovery of high-T_c superconductivity in 1986. Thus, we consider this course as a contribution the commemoratations of the 10th birthday of high-T_c superconductivity.

The upsurge of interest in strongly correlated magnetic and superconducting systems during these years has led to a great variety of research directions, both theoretical and numerical techniques, which more or less are intended to explain the intriguing normal and superconducting properties of cuprate materials.

In these proceedings the reader (not necessarily specialized) may find a good treatment of a variety of modern computational methods currently employed to deal with non-perturbative effects in strongly correlated systems. These proceedings have some distinctive features with respect to other meetings on strongly correlated systems. This field has evolved rapidly in recent years and there is a necessity to update the recent literature. In this regard, the content of this volume has mainly two novelties: a large presentation of real-space renormalization group methods, on the one hand, and likewise for ladder systems, on the other. The subject of ladders has become one of the most active fields in the study of strong correlations in cuprate materials. They are more manageable to both theoretical and numerical studies.

More explicitely, the contents of these proceedings can be summarizes as follows. The subject of real-space RG methods is extensively covered by S. White, including the most recent applications. T. Nishino deals with the applications of DMRG to classical statistical systems. M.A. Martín-Delgado covers old and recent treatments of BRG methods from a more theoretical viewpoint and J. Pérez-Conde presents a critical review of these methods for the Hubbard model.

The subject of ladder systems, including holes, is also covered by S. White, H.J. Schulz deals with coupled spin and charged chains using bosonization. G. Sierra presents the non-linear sigma model as a powerful tool to study ladders theoretically, and S. Haas also covers this subject in his study of photoemission bands in systems of strongly correlated electrons.

Field theory methods is another subject extensively studied in this school, by S. Sachdev with a nice introduction to quantum phase transitions and the recent developments in this field. G. Morandi covers the Hubbard model in two dimensions at an introductory level. J. González presents a field theory treatment of the Van Hove singularity in the cuprate materials and its relation to superconductivity, and L. Brey studies the role of Skyrmions in the quantum Hall effect.

The physics of one-dimensional systems is one of the most studied examples of strongly correlated systems. F. Guinea presents an introduction and recent results under the topic of quantum dissipative systems and G. Gómez-Santos deals with the problem of impurity effects in Luttinger liquids.

Finally, another standard and powerful numerical method, the quantum Monte Carlo method is presented with an introduction by A. Sandvik who also covers recent applications to cuprates.

It is a great pleasure to thank the Fundación General Universidad Complutense de Madrid and the sponsor Banco Central Hispano for giving us the opportunity to conduct and celebrate this summer school in the attractive venue of El Escorial and for providing us with so much help. Special thanks go to Prof. M.A. Alario (Director Cursos de Verano) and Prof. Antonio Fdz-Ranada (Director Científico) for their collaboration. We thank Fátima Esquivel for being so helpful and patient as our scientific secretary along with the rest of the secretaries and staff for their help with managing affairs, travels, photocopies, and related matters that are essential in the development of a conference. We acknowledge partial financial support from CICYT under contract AEN93-0776.

Last but not least, we are deeply grateful to the lecturers and to the participants for their attention, questions and enthusiasm, which produced a friendly atmosphere at El Escorial.

Madrid, May 1997 G. Sierra

M.A. Martín-Delgado

Contents

An Introduction to the Hubbard Model

E. Ercolessi[1], G. Morandi[1,2] and P. Pieri[1,2]

[1]Dipartimento di Fisica and INFM , Universita' di Bologna,
46 Via Irnerio, I-40126, Bologna, Italy.
[2] INFN, Universita' di Bologna.

Abstract. In these notes we review some of the basic features of the 2D Hubbard model, thought of as the appropriate model for the description of the $Cu - O$ planes in the cuprate superconductors. We discuss breifly the weak-coupling regime of the model and, in the opposite limit, the mapping of the one band Hubbard model onto an AFM Heisenberg model at half filling and onto the $t - J$ model below half filling. We discuss next Emery's three band model and its mapping onto the so-called "spin-fermion" model. Its continuum limit is discussed by making use of an adiabatic followed by a gradient expansion. We review briefly how the model maps onto a nonlinear sigma model and some of the features of the latter.

I   Essentials of the one-band Hubbard model

In its simpler, one-band, version the Hubbard model [1] is described by the many-body Hamiltonian:

$$H = -\sum_{ij}\sum_{\sigma} t_{ij}c_{i\sigma}^{\dagger}c_{j\sigma} + \frac{1}{2}\sum_{ijkl}\sum_{\sigma\sigma'} <ij|v|kl> c_{i\sigma}^{\dagger}c_{j\sigma'}^{\dagger}c_{l\sigma'}c_{k\sigma} \qquad (1)$$

where Latin indices label sites on an arbitrary lattice, $\sigma =\uparrow, \downarrow$ is a spin index (we consider here only spin-1/2 fermions) and the $c_{i\sigma}$'s are fermion annihilation operators. In (1), t_{ij} represents the hopping integral between sites i and j, while v is a two-body (e.g. Coulomb) potential. Usually one takes the approximation of nearest-neighbor (n.n.) hopping only:

$$t_{ij} = \begin{cases} t & (i,j) \text{ n.n.} \\ 0 & \text{otherwise} \end{cases} \qquad (2)$$

and of screened interactions:

$$<ij|v|kl> = \begin{cases} U & i = j = k = l \\ 0 & \text{otherwise} \end{cases} \qquad (3)$$

For a single band, this implies $\sigma' = -\sigma \equiv \bar{\sigma}$, so that one obtains the simplest version of the model, which we will refer to as the One Band Hubbard Hamiltonian

(OBH):

$$H = -t \sum_{<ij>} \sum_{\sigma} c_{i\sigma}^{\dagger} c_{j\sigma} + U \sum_{i} n_{i\uparrow} n_{i\downarrow} , \qquad (4)$$

where $n_{i\sigma} = c_{i\sigma}^{\dagger} c_{i\sigma}$ and $< ij >$ denotes a sum over n.n. ordered pairs, i.e. such that the couple (i,j) is counted only once.

The OBH is solvable in the two (almost trivial) following limits:

i) Band limit: $U = 0$.

In this case (4) reduces to the standard tight-binding hamiltonian [2]:

$$H_{U=0} = \sum_{\vec{k}\sigma} \epsilon(\vec{k}) c_{\vec{k}\sigma}^{\dagger} c_{\vec{k}\sigma} \qquad (5)$$

$$\epsilon(\vec{k}) = -\sum_{i0} t_{i0} e^{i\vec{k}\cdot\vec{R}_i} ,$$

where the vector $\vec{k}$ belongs to the Brillouin Zone (BZ). For example, for a 2D square lattice with spacing a and n.n. hopping only, one gets:

$$\epsilon(\vec{k}) = -t\eta(\vec{k}) \ , \ |k_x|, |k_y| \leq \frac{\pi}{a} \ ; \qquad (6)$$

$$\eta(\vec{k}) = 2[\cos(k_x a) + \cos(k_y a)] . \qquad (7)$$

Notice that $\epsilon(\vec{k} + \vec{\pi}) = -\epsilon(\vec{k})$, where $\vec{\pi} = (\frac{\pi}{a}, \frac{\pi}{a})$, for $\mu = 0$. Thus at half-filling when $\mu = 0$, the Fermi surface ($\epsilon(\vec{k}) = 0$) nests with a nesting vector $\vec{\pi}$).

ii) Atomic limit: $t_{ij} = 0$ for all i, j.

Now (4) reduces to the hamiltonian

$$H_{t=0} = U \sum_{i} n_{i\uparrow} n_{i\downarrow} \qquad (8)$$

and the Hilbert space decomposes in the direct sum

$$\mathcal{H} = \bigoplus_{\ell=0}^{M} \mathcal{H}_{\ell} , \qquad (9)$$

where M is the number of sites in the lattice and $\mathcal{H}_{\ell}$ is the subspace with exactly ℓ doubly occupied sites. Clearly, $\mathcal{H}_{\ell}$ is an eigenspace of (8) with eigenvalue:

$$E_{\ell} = \ell U . \qquad (10)$$

For N electrons ($0 \leq N \leq 2M$), the ground state manifold $\mathcal{D}_0$ is:

$$\mathcal{D}_0 = \begin{cases} \mathcal{H}_0 \\ \mathcal{H}_{N-M} \end{cases} \text{ with degeneracy } \begin{cases} 2^N \begin{pmatrix} M \\ N \end{pmatrix} & , N \leq M \\ 2^{2M-N} \begin{pmatrix} M \\ 2M-N \end{pmatrix} & , N \geq M \end{cases} . \qquad (11)$$

Notice that this could be a starting point for a perturbative expansion of the Hubbard hamiltonian in t/U. However, the degeneracy of the ground as well as of the excited states poses serious problems to a straightforward perturbative approach.

II Symmetries of the Hubbard model

Let us start by noticing that

$$n_{i\uparrow}n_{i\downarrow} = \frac{1}{2}n_i - \frac{2}{3}\vec{S}_i^2 \,, \tag{12}$$

where the spin $\vec{S}_i$ operator is given in the fermionic representation by

$$\vec{S}_i = \frac{1}{2}\sum_{\alpha\beta} c_{i\alpha}^\dagger [\vec{\sigma}]_{\alpha\beta} c_{i\beta} \,, \tag{13}$$

with $\vec{\sigma} = (\sigma_1, \sigma_2, \sigma_3)$ being Pauli matrices. Thus, on the Hilbert space of a single site, the spin operators act irreducibly with $S = 1/2$ on singly occupied states and trivially (S=0) on empty and doubly occupied states separately[3]:

$$\begin{aligned} \vec{S}_i^2|\uparrow\rangle = \tfrac{3}{4}|\uparrow\rangle \,, \ \vec{S}_i^2|\downarrow\rangle = \tfrac{3}{4}|\downarrow\rangle \\ \vec{S}_i^2|0\rangle = \vec{S}_i^2|\uparrow\downarrow\rangle = 0 \end{aligned} \tag{14}$$

It turns out then that the "U-term" of the Hubbard hamiltonian can be rewritten in a manifestly rotationally invariant form.

Another, perhaps more fundamental, way of looking at the same result is the following. Let us introduce the doublet:

$$\chi_i = \begin{pmatrix} c_{i\uparrow} \\ c_{i\downarrow} \end{pmatrix} \,, \tag{15}$$

so that

$$\vec{S}_i = \frac{1}{2}\chi_i^\dagger \vec{\sigma}\chi_i \,, \ n_i = \chi_i^\dagger \chi_i \tag{16}$$

and

$$\sum_\sigma c_{i\sigma}^\dagger c_{j\sigma} = \chi_i^\dagger \chi_j \,. \tag{17}$$

Then the OBH (4) can be rewritten as

$$H = -t\sum_{<ij>} \chi_i^\dagger \chi_j + U\sum_i \left(\frac{1}{2}\hat{n}_i - \frac{2}{3}\vec{S}_i^2 \right) \tag{18}$$

and is therefore invariant under a global $SU(2)$ action:

$$g : \chi_i \mapsto g\chi_i \quad \text{if } g \in SU(2) \,. \tag{19}$$

Indeed, this action leaves both the hopping term and the n_i's invariant in an obvious way. As for the spins, it is enough to recall the relationship between $SU(2)$ and $SO(3)$:

$$\text{if } g \in SU(2) \text{ then } g^{-1}\sigma^\alpha g = \sigma^\beta R_\beta^\alpha \text{ with } \|R_\beta^\alpha\| \in SO(3) , \qquad (20)$$

in order to show that the action of the transformation (19) amounts to a *rotation* of the spins, leaving scalar products (such as $\vec{S}_i^2$) invariant:

$$g : S_i^\alpha = \frac{1}{2}\chi_i^\dagger\sigma^\alpha\chi_i \mapsto \frac{1}{2}\chi_i^\dagger(g^{-1}\sigma^\alpha g)\chi_i = \frac{1}{2}\chi_i^\dagger(\sigma^\beta R_\beta^\alpha)\chi_i = R_\beta^\alpha S_i^\beta . \qquad (21)$$

Dropping the site index, the spin operators S^α that close on the Lie algebra of $SU(2)$ are explicitely given by:

$$S^z = \frac{1}{2}(n_\uparrow - n_\downarrow) , \ S^+ = c_\uparrow^\dagger c_\downarrow , \ S^- = (S^+)^\dagger \qquad (22)$$

But there is another realization of the same algebra, which is provided by:

$$J^z = \frac{1}{2}(n - 1) , \ J^+ = c_\uparrow^\dagger c_\downarrow^\dagger , \ J^- = (J^+)^\dagger . \qquad (23)$$

Also, the generators of the two algebras commute:

$$[J^I, S^K] = 0 \ \forall I, K . \qquad (24)$$

We can represent both algebras in a compact way by introducing the "super-spinor":

$$\widehat{\Psi} = \begin{pmatrix} c_\uparrow & c_\downarrow \\ c_\downarrow^\dagger & -c_\uparrow^\dagger \end{pmatrix} \qquad (25)$$

which implies

$$\begin{aligned} \vec{S} &= \tfrac{1}{4}\text{Tr}\left\{\widehat{\Psi}^\dagger\widehat{\Psi}\vec{\sigma}^T\right\} , \\ \vec{J} &= \tfrac{1}{4}\text{Tr}\left\{\widehat{\Psi}^\dagger\vec{\sigma}\widehat{\Psi}\right\} . \end{aligned} \qquad (26)$$

We will call $SU(2)_R$ the group generated by $\vec{S}$ and $SU(2)_L$ the one generated by $\vec{J}$, since for any $\mathcal{U}_R$ ($\mathcal{U}_L$) acting on the fermionic operators in the standard unitary way:

$$\mathcal{U}_{L,R} : c_\sigma \mapsto \mathcal{U}_{L,R}\, c_\sigma\, \mathcal{U}_{L,R}^\dagger , \qquad (27)$$

there is a unique element $g(h) \in SU(2)$, in the standard representation by 2×2 unitary unimodular matrices, such that

$$\begin{aligned} \mathcal{U}_L &: \Psi \mapsto g\Psi \\ \mathcal{U}_R &: \Psi \mapsto \Psi h . \end{aligned} \qquad (28)$$

It is not difficult to check [4] that the two Lie algebras are transformed one into the other by the following automorphism (particle-hole transformation in one spin channel):

$$\mathcal{A} \quad : \quad c_\uparrow \mapsto c_\uparrow , \tag{29}$$
$$c_\downarrow \mapsto c_\downarrow^\dagger ;$$
$$\mathcal{A}^2 \quad = \quad 1 ;$$

that is unitarily implemented as

$$U_\mathcal{A} c_\uparrow U_\mathcal{A}^\dagger = c_\uparrow \;\; , \;\; U_\mathcal{A} c_\downarrow U_\mathcal{A}^\dagger = c_\downarrow^\dagger \;\; , \tag{30}$$
$$U_\mathcal{A} = U_\mathcal{A}^\dagger = (1 - 2n_\uparrow)(c_\downarrow^\dagger + c_\downarrow) \;\; . \tag{31}$$

and acts on the superspinor by transposition:

$$\mathcal{A} : \Psi \mapsto \Psi^T \;\; . \tag{32}$$

Because of the existence of (29) the group acting on the fermionic operators generated by the two Lie algebras is

$$\mathcal{G} = \frac{SU(2)_L \otimes SU(2)_R}{Z_2} \simeq SO(4) \;\; . \tag{33}$$

If we consider now the single-site atomic limit hamiltonian

$$h = U n_\uparrow n_\downarrow - \mu \, , n \tag{34}$$

where we have introduced a chemical potential μ, it is easy to see that

$$[h, \vec{S}] = 0 \;\; , \tag{35}$$

while

$$[h, J^z] = 0 \;\; , \;\; [h, J^\pm] = \pm(U - 2\mu) \;\; , \tag{36}$$

so that $SU(2)_R$ is a symmetry only at half-filling when

$$\mu = \frac{U}{2} \;\; . \tag{37}$$

Going back to the full OBH and setting $H = T + V$ where T is the hopping part while $V = \sum_i h_i$, we can define a *global* $SU(2)_L$ via:

$$\vec{S} \equiv \sum_i \vec{S}_i \;\; , \tag{38}$$

which is easily checked to be a symmetry for H, since

$$[\vec{S}, H] = 0 \;\; . \tag{39}$$

We can also define a global $SU(2)_R$ by setting:

$$J^z = \sum_i J_i^z \ , \ J^+ = \sum_i \lambda_i J_i^+ \ , \ \lambda_i \in U(1) \ , \tag{40}$$

but now the commutation relations with H yield:

$$[J^z, H] = 0 \tag{41}$$
$$[J^\pm, V] = \pm(U - 2\mu)J^\pm$$
$$[J^\pm, T] = \sum_{ij}(t_{ij}\lambda_j + t_{ji}\lambda_j)c_{i\uparrow}^\dagger c_{j\downarrow}^\dagger$$

Thus at half-filling, see (37), and whenever

$$t_{ij}\lambda_j + t_{ji}\lambda_j = 0 \tag{42}$$

the Hubbard model has an additional global $SU(2)$ symmetry.

Taking into account also (29), we can conlude that in this case we have an $SO(4)$ symmetry. This symmetry was uncovered by Yang and Zhang [5], who showed that (42) can be solved for bipartite lattices and real hopping coefficients by putting

$$\lambda_i = (-1)^{|i|} \ . \tag{43}$$

Later, it has been proved [4] that this result can been generalized to non-bipartite lattices and complex hopping coefficients as long as the total "flux" per plaquette is quantized.

III The weak-coupling regime

In view of the implications of the Hubbard model for the understanding of the superconductivity of the cuprates, we will consider a 2D model on a square lattice with lattice size a. We are interested now in the weak coupling regime of the OBH and thus we start by examining again the unberturbed tight-binding model (5). We have already noticed that at half-filling there is perfect nesting. On the other hand, the density of states $g(\epsilon) = \sum_{\vec{k}} \delta(\epsilon - \epsilon(\vec{k}))$ is symmetric around $\epsilon = 0$ and has a van Hove singularity near the Fermi surface:

$$g(\epsilon) \approx \ln\left(\frac{4t}{|\epsilon|}\right) \quad \text{for} \quad \epsilon \approx 0 \ . \tag{44}$$

For non-zero U, and in the RPA, the static magnetic susceptibility is given by:

$$\chi(\vec{q}) = \frac{\chi^{(0)}(\vec{q})}{1 - U\,\chi^{(0)}(\vec{q})} \ , \ \text{with} \ \chi^{(0)}(\vec{q}) = \frac{1}{N}\sum_{\vec{k}} \frac{f(\epsilon(\vec{k} + \vec{q})) - f(\epsilon(\vec{k}))}{\epsilon(\vec{k}) - \epsilon(\vec{k} + \vec{q})} \ , \tag{45}$$

where $\chi^{(0)}(\vec{q})$ is the noninteracting static susceptibility and $f(\epsilon)$ is the Fermi function. Thus at half-filling:

i) for $\vec{q} \mapsto 0$:

$$\chi^{(0)}(\vec{q}) \propto \ln \left| \frac{t}{T} \right| , \tag{46}$$

as a consequence of the van Hove singularity at the Fermi surface, leading to a *ferro*magnetic instability. However

ii) for $\vec{q} \mapsto \vec{\pi}$:

$$\chi^{(0)}(\vec{\pi}) = \int d\epsilon \, g(\epsilon) \, \frac{f(-\epsilon) - f(\epsilon)}{2\,\epsilon} , \tag{47}$$

so that the low-temperature logarithmic divergence of the second factor in (45) piles up with the van Hove singularity, yielding an *antiferro*magnetic and stronger instability:

$$\chi^{(0)}(\vec{\pi}) \propto \left(\ln \left| \frac{T}{t} \right| \right)^2 . \tag{48}$$

So the model exhibits magnetic instabilities for arbitrarily small values of U and, as the antiferromagnetic (AFM) one is stronger, the Hubbard model at half-filling will make a transition to an AFM state. This is true at the RPA level, but, as we know, the Mermin-Wagner theorem [6] prohibits the transition at $T \neq 0$. To avoid this, let us study the magnetic ground states at the mean field level directly at $T = 0$. We make the ansatz:

$$< n_{i\sigma} >= \frac{n}{2} + (-1)^\sigma m \, \cos(\vec{Q} \cdot \vec{R}_i) , \tag{49}$$

where $m = m(T)$ and $\vec{Q} = 0, \vec{\pi}$ for the ferromagnetic and antiferromagnetic case respectively. In particular for $\vec{Q} = \vec{\pi}$, we can introduce the spinor:

$$\Psi_{\vec{k}\sigma} = \begin{bmatrix} c_{\vec{k}\sigma} \\ c_{\vec{k}+\vec{Q}\sigma} \end{bmatrix} \tag{50}$$

with $\vec{k}$ ranging over half of the BZ, so that a Hartree-Fock factorization yields the hamiltonian:

$$H_{HF} = \sum_{\vec{k}\sigma\sigma'} \sum_{aa'} \Psi^\dagger_{\vec{k}\sigma;a} E_{\vec{k},\sigma\sigma';aa'} \Psi_{\vec{k}\sigma';a'} , \tag{51}$$

where h is the 4×4 matrix:

$$E = \begin{bmatrix} \epsilon(\vec{k})\,\hat{1} & -\frac{1}{2}\Delta\,\sigma_3 \\ \frac{1}{2}\Delta\,\sigma_3 & \epsilon(\vec{k})\,\hat{1} \end{bmatrix} \tag{52}$$

with $\Delta = Um$ and $\hat{1}$ the unit matrix in spin space. It is then straightforward to diagonalize (51) and get:

- the single-particle spectrum

$$\pm E(\vec{k}) = \sqrt{\epsilon^2(\vec{k}) + \frac{\Delta^2}{4}} ; \tag{53}$$

- the gap equation

$$1 = \frac{U}{2N} \sum_{\vec{k}} \frac{1}{E(\vec{k})} \ . \tag{54}$$

Thus a gap Δ opens up at the Fermi surface and, in the mean field approximation, the half-filled Hubbard model is an *insulating antiferromagnet*.

Hirsch [7] has computed a phase diagram (U/t vs. filling) at $T = 0$ in which only uniform phases are considered. However, it as been shown later [8] that away from half filling uniform phases are unstable towards vortex and domain wall formation.

IV   The strong-coupling limit

We turn now to a more accurate analysis of the OBH in the strong-coupling limit $U \gg t$. We write (4) as:

$$H = H_0 + V \ , \tag{55}$$

where H_0 is the hopping integral and $V = U \sum_i n_{i\uparrow} n_{i\downarrow}$.

Let $\mathcal{H}_\ell$ be the eigenspace of V with exactly ℓ doubly occupied sites, corresponding therefore to the eigenvalue:

$$E_\ell = \ell U \ . \tag{56}$$

The projectors P_ℓ onto $\mathcal{H}_\ell$ can be generated by expanding:

$$\Pi(x) = \prod_{i=1}^{M} [1 - (1 - x)\hat{\nu}_i] \tag{57}$$

$$= \sum_{\ell=0}^{M} x^\ell P_\ell \ , \tag{58}$$

where $\hat{\nu}_i = n_{i\uparrow} n_{i\downarrow}$; $0 \leq x \leq 1$ and M is the number of sites in the lattice. In particular the Gutzwiller projector [9]

$$P_0 = \prod_{i=1}^{M} (1 - \hat{\nu}_i) \tag{59}$$

selects the subspace containing no doubly-occupied sites at all, i.e. $n_i \leq 1$ as an operator inequality.

We can also define the operator projecting onto the subspace containing at least one doubly-occupied site:

$$P_\eta \equiv \sum_{\ell>0} P_\ell \tag{60}$$

and of course:

$$P_0 + P_\eta = \hat{1} \ . \tag{61}$$

Using the decomposition of the identity (61), we can write:

$$H_0 \equiv = P_0 \, H_0 \, P_0 + P_\eta \, H_0 \, P_\eta + P_0 \, H_0 \, P_\eta + P_\eta \, H_0 \, P_0 \; , \tag{62}$$

while

$$V \equiv P_\eta \, V \, P_\eta \; . \tag{63}$$

Notice that, actually: $P_0 \, H_0 \, P_\eta = P_0 \, H_0 \, P_1$ and $P_\eta \, H_0 \, P_\eta = P_1 \, H_0 \, P_0$.

By using the (trivial) identity

$$c_{i\sigma} \equiv c_{i\sigma} \left[(1 - n_{i\bar{\sigma}}) + n_{i\bar{\sigma}} \right] \; , \tag{64}$$

we can rewrite H_0 as:

$$H_0 \equiv T_h + T_d + T_{mix} \; , \tag{65}$$

where:

$$T_h = -\sum_{\langle ij\rangle\sigma} t_{ij} \, (1 - n_{i\bar{\sigma}}) \, c^\dagger_{i\sigma} \, c_{j\sigma} (1 - n_{j\bar{\sigma}}) \; ; \; \bar{\sigma} = -\sigma \tag{66}$$

$$T_d = -\sum_{\langle ij\rangle\sigma} t_{ij} \, n_{i\bar{\sigma}} \, c^\dagger_{i\sigma} \, c_{j\sigma} \, n_{j\bar{\sigma}} \tag{67}$$

$$T_{mix} = -\sum_{\langle ij\rangle\sigma} \left\{ t_{ij} \, n_{i\bar{\sigma}} \, c^\dagger_{i\sigma} \, c_{j\sigma} \left(1 - n_{j\bar{\sigma}} \right) + h.c. \right\} \; . \tag{68}$$

One can easily check that:

i) Both T_h and T_d preserve the number of doubly-occupied sites, so that

$$P_0 \, T_h \, P_\eta = P_0 \, T_d \, P_\eta = 0 \; . \tag{69}$$

Moreover T_d vanishes in the subspace spanned by the Gutzwiller projection:

$$P_0 \, T_d \, P_0 = 0 \; . \tag{70}$$

ii) T_{mix} mixes eigenspaces with $\Delta \ell = \pm 1$.

Then the OBH is decomposed into the sum of a "diagonal" part $\tilde{H}_0$, and an "off-diagonal" part H_η coupling the subspaces spanned by P_0 and P_η:

$$H = \tilde{H}_0 + H_\eta \; , \tag{71}$$
$$\tilde{H}_0 = P_0 \, H_0 \, P_0 + P_\eta \, H_0 \, P_\eta + V \; , \tag{72}$$
$$H_\eta = P_0 \, H_0 \, P_\eta + P_\eta \, H_0 \, P_0 \; . \tag{73}$$

We now seek a canonical transformation eliminating the effect of H_η to lowest order [10], i.e. such that the transformed Hamiltonian H_{eff} satisfies:

$$P_0 \, H_{eff} \, P_\eta = 0 \tag{74}$$

to the required order.

Let us proceed formally and define:

$$H(\lambda) = \widetilde{H}_0 + \lambda H_\eta , \tag{75}$$

while seeking a canonical transformation of the form:

$$\mathcal{U}(\lambda) = e^{i\lambda S} \; ; \; S = S^\dagger , \tag{76}$$

where S has to be such that the transformed Hamiltonian $H_{eff}(\lambda)$ obeys:

$$H_{eff} = e^{i\lambda S} H(\lambda) e^{-i\lambda S} = \widetilde{H}_0 + \mathcal{O}(\lambda^2) . \tag{77}$$

As, at the end of the calculations, we have to set $\lambda = 1$, it is clear that λ is not an expansion parameter of any sort, but simply a bookkeeping device. Expanding (77), we find:

$$H_{eff}(\lambda) = \widetilde{H}_0 + \lambda \left(H_\eta + i[S, \widetilde{H}_0] \right) + \lambda^2 \left(i[S, H_\eta] + \frac{1}{2}[S, [\widetilde{H}_0, S]] \right) + \mathcal{O}(\lambda^3) , \tag{78}$$

while S is determined by

$$[\widetilde{H}_0, S] + i H_\eta = 0 . \tag{79}$$

Setting $\lambda = 1$, we eventually find:

$$H_{eff} = \widetilde{H}_0 + \frac{i}{2} [S, H_\eta] , \tag{80}$$

up to terms of order t^2.

The explicit calculations of H_{eff} can be found in [3, 11]. We report here only the final result:

$$P_0 H_{eff} P_0 = P_0 H P_0 - \frac{1}{U} P_0 H P_\eta H P_0 , \tag{81}$$

$$P_\eta H_{eff} P_\eta = P_\eta H P_\eta + \frac{1}{U} P_\eta H P_0 H P_\eta . \tag{82}$$

We proceed now to study the effective hamiltonian in the low-energy sector with no double-occupancy spanned by the Gutzwiller projector P_0. In this case H_{eff} reduces to a $\widehat{H}_{eff}$ given by:

$$P_0 H_{eff} P_0 = P_0 \widehat{H}_{eff} P_0$$
$$\widehat{H}_{eff} = T_h + H^{(1)} + H^{(2)} \tag{83}$$

where

1) T_h is the hopping term;

2) $H^{(1)} = 2 \sum_{<ij>} \sum_{\sigma\tau} |t_{ij}|^2 (1 - n_{i\bar\sigma}) c_{i\sigma}^\dagger c_{j\sigma} n_{j\bar\sigma} n_{j\bar\tau} c_{j\tau}^\dagger c_{i\tau} (1 - n_{i\bar\tau}) , \tag{84}$

3) $H^{(2)} = \sum_{<ijl>} \sum_{\sigma\tau} t_{ij} t_{jl} (1 - n_{i\bar\sigma}) c_{i\sigma}^\dagger c_{j\sigma} n_{j\bar\sigma} n_{j\bar\tau} c_{j\tau}^\dagger c_{l\tau} (1 - n_{l\bar\tau}) \tag{85}$

and the symbols $< ij >$, $< ijl >$ respectively denote a summation over n.n. and in which $i \neq l$, and both i and l are n.n. to j. Eq. (84) can be recast in a particularly interesting form. Indeed, rearranging terms, the following result has been obtained by various authors [10]:

$$H^{(1)} = -\sum_{(ij)} J_{ij} \left\{ \vec{S}_i \cdot \vec{S}_j - \frac{1}{4} n_i n_j \right\} \quad , \quad J_{ij} = 4 \frac{|t_{ij}|^2}{U} \tag{86}$$

where (ij) stands for unordered pairs and the spin operators have been defined as in (13). Thus $H^{(1)}$ contains two terms:

$i)$ an AFM Heisenberg coupling between nearby (singly-occupied) sites,

$ii)$ an "$n_i n_j$" coupling favouring simultaneous occupation of nearby sites.

As for $H^{(2)}$, to our knowledge, there is no corresponding simplifications. Notice that while $H^{(1)}$ describes virtual processes involving a double occupancy in the intermediate state, $H^{(2)}$ represents a real hopping process in which an electron jumps from site i to site k with again a double occupancy on site j. For this reason we will denote $H^{(2)}$ as the "three site term".

Summarizing we can conclude that:

- *at exact half filling*, when $n_i \equiv 1$ and hopping dynamics is not permitted, we are left only with a spin dynamics described by the standard AFM Heisenberg model:

$$H_{AFM} = -\sum_{(ij)} J_{ij} \vec{S}_i \cdot \vec{S}_j \ ; \tag{87}$$

- *below half-filling* $(n_i \leq 1)$, if we denote with δ the fraction of empty sites, the hamiltonian contains:

  $i)$ A direct hopping term, giving nonvanishing contribution only if the electron jumps from site i to an empty n.n. site j. So we expect and effective hopping coefficient t to be reduced by a factor of order δ [13].

  $ii)$ A three-site hopping term, which gives nonzero contribution only for the situation in which i, j are occupied while k is empty. Thus again this term should be renormalized by a factor of order δ.

  $iii)$ The AFM Heisenberg hamiltonian, which involves only virtual processes and thus does not get renormalized.

If we accept this "poor man's" renormalization prescription, the three-site term remains smaller by a factor of δ as compared with the AFM term and smaller by a factor $\frac{|t|}{U}$ as compared with the direct hopping term. That is why it is usually ignored, and one is led to the celebrated "t-J" model [12]:

$$\begin{aligned} H_{tJ} = \ & -\sum_{\langle ij \rangle \sigma} t_{ij} \left(1 - n_{i\bar{\sigma}}\right) c_{i\sigma}^\dagger c_{j\sigma} \left(1 - n_{j\bar{\sigma}}\right) \\ & -\sum_{(ij)} J_{ij} \left\{ \vec{S}_i \cdot \vec{S}_j - \frac{1}{4} n_i n_j \right\} \ . \end{aligned} \tag{88}$$

V "Zoology" at half-filling

We have just seen that at "half-filling" ($\nu_i = 1 \;\; \forall i$ not merely $< n_i > = 1$) the OBH reduces to the Heisenberg hamiltonian (87).

Now, it can be proved rather easily that

$$\vec{S}_i \cdot \vec{S}_j = \frac{\hat{n}_i \hat{n}_j}{4} - b_{ij}^\dagger b_{ij} \tag{89}$$

where

$$b_{ij}^\dagger = b_{ji}^\dagger = \frac{1}{\sqrt{2}}[c_{i\uparrow}^\dagger c_{j\downarrow}^\dagger - c_{i\downarrow}^\dagger c_{j\uparrow}^\dagger] \tag{90}$$

so that (87) becomes:

$$H = -J \sum_{(ij)} b_{ij}^\dagger b_{ij}. \tag{91}$$

Using the "poor man's" renormalization scheme described in the previous sections, one can also extend this construction below half-filling by rewriting the t-J hamiltonian (88) as:

$$H \simeq -t\delta \sum_{(ij)\sigma} c_{i\sigma}^\dagger c_{j\sigma} - J \sum_{(ij)} b_{ij}^\dagger b_{ij} \;, \tag{92}$$

which is suitable for a Hartree-Fock factorization with order parameters:

$$p_{ij} = \sum_\sigma < c_{i\sigma}^\dagger c_{j\sigma} > \;, \tag{93}$$

$$\Delta_{ij} = \sqrt{2} < b_{ij} > \;. \tag{94}$$

The case of uniform order parameters:

$$p_{ij}, \Delta_{ij} = \begin{cases} \text{const.} = p, \Delta & (i,j) n.n. \\ 0 & \text{otherwise} \end{cases} \tag{95}$$

has been studied by Anderson and coworkers [13] obtaining, as their main results, that:

1. at exact half-filling and below a certain transition temperature, electrons are paired in singlet bonds. The ground state is disordered and insulator. In analogy with what found previously for frustrated 2D Heisenberg antiferromagnets [14] this was called an RVB (Resonating Valence Bond) state.

2. As soon as doping starts, the system opens up a gap and the singlet pairs become mobile. The state is therefore superconducting, with pairing in real space.

Unfortunately there is a severe mismatch, as pointed out by Huang and Manousakis [15] between the predicted dependence of T_c on doping and the existing data on the cuprates, so this approach is by now only of historical interest, its merit being of having shown how the Hubbard-Heisenberg hamiltonian can exhibit non-trivial non-magnetic correlated ground states (at the mean field level, at least).

Notice that, by construction, $\Delta_{ji} = \Delta_{ij}$ i.e. the "pairing" order parameter has no "directional character". If we look now for mean-field solutions that preserve the original translational symmetry of the square lattice, then we are left with only two independent order parameters, Δ_x and Δ_y. If we further assume that $|\Delta_x| = |\Delta_y|$, solutions are classified by the relative phase of Δ_x and Δ_y. This has been examined by Kotliar [16], who found that the mean field equations admit the following solutions:

  i) "s-wave": $\Delta_x = \Delta_y$. This coincides with Anderson's uniform phase ;

  ii) "d-wave": $\Delta_y = -\Delta_x$ (relative phase $= \pi$);

  iii) "mixed" phase: $\Delta_y = i\Delta_x$ (relative phase $= \pi/2$) .

The "s" and "d-wave" solutions turn out to be energetically equivalent, and higher in energy than the "mixed" phase, which is therefore the most stable one.

We discuss now a more flexible scheme, that was first proposed by Affleck and Marston [17], by reporting also on a joint work of the Bologna-Napoli-Trieste-Bangalore group [18]. Let us start again from the Heisenberg Hamiltonian (87). Recall that the half-filling condition can be implemented by the projector

$$P = \prod_i n_i(2 - n_i) . \tag{96}$$

As the total particle number is fixed, the canonical partition function should be written as:

$$\mathcal{Z} = \text{Tr}\left\{e^{-PHP}P\right\} \tag{97}$$

However, P and H commute, so the exact partition function is given by the restricted trace

$$\mathcal{Z} = \text{Tr}\left\{e^{-\beta H}P\right\} . \tag{98}$$

To implement the constraint we introduce the generating function:

$$\mathcal{Z}[z] =: \text{Tr}\left\{e^{-\beta\left(H-\beta^{-1}\sum_i z_i n_i\right)}\right\} , \tag{99}$$

from which one finds:

$$\begin{aligned} \mathcal{Z} &\equiv [P(z)\mathcal{Z}[z]]_{z=0} \\ P(z) &= \prod_i \frac{\partial}{\partial z_i}\left(2 - \frac{\partial}{\partial z_i}\right). \end{aligned} \tag{100}$$

The generating function (99) is defined through an unrestricted trace which is easier to handle, since we can use the full (standard) machinery of many-body technology. We will show in particular how, with the help of this formulation, both magnetic and nonmagnetic phases can be studied within the same framework.

Let us start by noticing that, up to a constant, the Heisenberg hamiltonian can be recast in the form:

$$H = -\frac{1}{2}J \sum_{(ij)} \chi_{ij}^\dagger \chi_{ij} \tag{101}$$

where χ_{ij} is defined by

$$\chi_{ij} = \sum_\alpha c_{i\alpha}^\dagger c_{j\alpha} \ . \tag{102}$$

Notice that the operators χ_{ij} have a directional character since $\chi_{ji} = \chi_{ij}^\dagger \neq \chi_{ij}$.

At the same time, the hamiltonian (87) can also be rewritten in another useful form:

$$H = J \sum_{\vec{q}} \eta(\vec{q}) \vec{S}_{\vec{q}} \cdot \vec{S}_{-\vec{q}} \tag{103}$$

where $\eta(\vec{q})$ is given in (7) and

$$\vec{S}_{\vec{q}} = \frac{1}{\sqrt{N}} \sum_i e^{-i\vec{q}\cdot\vec{R}_i} \vec{S}_i \ . \tag{104}$$

While form (101) is useful for studying nonmagnetic states, form (103) is useful for studying antiferromagnetic states. We can however use the same general strategy to analyze both cases. It is as follows:

1. write down the partition function as a functional integral over anticommuting (Grassmann) variables [19]:

$$\mathcal{Z}[z] = \int [\mathcal{D}\psi_{i\alpha}^*(\tau)\mathcal{D}\psi_{i\alpha}(\tau)] \exp\left\{-\int_0^1 d\tau \sum_{i\alpha} \psi_{i\alpha}^*(\tau)(\partial\tau - z_i)\psi_{i\alpha}(\tau)\right\} \times$$
$$\times \quad \exp\{-\beta \int_0^1 d\tau H(\tau)\} \ ; \tag{105}$$

2. decouple the quartic interaction term in $H(\tau)$ using a Hubbard-Stratonovich transformation (which is basically a clever way of using the gaussian identity $\exp(x^2) = \int_{-\infty}^{+\infty} dy \exp\{-\pi y^2 + 2\sqrt{\pi}xy\}$). Clearly there are many different Hubbard-Stratonovich decouplings one can choose, depending on which kind of correlations one is interested in. After that, the Grassmann integrations are gaussian and one can proceed to:

3. integrate out the fermions, to get an effective action which is essentially the logarithm of the determinant of the fermionic operator;

4. implement the constraint by means of (100). At this stage $\mathcal{Z}$ is a functional integral over Hubbard-Stratonovich fields, with an effective action that is a highly nonlinear functional of the fields themselves. All one can do now is to:

5. develop approximations that are physically reasonable, sufficiently manageable and that yield sensible (and nontrivial!) results. In this sense, we have
considered two possible schemes:

5-a) saddle-point plus one-loop corrections of the effective action,

5-b) adiabatic expansion of the fermionic determinant arising from step (3).

Notice that steps (4) and (5) can be interchanged.
We will discuss now approximation (5-a), deferring (5-b) to Sec. (VII).

- Nonmagnetic states.

Skipping all calculations which can be found in [18], after step (4) we have
are left with a partition function which is written in terms of the auxiliary
Hubbard-Stratonovich fields

$$\mathcal{U}_{ij}(\tau) = \sum_\sigma \psi_{j\sigma}^*(\tau)\psi_{i\sigma}(\tau) \tag{106}$$

as:

$$\mathcal{Z} = \int [\mathcal{D}\mathcal{U}_{ij}^* \mathcal{D}\mathcal{U}_{ij}] \exp\left\{ -\pi \sum_{(ij)} |\mathcal{U}_{ij}|^2 - S_{eff}[\mathcal{U}] + \sum_i \log(4B_i[\mathcal{U}]) \right\}, \tag{107}$$

where $S_{eff}[\mathcal{U}]$ is the effective action deriving from the integration of the
Grassmann variables which is therefore given by:

$$S_{eff}[\mathcal{U}] = -2Tr \log\left\{ \hat{1} + \hat{G}\hat{U} \right\}|_{z=0} \tag{108}$$

$$(\hat{G}_0)_{ij}^{nn'} \equiv (i\omega_n)^{-1}\delta_{ij}\delta_{nn'}$$

$$\hat{U}_{ij}^{nn'} \equiv -\sqrt{\frac{\pi\beta J}{2}} U_{ij}(\omega_n - \omega_{n'}) + z_i\delta_{ij}\delta_{nn'}$$

and $B_i[\mathcal{U}]$ is the contribution of the constraint:

$$B_i[\mathcal{U}] \equiv \left[\frac{\partial^2 S_{eff}}{\partial z_i^2} - 2\frac{\partial S_{eff}}{\partial z_i} - \left(\frac{\partial S_{eff}}{\partial z_i} \right)^2 \right]_{z=0}. \tag{109}$$

For a square lattice, Affleck and Marston [17] considered an array of bonds
such as that depicted in Fig. (1). In the (static) saddle point approximation
one finds the following possible phases:

1. A *staggered dimer* (Peierls) solution with:

$$\chi_2 = \chi_3 = \chi_4 = 0 , \quad \chi_1 > 0. \tag{110}$$

This solution corresponds to a regular pattern of staggered singlet
bonds, as shown in Fig. (2)(a). This yields a transition temperature $T_c = 5J/12$ to be compared with the value $T_c = J/4$ one would
have obtained without the constraint, and, at $T = 0$, a free energy per
site $f_{stag} = -\frac{3}{4}$.

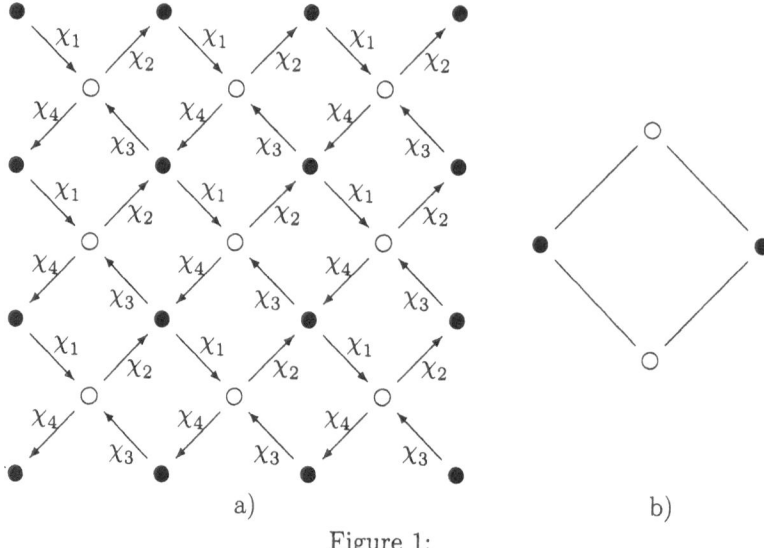

Figure 1:

a) The array of n.n. bonds corresponding to the Affleck and Marston solution.

b) The original 2D square lattice unit cell.

2. A *uniform phase* with

$$\chi_k = \chi > 0 \; ; \; k = 1, \ldots, 4 \; , \tag{111}$$

which exists at all temperatures.

3. The so-called *flux phase*, corresponding to:

$$\chi_k = \chi \, e^{i\frac{\pi}{4}} \; ; \; k = 1, \ldots, 4 \; , \tag{112}$$

with the same transition temperature as the dimer phase.

Besides the "staggered dimer" one can study other dimer solutions, like the *columnar dimer* discussed, e.g., by Read and Sachdev [20] using the $1/N$ expansion. The patterns of singlet bonds in the two cases are displayed in Fig. (2)(b). The columnar dimer corresponds to a doubling of the unit cell along one of the coordinate axes, and hence to a different residual translational symmetry w.r.t. the staggered dimer. At the pure MF-level and at half-filling, the two are degenerate in energy and both are slightly lower in energy (and hence more stable) than the flux phase.

As for the implementation of the half-filling constraint, one can check [18] that, already at this level of approximation, $< n_i >= 1$. However, $< n_i^2 > \neq 1$, with deviationns by at most the 10%. As we have noted, the constraint suppresses the fluctuations and thus raises the critical temperature.

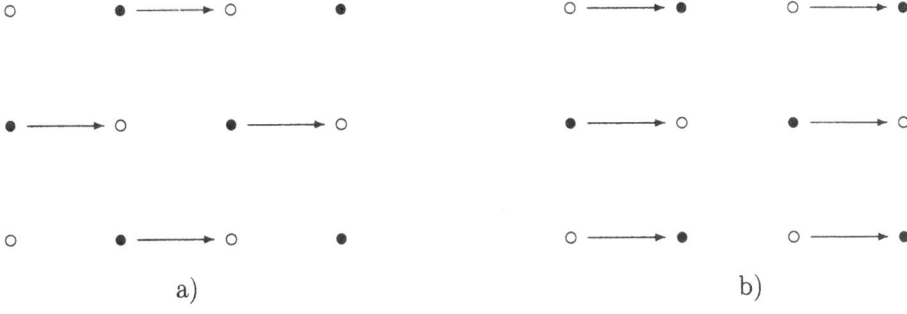

Figure 2:
The pattern of singlet bonds in
a) the staggered dimer phase, b) the columnar dimer phase.

Some remarks on the gauge invariance of the Heisenberg hamiltonian are in order at this stage:

i) (87) acquires an additional invariance that is not present in the full Hubbard Hamiltonian, namely a local *gauge* invariance:

$$c_{i\alpha} \mapsto e^{i\theta_i} \, c_{i\alpha} \, . \tag{113}$$

This reflects the fact that it is not only the total particle number that is conserved, but separately the particle number at each lattice site:

$$[H_{AFM}, n_i] = 0 \ \forall i \, . \tag{114}$$

It is then not difficult to check that, under this gauge transformation, the phase of the order parameter $\mathcal{U}_{ij} = |\mathcal{U}_{ij}| e^{i\theta_{ij}}$ transform as:

$$\theta_{ij} \mapsto \theta_{ij} + \theta_j - \theta_i \, . \tag{115}$$

i.e. as a $U(1)$ lattice gauge field [21]. This implies that the phases θ_j ($j = 1, \ldots, 4$) of the order parameters $\mathcal{U}_j$ are not separately observables: only the total phase per plaquette $\Theta \equiv \sum_{j=1}^4 \theta_j$ is gauge invariant and hence represents a physical observable: the flux Φ threading an elementary cell (in units of the elementary fluxon).

ii) For the flux phase, the flux across a plaquette is: $\Phi = \pi$. In this regard we simply recall that the existence of this phase has led to speculations that low-lying elementary excitations in high-T_c materials could obey fractional statistics induced by a Chern-Simons term in the effective low-energy action, and that superconductivity in these materials could be of "anyonic" type. For a more complete discussion of this problem we refer to [3] and references therein.

iii) The approaches of Anderson and coworkers and of Kotliar on one hand and of Affleck and Marston on the other seem at first sight to be totally uncorrelated. That this is not so was shown shortly after by Affleck, Zhou, Hsu and Anderson [22]. They showed that the $U(1)$ gauge symmetry described in (*i*) is actually a subgroup of a larger $SU(2)$ gauge symmetry of the Hubbard model at half-filling, and that the latter converts the order parameter of Anderson et al. into that of Affleck and Marston (and of course viceversa).

Going back to (107), to study one-loop corrections, one has to expand the free energy F_{eff} defined by $\mathcal{Z} = \exp\{-\beta F_{eff}\}$, around the saddle point solutions to second order in $\delta\mathcal{U}$. After some algebra, one finds [18]:

$$F_{eff} = F_{eff}^{(MF)} + F^{(2)} \tag{116}$$

$$\beta F^{(2)} = \pi Tr\{|\delta\mathcal{U}|^2\} + \delta^2 S_{eff} - \frac{1}{2}\sum_i\left\{\frac{\delta^2 B_i}{B_i} - \left(\frac{\delta B_i}{B_i}\right)^2\right\}.$$

Inclusion of quantum fluctuations around the MF solutions in the calculation of the free energy lifts the degeneracy in favor of the columnar dimer phase [18]: $f_{stag} = -0.668$, $f_{col} = -0.771$, a result that is is agreement with those of Read and Sachdev [20].

- Antiferromagnetic state.

A similar analysis can be performed for AFM correlations, by choosing as Hubbard-Stratonovich field:

$$\vec{M}(\vec{q},\tau) = \frac{1}{2\sqrt{N}}\sum_{i,\alpha\beta}e^{-i\vec{q}\cdot\vec{R}_i}\psi_{i\alpha}^*(\tau)\vec{\sigma}_{\alpha\beta}\psi_{i\beta}(\tau). \tag{117}$$

The MF approximation yields a Néel ground state, with a a free energy per site $f_{afm} = -\frac{1}{2}$, for $T = 0$. The one-loop approximations add the standard spin-wave corrections, leading to a total free energy $f_{afm} = -0.658$. In this case $< n_i >= 1$ and $< n_i^2 >= 1$ already at the MF level.

Notice that, at MF plus one-loop corrections, it appears that nonmagnetic (columnar dimer) phases are energetically favoured with respect to the AFM one. However, although no rigorous proof exists for D=1 and $S = \frac{1}{2}$, it is generally believed that Néel order should exist at $T = 0$, as follows mainly from numerical studies [23]. We can understand this apparent contraddiction, by remembering that the constraint is not exactly implemented for nonmagnetic phases ($< n_i^2 >\neq 1$), leaving more room for the fluctuation of the fields $\mathcal{U}_{ij}$ and hence lowering the free energy below that of the AFM phase.

VI The three-band Hubbard model and the spin fermion model

It is well known that a common feature of the copper oxyde superconductors are the CuO_2 planes, which have the structure shown in Fig.3(a).

As first suggested by Emery [24] the electronic structure of the CuO_2 planes can be described by a Hubbard Hamiltonian built up on the 2D-lattice depicted in Fig. 3(a), having one Cu-site and two O-sites per unit cell. A single $3d_{x^2-y^2}$-orbital in each Cu-site is assumed to enter into the game, and to become hybridized with each of the four $2p_x$, $2p_y$ orbitals of the surrounding O-sites pointing toward the i-site, as shown in the hybridization scheme of Fig. (3)(b). Also, a strong Coulomb repulsion term U is assumed between two electrons when they happen to be both in the 3d orbital at the same site i. The "Three Band Hubbard Hamiltonian" (TBH) which describes the previous picture is thus of the form

$$H = H_0 + V \tag{118}$$

with

$$H_0 = \sum_{i\sigma} \epsilon_d \, d_{i\sigma}^\dagger d_{i\sigma} + U \sum_i n_{i\uparrow} n_{i\downarrow} + \sum_{\mu\sigma} \sum_\alpha \epsilon_p \, p_{\mu\sigma}^{\alpha\dagger} p_{\mu\sigma}^\alpha \tag{119}$$

and

$$V = \sum_{i\sigma} \sum_{\mu_i} \left(V_{i\mu} \, d_{i\sigma}^\dagger p_{\mu_i\sigma} + h.c. \right) . \tag{120}$$

Here i and μ denote, respectively, the Cu-sites and the O-sites; $d_{i\sigma}^\dagger$, $p_{\mu_i\sigma}^{\alpha\dagger}$ create electrons with spin σ in the $Cu(3d_{x^2-y^2})$ and $O(2p_x)$ or $O(2p_y)$ orbitals of energies ϵ_d and ϵ_p. In (120) the sum over $\mu_i = i \pm c_x$; $\mu_i = i \pm c_y$ runs over the four O-sites around the Cu-site i and, according to the hybridization scheme of Fig. (3)(b), it is understood that the $p_{\mu_i\sigma}$'s indicate $p_{\mu_i\sigma}^x$ and $p_{\mu_i\sigma}^y$ for $\mu_i = i \pm c_x$ and $\mu_i - i \pm c_y$ respectively.

The hybridization matrix element $V_{i\mu}$ is assumed to be proportional to the overlap of the corresponding 3d and 2p orbitals and has then the form

$$V_{i\mu} = (-1)^{\alpha_{i\mu}} V \tag{121}$$

with

$$\alpha_{i\mu} = \begin{cases} 1 & \text{for } \mu_i = i + c_x \ ; \ \mu_i = i + c_y \\ 0 & \text{for } \mu_i = i - c_x \ ; \ \mu_i = i - c_y \end{cases} . \tag{122}$$

We shall concentrate in the following to the case of La_2CuO_4, in which substitution of the trivalent La with a divalent element $M = Ca$, Ba, Sr ...) creates holes with concentration x per unit cell in the CuO_2 planes of the "doped" system $La_{2-x}M_xCuO_4$. Experimental data [25] show that we are in the regime with $U > \Delta$ and $\Delta > 0$. A reasonable estimate of the parameters of the TBH gives $U \approx 10 \ eV$, $\Delta \approx 3 - 4 \ eV$, $V \approx 1.3 - 1.6 \ eV$ [26].

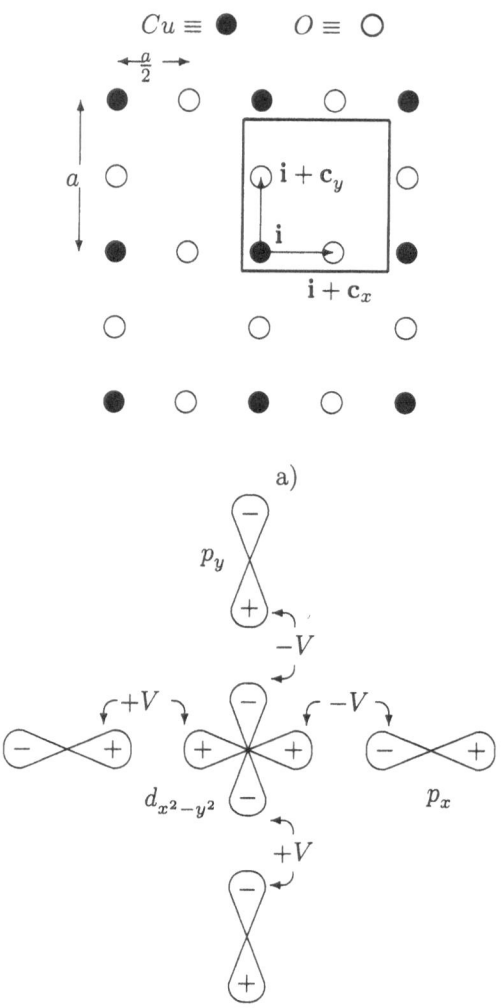

$Cu \equiv$ ● $O \equiv$ ○

a)

b)
Figure 3:
a) The lattice in the CuO_2 plane b) Scheme of hybridization between Cu-$3d_{x^2-y^2}$ and O-p_x, p_y orbitals.

Let us first consider the *pure La₂CuO₄* system. In absence of the Coulomb repulsion U, the hybridization term V would produce hopping processes of electrons from the O-sites to the Cu-sites and this would give rise to three hybridized $3d - 2p$ bands, the highest of which would be half-filled, since we started with an odd number (five, in fact) electrons per unit cell. The system should therefore be a conductor. But, as U is by far the largest parameter of the theory, real hopping processes, which should now take place between an initially doubly occupied O-site and a singly occupied Cu-site, thus implying a doubly occupied Cu-site in the final state, are strongly inhibited. The system behaves then as an insulator, of the kind that is often called of Mott-Hubbard. We thus recover for the pure system a situation showing a close analogy with that of the OBH in the half-filled case.

Since, as we have seen, V is the smallest parameter of the teory we would like to treat it as a perturbation. Again, as for the OBH, we expect that the effects produced by such a perturbation will be twofold, namely:

i) *in the pure case*, the resolution of the vaste degeneracy that the ground state of the pure system exhibits for $V = 0$, as a consequence of the two possible spin orientations of each of the singly occupied $3d$ orbitals at the Cu-sites;

ii) *in the doped system*, the activation of real hopping processes, even for large U. Such a term will arise in the perturbation theory at the order V^2/Δ, as it is clear by considering the processes shown in Fig. (4) which produce as a net result the hopping of a (spin-up in the Figure) hole from the O-site 1 to the O-site 2, which share the same Cu-site as a n.n. Note that while process a) does not change the orientations of the two spins, process b) produces a spin-flip of both the Cu-spin and of the hole-spin. Also note that, since each O-site has two nearest neighbours Cu-sites in the CuO_2 plane, by repeated processes of kind a) and b) a hole is allowed to truly propagate in the plane. The doped system has then become a conductor, even for $U \to \infty$.

We proceed now to study the strong coupling limit of (118) via a canonical transformation. Since conceptually the calculations are similar to the OBH case, we only quote the result [27, 11], remarking however that in the present case one is forced to push the expansion up to fourth order in V. In the subspace of singly occupied Cu-sites one finds:

$$\widetilde{H} = \widetilde{H}^{(2)} + \widetilde{H}^{(4)} + O(V^6) \ . \tag{123}$$

The 2nd order term $\widetilde{H}^{(2)}$ is non-zero only in presence of holes on the oxygens (i.e. in the doped system) and is given by:

$$H^{(2)} = H_{hop} + H_{ex} \ , \tag{124}$$

where

$$H_{hop} = -\frac{t_1}{4} \sum_{i\mu\nu} (-1)^{\alpha_{i\mu} + \alpha_{i\nu}} \, p^{\dagger}_{\mu\sigma} p_{\nu\sigma} \tag{125}$$

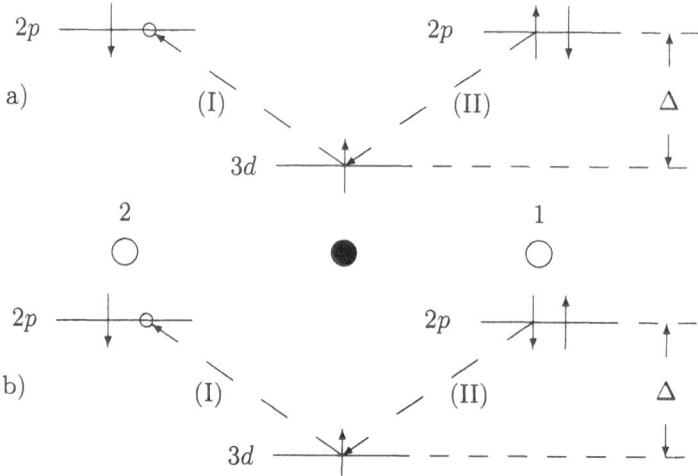

Figure 4:

Second order effective hopping processes of a hole between O-sites sharing the same Cu-site as a neighbour (no double occupancy of the Cu-site in the intermediate state).

represents an ordinary hopping between pairs (μ, ν) of O-sites sharing the same Cu-site as neighbour, and

$$H_{ex} = t_2 \sum_i \vec{S}_i \cdot \vec{R}_i \qquad (126)$$

represents an O-O hopping assisted by an exchange interaction between the Cu-spin $\vec{S}_i$ at the singly occupied site i

$$\vec{S}_i = \sum_{\alpha\beta} d_{i\alpha}^\dagger \, \vec{\sigma}_{\alpha\beta} \, d_{i\beta} \qquad (127)$$

and the oxygen hole "spin" $\vec{R}_i$

$$\vec{R}_i = \frac{1}{4} \sum_{<\mu\nu>_i} \sum_{\alpha\beta} (-1)^{\alpha_{i\mu} + \alpha_{i\nu}} \, p_{\mu\alpha}^\dagger \, \vec{\sigma}_{\alpha\beta} \, p_{\nu\beta} \; . \qquad (128)$$

The operator $\vec{R}_i$ is an appropriate combimation of the four oxygen orbitals surrounding the i copper site and represent a sort of "itinerant spin". The coupling constants in (125) and (126) have the values

$$t_1 \;\; = \;\; 2V^2 \left(\frac{1}{U - \Delta} - \frac{1}{\Delta} \right) , \qquad (129)$$

$$t_2 \;\; = \;\; 8V^2 \left(\frac{1}{U - \Delta} + \frac{1}{\Delta} \right) . \qquad (130)$$

The forth-order term in (123) turns out to be the more familiar AFM Heisenberg hamiltonian:

$$H^{(4)} = J \sum_{(ij)} \vec{S}_i \cdot \vec{S}_j \; , \qquad (131)$$

where

$$J = \frac{U V^4}{\Delta(U - \Delta)^3} \left(5 - \frac{U}{\Delta} \right) . \tag{132}$$

We are thus led to consider the so-called *spin-fermion model*:

$$H_{SF} = H_{hop} + H_{ex} + H_{AFM} . \tag{133}$$

The first term, H_{hop}, can very easily be diagonalized introducing appropriate Bloch operators $a^\dagger_{\vec{k}\sigma}$, $a_{\vec{k}\sigma}$: one gets

$$H_{hop} = \sum_{\vec{k}\sigma} \epsilon_{\vec{k}} a^\dagger_{\vec{k}\sigma} a_{\vec{k}\sigma} + (\text{ a flat band with } \epsilon = \epsilon_p = 0) , \tag{134}$$

representing two bands for the hybridized oxygen orbitals. One of these bands is dispersionless, and the other has dispersion given by

$$\epsilon_{\vec{k}} = -t_1 \Lambda_{\vec{k}} , \quad \Lambda_{\vec{k}} = \sin^2 \left(\frac{k_x a}{2} \right) + \sin^2 \left(\frac{k_y a}{2} \right) , \tag{135}$$

a being the Cu-Cu lattice constant in the CuO_2 planes The exchange term, given by (126), can also be expressed in term of the Bloch operators. One finds

$$H_{ex} = t_2 \sum_{\vec{q}} \vec{S}_{\vec{q}} \cdot \vec{R}_{-\vec{q}} , \tag{136}$$

with $\vec{S}_{\vec{q}}$ as in (104) and $\vec{R}_{-\vec{q}}$ given by

$$\vec{R}_{\vec{q}} = \frac{1}{\sqrt{N}} \sum_{\vec{k}} \sum_{\alpha\beta} \Lambda^{\frac{1}{2}}_{\vec{k}} \Lambda^{\frac{1}{2}}_{\vec{k}+\vec{q}} a^\dagger_{\vec{k}\alpha} \vec{\sigma}_{\alpha\beta} a_{\vec{k}+\vec{q}\beta} . \tag{137}$$

Also, the antiferromagnetic coupling H_{AFM} given by (131) can be written as in (103).

VII   The adiabatic expansion

The spin-fermion Hamiltonian (133) that we have derived at the end of the previous section is of course very complicated, because the term H_{ex} introduces a coupling of the dynamics of the (anti-ferromagnetically) coupled Cu-spins described by H_{AFM} with that of the fermions system consisting of the mobile holes propagating through the O-sites via the hopping term H_{hop}.

The strategy that has been widely adopted in literature [27, 28, 29, 11] to attack the problem is that of eliminating the spins by taking the trace of the equilibrium density matrix of the full system with respect to the spin degrees of freedom. This will in turn produce an effective interaction between the fermions which keeps into account the fact that the fermions "live" on top of the background of the Cu-spins to which they were coupled. Such a strategy can be easily accomplished, at least at a formal level, within the Feynman integral formalism, in which the fermion partition and correlation functions are expressed as path integrals over Grassman variables [19]. In our case the action is as a sum of two terms:

$$S_F = S_F^{(0)} + A_{eff} , \tag{138}$$

Here $S_F^{(0)}$ is the "free" action related to H_{hop} through

$$S_F^{(0)} = -i \int 01 d\tau \left[\sum_{\vec{k}\sigma} a^*_{\vec{k}\sigma} \partial_\tau a_{\vec{k}\sigma} + H_{hop}(a^*_{\vec{k}\sigma}, a_{\vec{k}\sigma}) \right] , \tag{139}$$

where $a^*_{\vec{k}\sigma}, a_{\vec{k}\sigma}$ are the Grassman variables corresponding to the fermion Bloch operators $a^\dagger_{\vec{k}\sigma}, a_{\vec{k}\sigma}$ and $H_{hop}(a^*_{\vec{k}\sigma}, a_{\vec{k}\sigma})$ has the same expression given by (134) with the operators $a^\dagger_{\vec{k}\sigma}, a_{\vec{k}\sigma}$ replaced by the Grassman variables $a^*_{\vec{k}\sigma}, a_{\vec{k}\sigma}$.

Furthermore S_{eff} is an "effective" action which takes origin from the thermal average (trace) operation over the spin degrees of freedom, so that:

$$e^{S_{eff}} = \frac{1}{Z_{AFM}} \left\{ Tr e^{-\beta (H_{AFM} + H_{ex})} \right\}_{spin} \tag{140}$$

where

$$Z_{AFM} = \left\{ Tr e^{-\beta H_{AFM}} \right\}_{spin} . \tag{141}$$

Here we will consider only the partition function (141) of the pure spin system, with H_{AFM} given by (87), while referring the reader to [11] for a more complete discussion of (140). We decouple the spin-spin interaction by introducing a set of auxiliary Hubbard-Stratonovich fields $\vec{\Phi}_i$ to obtain the following path-integral representation for the partition function:

$$\mathcal{Z}_{AFM} = \int [\mathcal{D}\Phi_i] \exp \left\{ \frac{\beta}{2} \int_0^1 \delta\tau \sum_{ij} (J^{-1})_{ij} \vec{\Phi}_i \cdot \vec{\Phi}_j + \sum_i \log \mathcal{L}[\vec{\Phi}_i] \right\} \tag{142}$$

$$\mathcal{L}[\vec{\Phi}] \equiv Tr \left\{ T_\tau \exp[\beta \int_0^1 d\tau \vec{\Phi}(\tau) \cdot \vec{S}] \right\} , \tag{143}$$

where T_τ denotes the (euclidean) time-ordering operator and the $\vec{\Phi}$'s satisfy bosonic-like boundary conditions

$$\vec{\Phi}_i(1) = \vec{\Phi}_i(0) . \tag{144}$$

Essentially we have traded the original spin-spin hamiltonian with an interaction of each spin with a *classical* τ-dependent field. It is not hard to imagine that restricting to static field configurations ($\vec{\Phi}_i(\tau) \equiv \vec{\Phi}_i$) yields to the standard mean field theory of the antiferromagnet, where the staggered magnetization

$$\vec{m}_i = \frac{(-1)^{|i|}}{2} m(T) \hat{n} . \tag{145}$$

is related to $\vec{\Phi}_i$ by:

$$\vec{\Phi}_i = \sum_j J_{ij} \vec{m}_j \ . \tag{146}$$

Let us go back now to the general form of the partition function (142), which contains the term (143):

$$\mathcal{L}[\vec{\Phi}] = TrU(1) \ , \tag{147}$$

where the operator

$$U(\tau) \equiv \mathrm{T}_\tau \exp[\beta \int_0^1 d\tau \vec{\Phi}(\tau) \cdot \vec{S}] \tag{148}$$

obeys the evolution equation (in euclidean time):

$$\frac{dU(\tau)}{d\tau} U^{-1}(\tau) = M(\tau) \ , \ U(0) = \hat{1} \tag{149}$$

$$M(\tau) = \beta \vec{\Phi}(\tau) \cdot \vec{S} \ , \ M(1) = M(0) \ . \tag{150}$$

From the fact that $M(\tau)$ is hermitian and traceless, it follows at once that $\det U(\tau) = \det U(0) = 1$, i.e. $U(\tau) \in SL(2,\mathbf{C}$. Denoting then the eigenvalues of $U(1)$ with $\exp(\pm\beta u)$, we find:

$$TrU(1) = 2\cosh\left(\frac{\beta u}{2}\right) \ . \tag{151}$$

In general the coefficients u depend on the element of the matrix U (i.e. on the fields $\vec{\Phi}$) in a nontrivial way. Our goal now is to find a *systematic expansion* which gives us $TrU(1)$ as a function of the $\vec{\Phi}$'s¿ This expansion, which has been originally devised by Berry [30], is an expansion in time derivatives and, as we shall see, in our case it is at the same time a low-temperature expansion. We will follow the derivation in [31].

The eigenvalues of the matrix $M(\tau)$ are given by $m\Gamma(\tau)$, where $m = \pm$ and $\Gamma(\tau) = |\vec{\Phi}(\tau)|/2$. Introducing then a unit vector $\hat{h}(\tau) = \vec{\Phi}(\tau)/|\vec{\Phi}(\tau)| \in S^2$ parametrized as $\hat{h}(\tau) \equiv (\sin\theta(\tau)\cos\varphi(\tau), \ \sin\theta(\tau)\sin\varphi(\tau), \ \cos\theta(\tau))$, the instantaneous eigenvectors of M are normalized two-component spinors

$$|m(\tau)\rangle \equiv (z_{1m}(\tau), z_{2m}(\tau)) \ , \ |z_{1m}|^2 + |z_{2m}|^2 \equiv 1 \ . \tag{152}$$

Explicitely:

$$\begin{aligned} |+(\tau)\rangle &= e^{i\chi_+} [\cos(\theta/2), \ e^{i\varphi} \sin(\theta/2)] \\ |-(\tau)\rangle &= e^{i\chi_-} [\sin(\theta/2), \ -e^{i\varphi} \cos(\theta/2)] \ . \end{aligned} \tag{153}$$

Notice that (152) $|m(\tau)\rangle \in S^3 \approx SU(2)$. The overall phases $\chi_\pm$ are of course not determined by the eigenvalue equation for M. In whichever way the phases are fixed, the instantaneous eigenvectors can be employed to construct a matrix $V(\tau)$ that diagonalizes $M(\tau)$, i.e. such that:

$$V(\tau) M(\tau) V^{-1}(\tau) = \Gamma(\tau) \sigma_3 \ . \tag{154}$$

Defining then a new matrix $S(\tau)$ as:

$$S(\tau) = V(\tau)\,U(\tau)\,, \tag{155}$$

the equation of motion obeyed by S turns ou to be:

$$\frac{dS}{d\tau}\,S^{-1} + \Gamma(\tau)\,\sigma_3 = -V\frac{dV^{-1}}{d\tau} \ ; \ \ S(0) = V(0)\,. \tag{156}$$

We may use now the freedom of fixing the overall phases $\chi_\pm$ to meet the requirement that the diagonal elements of the matrix

$$V\frac{dV^{-1}}{d\tau} = \langle m(\tau)|\frac{d}{d\tau}|m(\tau)\rangle\,, \tag{157}$$

in the right hand side of (156) vanish identically. We see then that a neat geometrical picture is emerging: in view of the (periodic) boundary conditions, the unit vector $\hat{h}(\tau)$ traces a closed path γ on the two-sphere S^2. For every choice of the initial phases, eq.ns (153) and (157) fix in a unique way the manner in which $\hat{h}$ "drives" the instantaneous eigenvectors on the three-sphere S^3. What we have achieved in this way is nothing but the well-known [32] *Hopf fibration*, i.e. the principal bundle:

$$U(1) \rightarrow S^3 \rightarrow S^2\,, \tag{158}$$

while the vanishing of $\langle m(\tau)|\frac{d}{d\tau}|m(\tau)\rangle$ is the parallel-transport condition on the Hopf bundle defined by the connection one-form:

$$\widetilde{\omega} = -i\langle u|\,d\,|u\rangle \ ; \ \ |u\rangle \in S^3 \tag{159}$$

Solving (157) and reconstructing $V(\tau)$, we find then:

$$V(\beta) = e^{\frac{i}{2}\,\omega_B(\gamma)\,\sigma_3}\,V(0)\,, \tag{160}$$

where $\omega_B(\gamma)$ is the solid angle associated with the closed path γ on the sphere (the usual 4π ambiguity in the definition of the latter is of no consequence here). In other words, $\omega_B(\gamma)$ is just the *Berry phase* [33] associated with the path γ.

We can solve now (156) by successive iterations. To lowest order, if we keep only diagonal matrix elements (i.e. neglect the r.h.s. altogether), we easily obtain:

$$U(\tau) = V(\tau)^{-1}\,e^{-\int_0^\tau d\tau'\Gamma(\tau')\,\sigma_3}\,V(0)\,, \tag{161}$$

and hence:

$$TrU(1) = 2\,\cosh(u_0) \tag{162}$$

$$u_0 = \frac{i}{2}\,\omega_B(\gamma) + \int_0^1 d\tau\Gamma(\tau)\,. \tag{163}$$

We can now iterate the scheme, writing (156) as

$$\frac{dS}{d\tau}\,S^{-1} = -M_1(\tau) \ ; \ \ M_1(\tau) \equiv \Gamma(\tau)\,\sigma_3 + V(\tau)\frac{V^{-1}}{d\tau}\,, \tag{164}$$

looking for the instantaneous eigenvectors of M_1, fixing phases via the parallel-transport condition and proceeding exactly as before. Clearly the procedure can be iterated in principle to yield a series expansion (actually, as discussed by Berry [30], an asymptotic series) for $\log U(\beta)$.

We quote here only the results that have been obtained in the second step of approximation and in the low temperature limit. One gets [31]:

$$TrU(1) = 2 \cosh(u_1) \tag{165}$$

where now:

$$u_1 = \int_0^1 d\tau \Gamma(\tau) + \frac{i}{2}\omega_B(\gamma) - \frac{1}{8}\int_0^1 d\frac{1}{\Gamma(\tau)}\left|\frac{d\hat{h}}{d\tau}\right|^2 + \mathcal{O}(\beta^{-2}) . \tag{166}$$

The second term in the r.h.s. of (166) is easily recognized to be $\mathcal{O}(\beta^{-1})$, which shows that what we have obtained is a systematic low temperature expansion of the auxiliary field action. Such an expansion is also the expansion in the time derivatives of the $\vec{\Phi}_i$'s we searched for.

Summarizing, we can rewrite the partition function (142) in terms of the effective action:

$$
\begin{aligned}
S_{eff}(\vec{\Phi}) &= S_0(\vec{\Phi}) + S_B(\vec{\Phi}) \\
S_0(\vec{\Phi}) &= \frac{\beta}{2}\int_0^1 \delta\tau \sum_{ij}(J^{-1})_{ij}\vec{\Phi}_i \cdot \vec{\Phi}_j + \sum_i \int_0^1 d\tau\Gamma_i(\tau) \\
&\quad -\frac{1}{8}\int_0^1 d\tau\frac{1}{\Gamma_i(\tau)}\left|\frac{d\hat{h}_i}{d\tau}\right|^2 + \mathcal{O}(\beta^{-2}) \\
S_B(\vec{\Phi}) &= \frac{i}{2}\omega_B(\gamma_i) ,
\end{aligned}
\tag{167}
$$

where γ_i is the closed path traced on S^2 by $\hat{h}_i(\tau)$ for $0 \leq \tau \leq 1$.

VIII   Continuum limit effective action: the nonlinear σ-model

We are ready now to obtain continuum limit action of (167), i.e. to study the low-energy, long-wavelength sector of our model. From now on we will work on a square 2D lattice and assume $J_{ij} \equiv J$ if (i,j) are n.n. and zero otherwise.

Since we want to consider smooth and long wavelength fluctuations around the Néel state (145),we assume

$$\vec{m}_i(\tau) = \frac{1}{2}(-1)^{|i|}\hat{n}_i(\tau) , \tag{168}$$

with the unit vector $\hat{n}_i(\tau)$ being a smooth and slow-varying function of both i and τ, so that

$$\vec{\Phi}_i(\tau) = J \sum_{\vec{\delta}} (-1)^{|i|+1} \hat{n}_{i+\vec{\delta}}(\tau) , \tag{169}$$

with $\vec{\delta} = \delta_x, \delta_y$ connecting a site i to its n.n.'s. We now substitute this ansatz in (167) and, setting also $\vec{\Delta} = \sum_{\vec{\delta}} \hat{n}_{i+\vec{\delta}} - 4\hat{n}_i$, after a long but straightforward calculation we obtain:

$$S_{eff} \approx -J \sum_i \int_0^\beta d\tau \left[\vec{n}_i^2 - 2|\vec{n}_i| \right] + \sum_i \int_0^\beta d\tau \frac{\vec{n}_i \cdot \vec{\Delta}_i}{4} \left(\frac{1}{2} - \frac{1}{|\vec{n}_i|} \right) + \tag{170}$$

$$+ \frac{i}{2} \sum_i \omega_B \left[(-1)^{|i|} \frac{\vec{n}_i}{|\vec{n}_i|} \right] - \frac{1}{16J} \sum_i \int_0^\beta d\tau \left[\frac{d}{d\tau} \left(\frac{\vec{n}_i}{|\vec{n}_i|} \right) \right]^2 \frac{1}{|\vec{n}_i|} .$$

We want now to go to the continuum limit, which amounts to sending the lattice spacing $a \to 0$, while $\sum_i \mapsto a^{(-2)} \int d^2x$ and $\Delta_i \mapsto a^2 \nabla^2 \vec{n}_i$. Notice that the first term in the right hand side of (171) becomes then an integral representation of the delta function $\delta(|\vec{n}_i| - 1)$, so that we can set $|\vec{n}_i| = 1$ in all the remaining terms. After doing so, we are left with a continuum action:

$$S_{lw} = S_{\sigma M} + T_2 \tag{171}$$

where

$$T_2 = \lim_{a \to 0} \left(\frac{i}{2} \right) \sum_i \omega_B \left[(-1)^{|i|} \vec{n}_i \right] \tag{172}$$

and

$$S_{\sigma M} = -\frac{J}{8\hbar} \int d^2r \int_0^\beta d\tau \left[(\nabla \vec{n})^2 + \frac{1}{c^2} (\partial_\tau \vec{n})^2 \right] , \tag{173}$$

is the standard action of the $O(3)$ nonlinear σ-model, with spin-wave velocity $c = \sqrt{2}Ja/\hbar$.

To end up, let us now briefly discuss the extra term T_2 that enters in the expression (172) of S_{lw}. For a one-dimensional lattice the corresponding term

$$T_1 = \lim_{a \to 0} \frac{i}{2} \sum_i \omega_B \left[(-1)^i \vec{n}_i \right] \tag{174}$$

has a simple topological meaning and plays an important role. By splitting the chain into even and odd sites, (175) can be written as

$$T_1 = \lim_{a \to 0} \left(\frac{i}{2} \right) \sum_i{}' \left(\omega_B[\vec{n}_i] - \omega_B[\vec{n}_{i+1}] \right) \approx -\frac{i}{4} \int dx \partial_x \omega_B , \tag{175}$$

where the sum $\sum_i{}'$ is performed over one of the two sublattices of the original (bipartite) one-dimensional lattice. On the other hand, recalling the geometrical meaning of $\omega_B[\vec{n}]$ as the solid angle singled out by the closed path described by $\vec{n}(x, \tau) \equiv (\sin\theta\cos\varphi, \sin\theta\sin\varphi, \cos\theta)$ on S^2, it is immediate to recognize that

$$\omega_B[\vec{n}] = \int_0^1 d\tau(\partial_\tau\varphi)(1 - \cos\theta) , \qquad (176)$$

and therefore:

$$T_1 = \frac{i}{4}\int dx\,d\tau\,\vec{n}\cdot(\partial_\tau\vec{n}\wedge\partial_x\vec{n}) . \qquad (177)$$

This is the famous Haldane "topological term" [34] in the case $S = 1/2$. The integral in the r.h.s. of (178) represents the area of the unit sphere "swept" by the closed path $\vec{n}(x, \tau)$ (at fixed x) when x increases along the 1D-lattice. Owing to the periodic boundary conditions that $\vec{n}(x, \tau)$ must satisfy with respect to both x and τ, it is easy to realize that this area can only be a multiple of the total area of the sphere, i.e. $4\pi N$ and that

$$T_1 = \left(\frac{i}{4}\right)4\pi N = i\pi N , \qquad (178)$$

where the integer N is the "winding number" (or "Pontrjagin index" [32]) of the order parameter field configuration $\vec{n}(x, \tau)$. (179) shows that different "paths" $\vec{n}(x, \tau)$ with different values of the winding number N contribute to the partition function with an extra-phase factor $e^{T_1} = \pm 1$, which is positive (or negative) according that N is even (or odd). This only happens for half-integer spins ($S = \frac{1}{2}$ in our case) and it plays a very important role in setting different behaviours of integer and half-integer quantum spin chains [35].

Turning back to our case of a 2D-square lattice, if we evaluate T_2 in the same way as we did for T_1, we obtain, in place of (176):

$$T_2 = \frac{i}{8}\int dx\,dy(\partial_y\partial_x\omega_B) = -\frac{1}{2}\int dy\partial_y T_1 . \qquad (179)$$

But we have just seen that T_1 is an integer-valued function, which, for slowly varying $\vec{n}(\vec{r}, \tau)$ must be also continuous [36]. We then conclude that $\partial_y T_1 = 0$, and therefore $T_2 = 0$.

Thus, in two-dimensions, the total action in the long wavelength limit is given by the pure non linear σ model term (174). The renormalization group analysis of the NLσM [37, 11] leads to a flow diagram for the renormalized coupling with a fixed point at $T = 0$ (see, e.g., Fig. 13 of [11]) separating a weak coupling "renormalized classical" regime from a strong-coupling "quantum disordered" regime. With some adjustment of the bare parameters (see the discussion in [37]) the theoretical predictions of the NLσM can be made to fit quite nicely the existing experimental data [38].

Acknowledgments. Two of us, G.M. and E.E., would like to thank the organizers of El Escorial School for their invitation and their warm hospitality.

References

[1] J. Hubbard. *Proc. Roy. Soc.* **A276** (1963) 238; **A277** (1964) 237; **A281** (1964) 401.

[2] N. W. Ashcroft and N. D. Mermin, *Solid State Physics*, (Saunders College, 1976).

[3] A. P. Balachandran, E. Ercolessi, G. Morandi, A. M. Srivastava, *The Hubbard Model and Anyon Superconductivity, Lect. Notes in Phys.* **38** (World Scientific, 1990).

[4] E. Ercolessi, G. Morandi and F. Ortolani, *Mod. Phys. Lett.* **B6** (1992) 77.

[5] C. N. Yang, *Phys. Rev. Lett* **63** (1989) 2144; C. N. Yang and S. C. Zhang, *Mod. Phys. Lett.* **B4** (1990) 759 ; S. C. Zhang, *Phys. Rev. Lett.* **65** (1990) 120.

[6] N. D. Mermin and H. Wagner, *Phys. Rev. Lett.* **17** (1966) 1133.

[7] J. E. Hirsch, *Phys. Rev.* **B28** (1983) 4059; **B31**, (1985) 4403.

[8] H. J. Schulz, *Phys. Rev. Lett.* **64** (1990) 1445;
A. R. Bishop, F. Guinea, P. S. Lomdhar, E. Louis and J. A. Vergès, *Europh. Lett.* **14** (1991) 157.

[9] M. Gutzwiller, *Phys. Rev.* A137 (1965) 1726.

[10] K. A. Chao. J. P. Spalek and A.M. Oles, *J. Phys.* **C 10** (1977) L271; *Phys. Rev.* **B 18** (1978) 3453.

[11] E. Galleani d'Agliano, G. Morandi and F. Napoli, in *High Temperature Superconductivity*, M¿ Acquarone ed. (World Scientific, 1996).

[12] J. E. Hirsch, *Phys. Rev. Lett* **54** (1985) 1317;
C. Gros, R. Joynt and T. M. Rice, *Phys. Rev.* **B36** (1987) 8190.

[13] G. Baskaran and P. W. Anderson, *Phys. Rev.* **B37** (1988) 580;
G. Baskaran, Z. Zou and P. W. Anderson, *Solid State Commun.* **63** (1987) 973.

[14] P. Fazekas and P. W. Anderson, *Phil. Mag.* **30** (1974) 432

[15] C. Y. Huang and E. Manousakis, *Phys. Rev.* **B36** (1987) 8302.

[16] G. Kotliar, *Phys. Rev.* **B37** (1988) 3664.

[17] I. Affleck and J. Brad Marston, *Phys. Rev.* **B37** (1988) 3744.

[18] M. Di Stasio, E. Ercolessi, G. Morandi and A. Tagliacozzo, *Phys. Rev.* **B49** (1994) 10908; *Int. J. Mod. Phys.* **B8** (1994) 757;
M. Di Stasio, E. Ercolessi, G. Morandi, A. Tagliacozzo and F. Ventriglia, *Phys. Rev.* **B45** (1992) 1939, *Int. J. of Mod. Phys.* **B7** (1993) 3281; *Phys. Rev.* **B49** (1994) 10908;
M. Di Stasio, E. Ercolessi, G. Morandi, J. Samuel, A. Tagliacozzo and G. P. Zucchelli, in *Superconductivity and Strongly Correlated Electron Systems*, A. Romano and G. Scapino ed.s (World Scientific, 1994).

[19] J. W. Negele and H. Orland, *Quantum Many-Particle Systems*, (Addison Wesley, 1988);
R. Cenni, E. Galleani d'Agliano, F. Napoli, P. Saracco and M. Sassetti, *Feynman Integrals in Theoretical, Nuclear and Statistical Physics*, (Bibliopolis, Napoli 1989).

[20] N. Read and S. Sachdev, *Nucl. Phys.* **B316** (1989) 609.

[21] J. Kogut, *Rev. Mod. Phys.* **51** (1979) 659.

[22] I. Affleck, Z. Zou, T. Hsu and P. W. Anderson, *Phys. Rev.* **B38** (1988) 745.

[23] S. Liang, *Phys. Rev.* **B42** (1990) 6555 and references therein.

[24] V. J. Emery, *Phys. Rev. Lett.* **58** (1987) 2794.

[25] A. W. Sleight, *High Temperature Superconductivity*, Eds. D. P. Tunstall and W. Barford (Adam Hilger, 1991) p. 97;
J. M. Tranquada, D. E. Cox, W. Kunnmann, H. Moudden, G. Shirane, M. Suenage and P. Zolliker, *Phys. Rev. Lett.* **60** (1988) 156.

[26] M. S. Hybertsen et al., *Phys. Rev.* **B41** (1994) 11068;
G. Dopf, A. Muramatsu and W. Hanke, *Phys. Rev.* **B41** (1990) 9264.

[27] A. Muramatsu, R. Zeyher and D. Schmelzer, *Europhys. Lett.* **7** (1988) 473.

[28] T. Aste, E. Galleani d'Agliano and F. Napoli, *Physica* **C182** (1991) 307.

[29] Z. B. Su, L. Yu, J. M. Dong and E. Tosatti, *Z. Physik* **B70** (1988) 131.

[30] M. V. Berry, *Proc. Roy. Soc.* **A414** (1987) 31.

[31] M. Di Stasio, E. Ercolessi, G. Morandi, R. Righi, A. Tagliacozzo and G. P. Zucchelli, *Int. J. of Mod. Phys.* **B8** (1994) 1391.

[32] R. Bott and L. Tu, *Differential Forms in Algebraic Topology* (Springer, 1982);
G. Morandi, *The Role of Topology in Classical and Quantum Physics* (Springer, 1992).

[33] M. V. Berry, *Proc. Roy. Soc.* **A392** (1984) 45;
A. Shapere and F. Wilczek, Eds, *Geometric Phases in Physics*, World Scientific (1989).

[34] E. Fradkin, *Field Theories of Condensed Matter Systems*, (Addison Wesley, 1991);
F. D. M. Haldane, *Phys. Lett.* **93A** (1983) 464; *Phys. Rev. Lett.* **50** (1983);
I. Affleck, *Nucl. Phys.* **B257** (1985) 397.

[35] I. Affleck, *Strings, Fields and Critical Phenomena*, Les Houches Summer School 1988, Eds. E. Brezin and J. Zinn-Justin, (North-Holland, 1990), *J. Phys. Condens. Matter* **1** (1989) 3047.

[36] F. D. M. Haldane, *Phys. Rev. Lett.* **61** (1988) 1029;
T. Dombre and N. Read, *Phys. Rev.* **B38** (1988) 7181.

[37] S. Chakravarty, B. I. Halperin and D. R. Nelson, *Phys. Rev. Lett.*, **60** (1988) 1957; *Phys. Rev.* **B39** (1989) 2344.

[38] R. J. Birgeneau and G. Shirane, *Physical Properties of High Temperature Superconductors*, D. M. Ginsberg ed. (World Scientific, 1989).

A Quantum Critical Trio:
Solvable Models of Finite Temperature Crossovers Near Quantum Phase Transitions

Subir Sachdev

Department of Physics, P.O. Box 208120, Yale University, New Haven, CT 06520-8120
(May 14, 1997)

The physics of three simple solvable quantum models is reviewed in some detail: the spin $1/2$ XX chain (and the related dilute spinless Fermi gas), the Ising chain in a transverse field, and the large N limit of the $O(N)$ quantum rotor model. A unified scaling description is presented, and the many common features among the models are highlighted. The crossovers in these systems are the simplest paradigms of finite temperature physics near quantum critical points.

I. INTRODUCTION

Consider a quantum system on an infinite lattice described by the Hamiltonian $\mathcal{H}(g)$, with g a dimensionless coupling constant. For any reasonable g, all observable properties of the *ground state* of $\mathcal{H}$ will vary smoothly as g is varied. However, there may be special points, like $g = g_c$, where there is a non-analyticity in some property of the ground state: we identify g_c as the position of a quantum phase transition. In finite lattices, non-analyticities can only occur at level crossings; the possibilities in infinite systems are richer as avoided level crossings can become sharp in the thermodynamic limit. In this paper, I will restrict my discussion to second order quantum transitions, or transitions in which the length and time scales over which the degrees of freedom are correlated diverge as g approaches g_c. As I will discuss below, any such quantum transition can be used to define a continuum quantum field theory (CQFT): the CQFT has no intrinsic short-distance (or ultraviolet) cutoff. The main purpose of this paper is to discuss some properties of $\mathcal{H}(g)$ at *finite temperatures* (T) in the vicinity of $g = g_c$, in the context of three simple solvable, but illustrative, models. These studies are equivalent to a determination of the finite T crossovers of the associated CQFTs.

We begin by stating some basic concepts on the relationship between quantum critical points and CQFT's [1–3]; these will also be discussed in more detail in our explicit study of the three models. As correlations become long range in time in the vicinity of the critical point, every system must be characterized by an experimentally measureable energy scale, Δ which vanishes at $g = g_c$. Convenient choices are an energy gap, if one exists, or a stiffness of an ordered phase to changes in the orientation of an order parameter; we will meet several explicit examples later. (More precisely, there are two energy scales Δ_+, Δ_- corresponding to the phases with $g > g_c$, $g < g_c$.) It should be emphasizes that Δ is a dimensionful parameter, expressed in the laboratory units of energy, and directly measurable in an experiment. In all the models we shall consider here, Δ vanishes as a power-law as g approaches g_c:

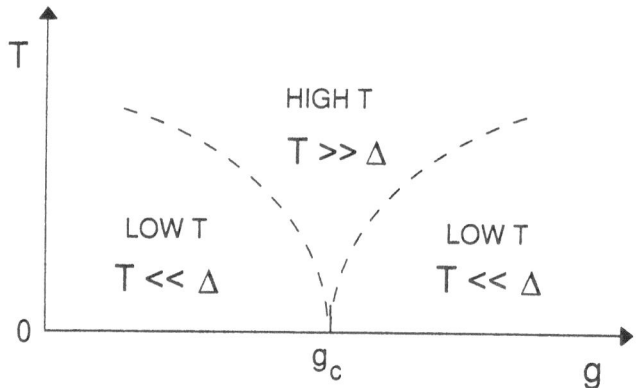

FIG. 1. Schematic phase diagram as a function of the coupling constant of the quantum Hamiltonian g, and the temperature T. The quantum critical point is at $T = 0$, $g = g_c$. The dashed lines indicate crossovers. There may finite temperature phase transitions in either of the two low temperature regimes. The symbol Δ represents a characteristic energy scale of the $T = 0$ theory which vanishes at $g = g_c$ according to 1.1. We will sometime use the symbols Δ_+, Δ_- to distinguish the ground states on either side of the $g = g_c$ point.

$$\Delta \sim \Lambda|g - g_c|^{z\nu} \tag{1.1}$$

where Λ is an ultraviolet cutoff, measured for convenience in the units of energy too, and $z\nu > 0$ is a critical exponent; this exponent is intimately related to the behavior of the system under rescaling transformations, as will be discussed later. From the perspective of a field theorist, the CQFT associated with the quantum critical point is now defined by taking the limit $\Lambda \to \infty$ at fixed Δ; from (1.1) we see that, because $z\nu > 0$, it is possible to take this limit by tuning the bare coupling g closer and closer to the critical point as Λ increases. (A condensed matter physicist would take the complementary, but equivalent, perspective of keeping Λ fixed but moving closer to criticality by lowering his probe frequency $\omega \sim \Delta$). Assuming the $\Lambda \to \infty$ limit exits, the resulting CQFT then contains only the energy scale Δ. At finite temperatures, there is a second energy scale k_BT; its thermodynamic properties will then be a universal function of the only dimensionless ratio available—Δ/k_BT. This paper will describe the physical properties of these universal functions in some detail for three solvable models.

Our study will find two regimes with very different physical properties, as sketched in Fig 1:

(i) *The low temperature region $k_BT \ll \Delta$*

There are actually two regimes of this type, one on either side of g_c. Correlations in this region are similar to those of the $T = 0$ ground state. The low temperature creates a small density of excitations which can sometimes have significant effects at very long scales.

(ii) The high temperature region $\Delta \ll k_B T$

As we are discussing universal properties of the CQFT, it is implicitly assumed that all energy scales, including $k_B T$, are smaller than the upper cutoff Λ which has been sent to infinity; so we also require that $k_B T \ll \Lambda$. The thermal energy, $k_B T$, sets the scale for all physical phenomena in this region, and the system behaves as if it's couplings are at the $g = g_c$ critical point. We shall devote much attention to the unfamiliar and unusual properties of this region. It is perhaps worth noting explicitly why the high T limit of the CQFT can be non-trivial. A conventional high T expansions of the lattice model $\mathcal{H}$ proceeds with the series

$$\text{Tr} e^{-\mathcal{H}/k_B T} = \text{Tr} 1 - \frac{1}{k_B T} \text{Tr} \mathcal{H} + \frac{1}{2(k_B T)^2} \text{Tr} \mathcal{H}^2 + \dots \tag{1.2}$$

The successive terms in this series are well-defined and finite because of the ultraviolet cutoffs provided by the lattice. Further, the series is well-behaved provided T is larger than all other energy scales; in particular we need $k_B T \gg \Lambda$. In contrast, the CQFT was defined by the limit $\Lambda \to \infty$ at fixed $k_B T$, Δ, and, as already stated, the high T limit of the CQFT corresponds to the intermediate temperature range $\Delta \ll k_B T \ll \Lambda$ of the lattice model. It is not possible to access this temperature range by an expansion as simple as (1.2), and more sophisticated techniques, to be discussed here, are necessary.

The three solvable models to be considered are described in the following three sections. In Section II will consider a simple spin chain with nearest neighbor exchange acting equally on only two spin components: this model is also known as the XX chain. We will decribe its relationship to the one-dimensional spinless Fermi gas, and then study the quantum phase transition in an arbitrary dilute spinless Fermi gas in general spatial dimension d. Section III will consider the second solvable model, the Ising chain in a transverse field, which is also in $d = 1$. Section IV will consider quantum transitions solely in $d > 1$: we will describe the $O(N)$ quantum rotor model, which is solvable in the $N = \infty$ limit.

II. THE QUANTUM XX CHAIN AND THE DILUTE SPINLESS FERMI GAS

We will begin our study of quantum phase transitions by examining a simple class of models, most of whose critical properties can be exactly determined. As will become clear, in many ways these models realize the "simplest" quantum phase transition; nevertheless, many of their properties are quite non-trivial.

The main model of interest here is the "XX chain", which is a particular Hamiltonian describing the interactions of spin-$1/2$ degrees of freedom in one dimension. As shown below, it can be solved by an exact mapping to a gas of non-interacting spinless fermions in one dimension. This will lead us to a study of a quantum phase transition exhibited by dilute spinless fermions in d dimensions [3]: its critical properties can determined exactly for *all* d. We emphasize that the two problems being studied here, the XX chain and the dilute spinless Fermi gas, are related only in $d = 1$; the critical properties of the XX model for $d > 1$ are different, and will not be discussed here.

A. The quantum XX chain

The quantum XX chain is described by the Hamiltonian

$$H_{XX} = -J \sum_i \left(\sigma_i^x \sigma_{i+1}^x + \sigma_i^y \sigma_{i+1}^y \right) - h \sum_i \sigma_i^z \tag{2.1}$$

where σ_i^α ($\alpha = x, y, z$) are Pauli matrices describing spin-1/2 degrees of freedom on the sites, i, of an infinite chain. The spins are placed in a uniform magnetic field, $h > 0$, which couples to their z component, and interact with a ferromagnetic exchange, $J > 0$, which couples only their x and y components. This peculiar interaction is designed to ensure solvability of the model; later, we will drop the restriction on the absence of an exchange coupling between z components, and discuss universal properties of a general class of models.

1. Classical Limit

It is useful to begin by examining H_{XX} in the classical limit, in which the σ_i^α are treated as commuting vectors of unit length ($\sum_\alpha \sigma_i^{\alpha 2} = 1$); this will give us a feel for the possible ground states. At zero temperature in such a limit, there are neither quantum nor thermal fluctuations, and the ground state is specified by the static orientations of the spins on each site. As there is no frustration in H_{XX}, it is clear that the lowest energy configuration has all the spins oriented in the same direction, and the ground state always has ferromagnetic long-range order. Let the optimum spin orientation be

$$\vec{\sigma} = (\sin \theta, 0, \cos \theta), \tag{2.2}$$

where we have used the rotation invariance in the x,y plane to set the y-component to zero, and recall that we are momentarily treating the $\vec{\sigma}$ as commuting variables. The ground state energy is then

$$E_{XX} = -N(J \sin^2 \theta + h \cos \theta) \tag{2.3}$$

where N is the number of sites. This is easily minimized at $\theta = \theta_0$, and gives for the expectation value of the z polarization

$$\langle \sigma^z \rangle = \cos \theta_0 = \begin{cases} h/2J & h \leq 2J \\ 1 & h \geq 2J \end{cases} . \tag{2.4}$$

Thus the spins lie in the x,y plane at $h = 0$, gradually tilt upwards into the z direction with increasing field, until they lock onto a complete z polarization for $h \geq 2J$. Ground state properties of the system, when considered as functions of h at fixed J, have a *non-analyticity* at $h = 2J$. This signals the occurence of a critical point with a phase transition. As we will see shortly, the exact solution of the full quantum theory of H_{XX} has a closely related quantum phase transition, whose critical singularities are however different from those of the classical model.

2. Solution of quantum model

The essential tool in the solution of the spin-1/2 model H_{XX} is the Jordan-Wigner transformation [4]. This is a very powerful mapping between models with spin-1/2 degrees of freedom and spinless fermions. The central observation is that there is a simple mapping between the Hilbert space of a system with a spin-1/2 degree of freedom per site, and that of spinless fermions hopping between sites with single orbitals. We may associate the spin up state with an empty orbital on the site, and a spin-down state with an occupied orbital. If the canonical fermion operator c_i annhilates a spinless fermion on site i, then this simple mapping immediately implies the operator relation

$$\sigma_i^z = 1 - 2c_i^\dagger c_i \qquad (2.5)$$

It is also clear that the operation of c_i is equivalent to flipping the spin from down to up, or the operation of $\sigma_i^+ = (\sigma_i^x + i\sigma_i^y)/2$; similar creating a fermion by $c_i^\dagger$ is equivalent to lowering the spin by $\sigma_i^- = (\sigma_i^x - i\sigma_i^y)/2$. While this equivalence works for a single site, we cannot yet equate the fermion operators with the corresponding spin operators for the many site problem; this is because while two fermionic operators on different sites anticommute, two spin operators commute. The solution to this dilemma was found by Jordan and Wigner, who showed that the following representation satisfied both on-site and inter-site (anti)commutation relations:

$$\sigma_i^+ = \prod_{j<i} \left(1 - 2c_j^\dagger c_j\right) c_i$$

$$\sigma_i^- = \prod_{j<i} \left(1 - 2c_j^\dagger c_j\right) c_i^\dagger. \qquad (2.6)$$

The naive single-site correspondence has been modified by a 'string' of operators, whose value is $|1$ (1) if the total number of fermions on the sites to the left of site i are even (odd). Notice that the spin operators have a highly non-local representation in terms of the fermion operators. This feature is also found in the inverse of (2.6)

$$c_i = \left(\prod_{j<i} \sigma_j^z\right) \sigma_i^+$$

$$c_i^\dagger = \left(\prod_{j<i} \sigma_j^z\right) \sigma_i^-. \qquad (2.7)$$

It can now be verified that (2.5,2.6,2.7) are consistent with the relations

$$\left\{c_i, c_j^\dagger\right\} = \delta_{ij} \quad \{c_i, c_j\} = \left\{c_i^\dagger, c_j^\dagger\right\} = 0$$

$$[\sigma_i^+, \sigma_j^-] = \delta_{ij}\sigma_i^z \quad [\sigma_i^z, \sigma_j^\pm] = \pm 2\delta_{ij}\sigma_i^\pm, \qquad (2.8)$$

where the curly brackets represent anticommutators, and square brackets are commutators.

Let us now perform the Jordan-Wigner transformation on the Hamiltonian H_{XX}. Inserting (2.5,2.6) into (2.1), we get

$$H_{XX} = 2 \sum_i \left(-J(c_{i+1}^\dagger c_i + c_i^\dagger c_{i+1}) + h c_i^\dagger c_i \right) - Nh \qquad (2.9)$$

where N is the number of sites. Notice that H_{XX} is simply a free spinless fermion Hamiltonian and its spectrum can therefore be easily determined; indeed, the original form of H_{XX} was carefully chosen to ensure this solvability. The Jordan-Wigner transformation on a general spin Hamiltonian will also lead to terms describing interactions among fermions, which usually makes exact determination of the spectrum impossible; we will discuss the consequences of such terms later. The Hamiltonian H_{XX} is diagonalized by transforming to fermions, c_k, with momentum k

$$c_k = \frac{1}{\sqrt{N}} \sum_i c_i e^{-ikr_i}, \qquad (2.10)$$

where r_i is the spatial coordinate of site i, and H_{XX} becomes

$$H_{XX} = \sum_k (2h - 4J\cos(ka)) c_k^\dagger c_k - Nh, \qquad (2.11)$$

where a is the lattice spacing. The ground state of H_{XX} is obtained by filling all fermion states with negative single particle energy. For $h > 2J$, all single particle states have positive energy, so the ground state is the empty state with no fermions. For $h < 2J$ the fermions occupy states with momenta, k, which satisfy $2h - 4J\cos(ka) < 0$. Using (2.5), we can easily obtain now the exact result for the $T = 0$ expectation value of σ^z:

$$\langle \sigma^z \rangle = \begin{cases} 1 - (2/\pi)\cos^{-1}(h/2J) & h \leq 2J \\ 1 & h \geq 2J \end{cases} . \qquad (2.12)$$

Compare this result with the classical result (2.4), as has been done in Fig 2. There is now a quantum phase transition at $h = 2J$ where the spins first move away from a saturated polarization along the z direction. For h close to $2J$, we now find for the deviation from saturated polarization that $1 - \langle \sigma^z \rangle \sim (1 - h/2J)^{1/2}$, in contrast to the linear dependence $\sim (1 - h/2J)$ in the classical model. We identify the power of $1/2$ as a critical exponent of the quantum critical point at $h = 2J$.

Having identified a quantum critical point and determined one of its exponents in the Hamiltonian H_{XX}, it is now natural to ask how general these results are. In particular, we would like to know if the critical exponent would be modified by exchange interactions among the z components, or by non-nearest neighbor interactions. These questions are addressed in the next section, where we study the vicinity of the critical point in more detail by formulating the appropriate continuum theory.

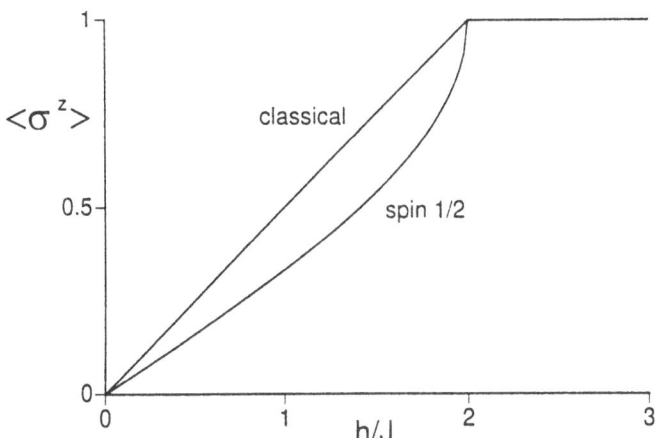

FIG. 2. The $T = 0$ value of $\langle \sigma^z \rangle$ for the Hamiltonian H_{XX} for the classical and spin-1/2 cases. We have $1 - \langle \sigma^z \rangle \sim (1 - h/2J)^{1/2}$ near the $h = 2J$ critical point for all spins less than infinity.

B. The dilute spinless Fermi gas

This section will focus on spinless fermion models described by generalizations of (2.9). We will temporarily not make any reference to the spin system, but instead discuss properties of the quantum phase transition using the fermionic degrees of freedom. We will find it useful to consider the generalization of the transition to spinless fermions moving in an arbitrary number (d) of spatial dimensions [3]; it must be emphasized, however, that the mapping of spinless fermions to the spin chain is valid only in $d = 1$.

Notice that near the quantum critical point $h = 2J$ the fermions only occupy states with momenta near $k = 0$. This correctly suggests that a naive long wavelength expansion in spatial gradients will yield the continuum theory characterizing the critical point. We introduce the continuum spinless fermion field $\Psi_F(x)$ by

$$\Psi(x_i) = a^{-d/2} c_i. \tag{2.13}$$

The factor of $a^{-d/2}$ is in a form involving arbitrary d dimensions, anticipating our discussion of spinless fermions in d dimenions; it ensures that the anticommutation relation

$$\{\Psi(x), \Psi^\dagger(x')\} = \delta^d(x - x') \tag{2.14}$$

is satisfied, where the right hand side is a Dirac delta function in $d = 1$ spatial dimensions. We insert (2.13) into (2.9), perform an expansion in spatial gradients, and obtain the continuum fermion Hamiltonian

$$H_F = E_0 + \int d^d x \left(-\frac{\hbar^2}{2m} \Psi^\dagger(x) \nabla^2 \Psi(x) - \mu \Psi^\dagger(x) \Psi(x) \right) \tag{2.15}$$

where $E_0 = -Nh$ is the ground state energy, the chemical potential $\mu = -2h + 4J$, and the fermion mass $m = \hbar^2/(4Ja^2)$ in the mapping to the $d = 1$ model (2.9). Notice that the quantum critical point is now at $\mu = 0$. The deviation from the saturated magnetization of the quantum XX chain, $1 - \langle \sigma^z \rangle$ is proportional to the fermion density $\langle \Psi^\dagger \Psi \rangle$. At $T = 0$, this is non-zero only for $\mu > 0$, when the fermions all states with momenta less than the Fermi momentum $k_F = (2m\mu/\hbar^2)^{1/2}$. The fermion density is therefore

$$\langle \Psi^\dagger \Psi \rangle = \begin{cases} \mathcal{C}_d (2m\mu/\hbar^2)^{d/2} & \mu > 0 \\ 0 & \mu < 0 \end{cases} \tag{2.16}$$

where $\mathcal{C}_d$ is a dimensionless *universal* number given by $\mathcal{C}_d =$. In $d = 1$, this result is clearly the continuum limit of (2.12), and the sense in which it is universal will be made clearer below.

It is convenient to perform our subsequent scaling analysis in a Lagrangean path integral representation of the dynamic of H_F. Using the standard Grassman path integral of canonical Fermi operators we obtain for the partition function $Z = \mathrm{Tr} e^{-H_F/k_B T}$

$$Z = \int \mathcal{D}\Psi \mathcal{D}\Psi^\dagger \exp\left(-\frac{1}{\hbar} \int_0^{\hbar/k_B T} d\tau d^d x \mathcal{L}_0 \right) \tag{2.17}$$

where the functional integral is over complex Grassman fields Ψ, $\Psi^\dagger$ in space (x) and imaginary time (τ), and the Lagrangean density $\mathcal{L}_0$ is

$$\mathcal{L}_0 = -\hbar \Psi^\dagger \frac{\partial \Psi}{\partial \tau} + \frac{\hbar^2}{2m} \Psi^\dagger \nabla^2 \Psi + \mu \Psi^\dagger \Psi, \tag{2.18}$$

The temperature T appears only in the length $(\hbar/k_B T)$ of the imaginary time interval, and the Grassman fields satisfy antiperiodic temporal boundary conditions $\Psi(x, \tau = \hbar/k_B T) = -\Psi(x, 0)$, and similarly for $\Psi^\dagger$.

The quantum critical point of $\mathcal{L}_0$ is at $\mu = 0$, $T = 0$. The central idea behind our analysis of this critical point is the examination of the behavior of $\mathcal{L}_0$ under a rescaling of length and time scales. In particular, let us coarse-grain the system and express its properties in terms of new length (x') and imaginary time (τ) co-ordinates [5]

$$x' = x e^{-\ell} \qquad \tau' = \tau e^{-z\ell} \tag{2.19}$$

Here $e^{-\ell}$ is the rescaling factor of length scales, and z is known as the *dynamic critical exponent* which determines the relative rescaling of time scales. The value of z will be chosen by us to ensure the scaling invariance of the quantum critical point. Now notice that if we rescale the field Ψ by

$$\Psi'(x',\tau') = e^{d\ell/2}\Psi(x,\tau), \tag{2.20}$$

and choose

$$z = 2, \tag{2.21}$$

then at $\mu = 0$, $\mathcal{L}_0$ is scale invariant; in other words it has the same form in both the primed and unprimed variables. The mass m does not change during this scale transformation: it merely plays the role of fixing the relative units of measurement of time and spatial scales.

This scale invariance has strong consequences for correlators of $\mathcal{L}_0$ which must respect the constraints imposed by (2.19) and (2.20). Although there are no particles in the ground state at $\mu = 0$, many of the correlators of $\mathcal{L}_0$ are non-zero; for instance we have at $T = \mu = 0$

$$G(x, -i\tau) \equiv \langle \mathcal{T}\Psi(x,\tau)\Psi^\dagger(0,0)\rangle = \int \frac{d\omega}{2\pi} \int \frac{d^d k}{(2\pi)^d} \frac{-e^{i(kx-\omega\tau)}}{i\omega - \hbar^2 k^2/(2m)}$$

$$= \theta(\tau) \left(\frac{m}{2\pi\hbar\tau}\right)^{d/2} \exp\left(-\frac{mx^2}{2\hbar\tau}\right) \tag{2.22}$$

where $\mathcal{T}$ is the time-ordering symbol. We have written the temporal argument of the Green's function G as $-i\tau$ to emphasize that it is a correlator in imaginary time. When the above correlator is interpreted as that of Grassmann fields under the functional integral (2.17), the $\mathcal{T}$ symbol can be ignored–it is however required when the Ψ's are treated as operators on the Hilbert space of spinless fermions. The function $\theta(\tau)$ is a step function, and so G vanishes for $\tau < 0$. Physically, (2.22) describes the quantum motion of a free particle, whose propagator has the form of the probability distribution of a diffusing particle in imaginary time.

The invariance of $\mathcal{L}_0$ at $\mu = 0$ under the scaling transformation implies that (2.22) also holds in terms of the primed variables. It can now be checked that these two forms of (2.22) are consistent with (2.19) and (2.20). Indeed the prefactor of the power of τ appearing in (2.22) is the only one consistent with the scaling transformations, and therefore could have been guessed a priori. The exponential depends only upon the combination x^2/τ which is invariant under the scaling transformation, which would also have been consistent with any function of this combination. The step function $\theta(\tau)$ is invariant under arbitrary rescalings of τ and its appearance is only determinable by an explicit calculation.

Let us now move away from the critical point $\mu = 0$. Under the rescaling (2.19,2.20) the action $\mathcal{L}_0$ remains invariant only if we introduce a new chemical potential μ',

$$\mu' = \mu e^{2\ell}. \tag{2.23}$$

Unlike at the critical point, it is now necessary to redefine a coupling constant in $\mathcal{L}_0$ to understand its behavior under scaling transformations. So at a fixed $\mu \neq 0$, the correlators of $\mathcal{L}_0$ are not scale invariant. Nevertheless, the simple behavior of μ under the rescaling transformation does place constraints on the allowed form of its correlators. We also find it useful to consider the consequences of repeated scaling transformations, in which case it is useful to define an ℓ-dependent $\mu(\ell)$ which now satisfies the differential equation

$$\frac{d\mu}{d\ell} = 2\mu \tag{2.24}$$

We see that μ grows indefinitely as one transforms to larger scales (larger ℓ), and such perturbations away from the scale-invariant quantum critical point are known as *relevant* perturbations. It is clear that they destroy scale-invariance at the largest scales and therefore must be included in any theory of the system. The parameter μ thus plays the role of the coupling g in the general discussion of Section I.

This is a convenient place to introduce the concept of the a **scaling dimension** of a coupling constant. This is simply the power to which the length rescaling factor e^ℓ must to raised to obtain its scaling transformation. We will denote the scaling dimension of μ by $\dim[\mu]$, and so clearly,

$$\dim[\mu] = 2 \tag{2.25}$$

It is conventional to define the exponent ν as the inverse of the scaling dimension of the most relevant perturbation about a quantum critical point; in the present case, this will turn out to be μ, and so

$$\nu = 1/2 \tag{2.26}$$

We can also talk about the scaling dimension of an operator, and clearly from (2.20) we have

$$\dim[\Psi] = d/2. \tag{2.27}$$

Finally, we may talk of scaling dimensions of space and time themselves, which are clearly

$$\dim[x] = -1 \qquad \dim[\tau] = -z \tag{2.28}$$

The knowledge of scaling dimensions allows to predict the dependences of correlators on coupling constants or space-time co-ordinates. For instance, the power of τ in (2.22) comes simply from the requirement that the right hand side have a scaling dimension equal to twice that of Ψ. Similarly, the power of μ in (2.16) follows immediately from the values of $\dim[\Psi]$ and $\dim[\mu]$.

Let us also consider the scaling dimension of the free energy density $\mathcal{F}$ of the system. From all subsequent results for $\mathcal{F}$, we will subtract out the ground state energy density at $\mu = 0$; this is always a non-universal number and not dominated by the effects of critical fluctuations (for the $d = 1$ XX chain it equals h/a, while for the continuum spinless Fermi gas it has implicitly been set equal to zero). The free energy density has dimensions of $(\text{length})^{-d}(\text{time})^{-1}$ and so

$$\dim[\mathcal{F}] = d + z \tag{2.29}$$

For $T = 0$, $\mu < 0$, there are no particles present, and so we must have $\mathcal{F} = 0$. However, for $T = 0$, $\mu > 0$, there is non-zero ground state energy, and using the scaling dimensions of $\mathcal{F}$ and μ we expect

$$\mathcal{F} \sim \mu^{(d+z)/2}\theta(\mu) \qquad \text{for } T = 0 \tag{2.30}$$

Indeed, at $T = 0$, the ground state energy can be obtained by adding up the kinetic energy of all the particles in the occupied states with $k < k_F$, and we get

$$\mathcal{F} = \int_{k<k_F} \frac{d^d k}{(2\pi)^d} \frac{\hbar^2 k^2}{2m}$$

$$= \frac{2}{(4\pi)^{d/2}(d+2)\Gamma(d/2)} \left(\frac{2m}{\hbar^2}\right)^{d/2} \mu^{(d+2)/2}\theta(\mu) \quad \text{for } T = 0 \tag{2.31}$$

in agreement with the predictions of the scaling analysis.

We have so far been considering quite a trivial non-interacting fermion problem, and this elaborate machinery of scaling transformation might seem like overkill to the reader. However, we will now address the question of the robustness of the above results to perturbations with interactions to $\mathcal{L}_0$, and show that the above formalism allows us to rapidly, and simply, deduce their consequences.

Let us examine the simplest two-body interaction term that can be added to $\mathcal{L}_0$. A contact interaction like $\int dx (\Psi^\dagger(x)\Psi(x))^2$ vanishes because of the fermion anti-commutation relation, and simplest allowed term is

$$\mathcal{L}_1 = \lambda \left(\Psi^\dagger(x,\tau)\nabla\Psi^\dagger(x,\tau)\Psi(x,\tau)\nabla\Psi(x,\tau) \right) \tag{2.32}$$

where λ is a coupling constant measuring the strength of the interaction. Under the scaling transformation (2.19,2.20) we now find that $\lambda' = e^{-d\ell}\lambda$, or, equivalently

$$\frac{d\lambda}{d\ell} = -d\lambda \qquad \dim[\lambda] = -d \tag{2.33}$$

Unlike μ, we now find that the coupling λ flows to 0 under repeated scaling transformations. This is a property of couplings with negative scaling dimensions, or *irrelevant couplings*, the name suggesting their unimportance for long distance or low energy properties. The analysis of quantum critical points is usually facilitated by setting such couplings to zero at an early stage, and will do this often. Here, let us discuss the consequences of λ a little more explicitly. It is useful to consider, for example, the effect λ would have on the result (2.16). Knowledge of the scaling dimensions allows us to deduce

$$\langle \Psi^\dagger\Psi \rangle \sim \theta(\mu)\mu^{d/2}(1 + c\lambda\mu^{d/2} + \ldots) \tag{2.34}$$

where c is a numerical constant, and the combination $\lambda\mu^{d/2}$ has net scaling dimension zero, as must be the case for consistency under scale transformations. Notice that λ does not modify the leading singularity as a function of μ, and therefore unimportant for small enough μ. A similar argument can be used to show the unimportance of λ for large distances, long times, or low temperatures. We can therefore safely set $\lambda = 0$ in all our subsequent discussion of critical properties of the $\mu = 0$ point.

It is not difficult to extend the above analysis to other perturbations of $\mathcal{L}_0$, and find that, in fact, all of them are irrelevant. Thus, the Lagrangean $\mathcal{L}_0$ is the complete theory of the quantum critical point of the dilute spinless Fermi gas and of all its relevant perturbations, in all spatial dimensions. The scale-invariant quantum critical point is at $\mu = 0$ and one moves away from adding to the action a relevant perturbation whose bare coupling strength is measured by μ. This bare coupling should be distinguished from the energy scale, called Δ in Section I, at which the consequences of the perturbation from the critical point are first apparent; this is the energy scale which will appear in arguments of universal scaling functions describing the deviation from the critical point. In general, this energy scale is a complicated and non-universal function of the bare coupling constant. However, the present model of spinless fermions has the potentially confusing property that this energy scale is also exactly μ. In other words, for this model, $g = \Delta = \mu$. Even the irrelevant couplings, like λ, do not modify the value of the energy scale μ. This remarkable identity between a bare coupling constant and a universal energy scale was dubbed "no scale-factor universality" [6], and is a consequence of the fact that the ground state at the critical point contains no particles. The other models considered later will have scale-factors (*i.e.* the factors appearing in the relatinship between g and Δ) which are non-universal.

Finally, it is worth noting here that the reader should not be misled by the deceptive simplicity of the above results and the apparent triviality of a quantum critical point with no particles. If we examine precisely the same situation for a spin-1/2 Fermi gas, the quantum critical point with no particles turns out to have rather non-trivial properties. In this latter situation it is possible to have a contact interaction $\Psi_\uparrow^\dagger \Psi_\uparrow \Psi_\downarrow^\dagger \Psi_\downarrow$ which is relevant about the non-interacting point for $d < 2$.

1. Non-zero temperatures

The inverse temperature appears as the "length" of the system along the imaginary time direction, and so the scaling dimension of temperature must be the negative of that of time

$$\dim[T] = z \qquad (2.35)$$

Clearly, $k_B T$ is an externally imposed energy scale which breaks scale invariance and must be included in a description of the correlators of the system. The deviation from the quantum critical point is measured, as discussed above, by the energy scale μ. Note that the scaling dimension of μ is also that of energy $\dim[\mu] = 2 = z$, as it must be. By dimensional analysis it now follows that all properties are determined by the dimensionless ratio $\mu/k_B T$, and we are then immediately able to draw the phase diagram shown in Fig 3. There are three limiting regimes, which are separated by smooth crossover denoted by the dashed lines. There is a high temperature regime ($k_B T \gg |\mu|$), and two low temperature regimes ($k_B T \ll \mu$ for

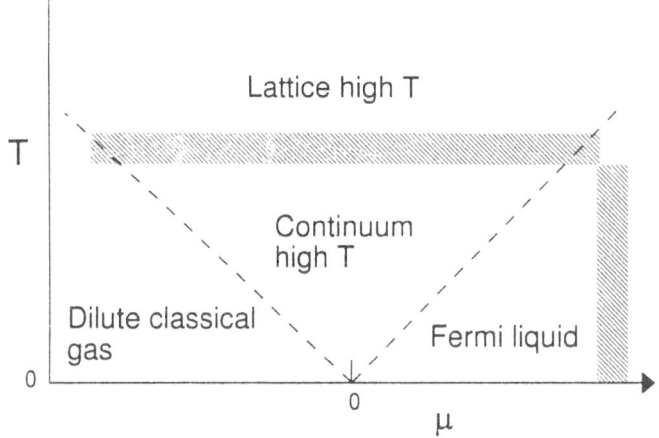

FIG. 3. Phase diagram of the dilute Fermi gas H_F (Eqn (2.15)) as a function of the chemical potential μ and the temperature T. The regions are separated by crossovers denoted by dashed lines, and their physical properties are discussed in Section II B 1. The hatched region marks the boundary of applicability of the continuum theory and occurs at $\mu, T \sim \Lambda$.

$\mu > 0$, and $k_B T \ll -\mu$ for $\mu < 0$). It should be clear that the two low T regimes can be described in terms of a weak thermal perturbation of the $\mu > 0$ and $\mu < 0$ ground states. We will argue below that the high T regime is controlled by the $\mu = 0$ quantum critical point.

Before discussing the nature of the regimes in Fig 3 more explicitly, let us write down the scaling forms which contain the necessary crossovers. We will meet many such scaling forms in our analyses, and they are all deduced from a simple set of rules:

(i) write down the power of T, which has the same scaling dimension as the observable, as a prefactor;

(ii) the scaling function is a dimensionless quantity, and all of its arguments have net scaling dimension 0; again this is ensured by inserting appropriate powers of T;

(iii) finally, insert powers of non-scaling variables like m, k_B, $\hbar$ to ensure consistency in ordinary physical units of measurements.

These rules use T as the primary scaling variable which sets the scale for all other quantities; the advantages of doing this will become clear as we proceed. For the case of the fermion two-point function G, defined in (2.22), these rules give

$$G(x, -i\tau) = \left(\frac{2mk_B T}{\hbar^2}\right)^{d/2} \Phi_G\left(\frac{(2mk_B T)^{1/2}}{\hbar}x, -i\frac{k_B T}{\hbar}\tau, \frac{\mu}{k_B T}\right) \qquad (2.36)$$

The function Φ_G is a scaling function, and everything about it is universal, including its overall scale and those of all its arguments. The reasons for this were discussed above: all corrections due to irrelevant couplings like λ will carry prefactors of higher powers of T and can therefore be neglected at low T near the quantum critical point. (Later we will meet more typical critical points where a non-universal renormalization of bare couplings, dependent on the strength of irrelevant couplings, is required before the scales of the scaling functions become universal).

It is, of course, quite straightforward to obtain an explicit expression for Φ_G from the propator of a single free fermion:

$$\Phi_G(\bar{x}, -i\bar{\tau}, y) = \int \frac{d^d w}{(2\pi)^d} e^{-(w^2 - y)\bar{\tau}} e^{iw\bar{x}} \left(\frac{\theta(\bar{\tau})}{1 + e^{-(w^2-y)}} - \frac{\theta(-\bar{\tau})}{e^{w^2-y}+1} \right) \qquad (2.37)$$

Similar scaling forms hold for all other observables. The fermion density $\langle \Psi^\dagger \Psi \rangle$ is given by $-G(x = 0, -i\tau = 0^-)$, and its scaling form follows from (2.36). Similarly, momentarily returning to the $d = 1$ XX model, let us note that the finite T behavior of the deviation from saturated magnetization $1 - \langle \sigma^z \rangle$ is given by

$$1 - \langle \sigma^z \rangle = -aG(x = 0, -i\tau = 0^-) \qquad (2.38)$$

and so its scaling form also follows from (2.36). The free energy density $\mathcal{F}$ obeys

$$\mathcal{F} = T^{d/2+1} \Phi_{\mathcal{F}} \left(\frac{\mu}{T} \right) \qquad (2.39)$$

where again the prefactor of T follows from the scaling dimension of $\mathcal{F}$ in (2.29). The scaling function $\Phi_{\mathcal{F}}$ is

$$\Phi_{\mathcal{F}}(y) = -\int \frac{d^d w}{(2\pi)^d} \ln(1 + e^{y-w^2}) \qquad (2.40)$$

Let us now discuss physical properties of the above correlators in the three regimes of Fig 3.

(a) Dilute Classical Gas, $k_B T \ll |\mu|$, $\mu < 0$

The ground state for $\mu < 0$ is simply the vacuum with no particles. Turning on a non-zero temperature produces particles with a small non-zero density $\sim e^{-|\mu|/T}$. The de Broglie wavelength of the particles is of order $T^{-1/2}$ which is significantly smaller than the mean spacing between the particles which diverges as $e^{|\mu|/dT}$ as $T \to 0$. This implies that the particles behave classically, obey Boltzmann statistics, and their fermionic nature is unimportant. Because of the rather trivial nature of the ground state in this case, the spacetime dependence of correlation functions like G is rather uninteresting; it does not display the any siginificant crossovers as a function of space or time, as will be the case in the corresponding regime of more typical quantum critical points. We will therefore not discuss it in any detail here.

(b) Fermi Liquid, $k_B T \ll \mu$, $\mu > 0$

The behavior in this regime is quite complex and rich, and illustrates many common, important features of quantum critical points. It is therefore quite useful to study it in some detail.

First it can be argued, *e.g.* by studying asymptotics of the integral in (2.37), that for very short times or distances, the correlators do not notice the consequences of a non-zero T or μ and are therefore given by the $T = \mu = 0$ result in (2.22). More precisely we have

$$G(x, -i\tau) \text{ is given by (2.22) for } |x| \ll \left(\frac{\hbar^2}{2m\mu}\right)^{1/2} \quad |\tau| \ll \frac{\hbar}{\mu} \qquad (2.41)$$

With increasing x or τ, the restrictions in (2.41) are eventually violated and the consequences of a non-zero μ become apparent. Notice that as μ is much larger than $k_B T$, it is the first energy scale to be noticed, and as a first approximation to understand the behavior at larger x we may ignore the effects of $k_B T$.

Let us therefore discuss the ground state for $\mu > 0$. It consists of a filled Fermi sea of particles (a Fermi liquid) with momenta $k < k_F = (2m\mu/\hbar^2)^{1/2}$. An important property of the this state that it permits excitations at arbitrarily low energies *i.e.* it is *gapless*. These low energy excitations correspond to changes in occupation number of fermions arbitrarily close to k_F. As a consequence of these gapless excitations, the line $\mu > 0$ ($T = 0$) is a *line of quantum critical points*, as will become apparent from our analysis below. Each point on the line exhibits an invariance under a scaling transformation, but this transformation is quite different from that describing the quantum-critical end point $\mu = 0$.

We now explain the statements above in some more detail. We are interested here only in x and τ which violate the constraints in (2.41), or alternatively, in excitations above the ground state with momenta near k_F. So let us parametrize, in $d = 1$

$$\Psi(x, \tau) = e^{ik_F x}\Psi_R(x, \tau) + e^{-ik_F x}\Psi_L(x, \tau) \qquad (2.42)$$

where $\Psi_{R,L}$ describe right and left moving fermions, and are fields which vary slowly on spatial scales $\sim 1/k_F = (\hbar^2/2m\mu)^{1/2}$ and temporal scales $\sim \hbar/\mu$. A similar parametrization can be used for $d > 1$ but we will not explicitly discuss it here; most of the results discussed below hold, with small modifications, in all d (see the work by Shankar [7] for more details on a renormalization group analysis of fermions in $d > 1$). Inserting the above parametrization in $\mathcal{L}_0$, and keeping only terms lowest order in spatial gradients, we obtain the "effective" Lagrangian for the Fermi liquid region, $\mathcal{L}_{FL}$:

$$\mathcal{L}_{FL} = \Psi_R^\dagger \left(-\frac{\partial}{\partial \tau} - iv_F \frac{\partial}{\partial x}\right)\Psi_R + \Psi_L^\dagger \left(-\frac{\partial}{\partial \tau} + iv_F \frac{\partial}{\partial x}\right)\Psi_L \qquad (2.43)$$

where $v_F = \hbar k_F/m = (2\mu/m)^{1/2}$ is the Fermi velocity. The Lagrangian $\mathcal{L}_{FL}$ also describes a massless Dirac field in one spatial dimension, and is invariant under relativistic and conformal transformations of spacetime: these facts shall be of some use to us later. Now notice that $\mathcal{L}_{FL}$ is invariant under a scaling transformation, which is rather different from that discussed earlier:

$$
\begin{aligned}
x' &= xe^{-\ell} \\
\tau' &= \tau e^{-\ell} \\
\Psi'_{R,L}(x',\tau') &= \Psi_{R,L}(x,\tau)e^{\ell/2} \\
v'_F &= v_F
\end{aligned}
\tag{2.44}
$$

Notice that this transformation implies scaling dimensions which are different from those discussed earlier for the $\mu = 0$ critical point. This highlights an important point: scaling dimensions of any operator or coupling constant are not its absolute attributes, but always refer to a particular quantum critical point. The above results imply

$$
z = 1,
\tag{2.45}
$$

unlike $z = 2$ (Eqn (2.21)) at the $\mu = 0$ critical point, and $\dim[\Psi] = 1/2$ which actually holds for all d and therefore differs from (2.27). Further notice that v_F, and therefore μ, are now *invariant* under rescaling, unlike the transformation (2.23) at the $\mu = 0$ critical point. Thus now v_F now plays a role rather analogous to that of m at the $\mu = 0$ critical point: it simply the physical units of spatial and length scales. The transformations (2.44) show that $\mathcal{L}_{LF}$ is scale invariant for each value of μ, and we therefore have a line of quantum critical points as claimed earlier. It should also be emphasized that the scaling dimension of interactions like λ will now also change; in particular not all interactions are irrelevant about the $\mu \neq 0$ critical points. We will ignore the ramifications of this complication here.

The action (2.43) and the scaling transformations (2.44) can be considered as defining scaling forms on their on right, independent of any derivation from the original $\mathcal{L}_0$. By complete analogy with the arguments presented earlier, we may now deduce that

$$
G_{R,L}(x,-i\tau) = \left(\frac{k_B T}{\hbar v_F}\right)\phi_{R,L}\left(\frac{k_B T}{\hbar v_F}x, -i\frac{k_B T}{\hbar}\tau\right)
\tag{2.46}
$$

where $G_{R,L}$ are two point correlators of $\Psi_{R,L}$ in an obvious notation, the powers of T follow from the scaling dimensions of G, x, and τ, the factors of v_F, k_B, $\hbar$ merely keep track of physical units, and $\phi_{R,L}$ are universal scaling functions.

What is now the relationship between the original scaling function Φ_G of the $\mu = 0$ critical point in (2.36), and the scaling functions $\phi_{R,L}$ introduced above ? The answer to this question requires introduction of the very useful concept of a **reduced scaling function**; such reduced scaling functions will appear in all three models considered in this article. So it will help the reader to fully absorb this somewhat subtle idea in the present simple setting. Notice that the regime of validity of Φ_G includes that of $\phi_{R,L}$. The latter functions require the additional restrictions $k_B T \ll \mu$, and also the converse of those in (2.41). So it must be the case that the $\phi_{R,L}$ are contained in Φ_G *i.e.* they are reduced scaling functions of Φ_G. Just this requirement allows us to place important restrictions on the functional form of Φ_G in the limit of large μ/k_BT. The scaling forms (2.36) and (2.46), and the mapping (2.42), imply that, for large μ/k_BT, the Greens function takes the form

$$G(x, -i\tau) = e^{ik_Fx}G_R(x, -i\tau) + e^{-ik_Fx}G_L(x, -i\tau), \qquad (2.47)$$

or in terms of the scaling functions Φ_G, $\phi_{R,L}$ we have for large $y = \mu/k_BT$

$$\Phi_G(\bar{x}, -i\bar{\tau}, y) = e^{i\sqrt{y}\bar{x}}\Phi_{GR}(\bar{x}, -i\bar{\tau}, y) + e^{-i\sqrt{y}\bar{x}}\Phi_{GL}(\bar{x}, -i\bar{\tau}, y), \qquad (2.48)$$

and further that

$$\phi_{R,L}(\bar{x}, \bar{\tau}) = \lim_{y\to\infty} 2\sqrt{y}\Phi_{GR,L}(2\sqrt{y}\bar{x}, \bar{\tau}, y). \qquad (2.49)$$

The reasoning we have presented above implies that a non-zero limit of the right-hand-side exists. Notice that the reduced scaling function ϕ_{RL} has one less argument than the primary function Φ_G: this is a characteristic feature of the collapse of one scaling function into another.

Let us now show the above collapse into a reduced scaling function explicitly. For simplicity, we will limit ourselves in this paragraph to $\tau > 0$. Then the full expression for the original Greens function is

$$G(x, -i\tau) = \int_{-\infty}^{\infty} \frac{dk}{2\pi} \frac{e^{ikx}e^{-(\hbar^2k^2/2m-\mu)\tau}}{1 + e^{-(\hbar^2k^2/2m-\mu)/k_BT}} \qquad (2.50)$$

For $|x| \gg (\hbar^2/2m\mu)^{1/2}$, $\tau \gg \hbar/\mu$, and $k_BT \ll \mu$, this integral is dominated by contributions near the Fermi points $k = \pm k_F$. So near k_F let us parametrize $k = k_F + p$, expand terms in the integrand to linear order in p, and to leading order let the integral extend over all real p; a similar procedure can be carried out near $-k_F$. In this manner the above expression for G reduces to

$$G(x, \tau) = e^{ik_Fx} \int_{-\infty}^{\infty} \frac{dp}{2\pi} \frac{e^{p(ix-v_F\tau)}}{1 + e^{-\hbar v_F p/k_BT}} + e^{-ik_Fx} \int_{-\infty}^{\infty} \frac{dp}{2\pi} \frac{e^{p(-ix-v_F\tau)}}{1 + e^{-\hbar v_F p/k_BT}} \qquad (2.51)$$

This result could, of course, also have been obtained direction from $\mathcal{L}_{FL}$, combined with (2.42). The integrals over p can be evaluated exactly and we obtain

$$G_{R,L}(x, \tau) = \left(\frac{k_BT}{\hbar v_F}\right) \frac{1}{2\sin(\pi k_BT(\tau \mp ix/v_F)/\hbar)} \qquad (2.52)$$

This result is clearly consistent with the scaling form (2.46).

Let us now discuss the physical properties of the Green's functions obtained above for the region $k_BT \ll \mu$. The results are schematically indicated in Fig 4. Recall first that at the shortest scales $|x| \ll (\hbar^2/2m\mu)^{1/2}$, $\tau \ll \hbar/\mu$, we get the behavior (2.22) of the groundt state at the $\mu = 0$ critical point. At slightly larger scales, we crossover to behavior characterizing the $\mu > 0$ ground state. In particular for $(\hbar^2/2m\mu)^{1/2} \ll |x| \ll \hbar v_F/k_BT$ or $\hbar/\mu \ll \tau \ll \hbar/k_BT$, we have

Fermi liquid

High T

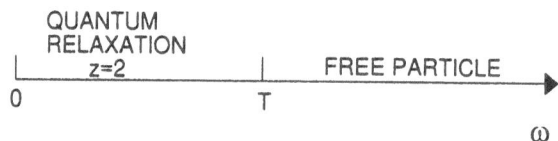

FIG. 4. Crossovers as a function of frequency for the spinless Fermi gas in two of the regimes of Fig 3. The dilute classical gas regime is not shown because it does not have any interesting crossovers in this simple model. The term 'quantum relaxational' refers to the exponential decay of correlations on a scale determined by T: the reason for this nomenclature will be discussed further in Section III D

$$G_{R,L}(x,\tau) = \frac{1}{2\pi\hbar v_F(\tau \mp ix/v_F)/\hbar} \tag{2.53}$$

This is the power law decay characteristic of the $\mu > 0$ critical ground state: note that it is consistent with the scaling transfromations (2.44). At the very largest space and time scales, $|x| \gg \hbar v_F/k_B T$ and real time $|t| \gg \hbar/k_B T$, the effects of a finite temperature became manifest. The gapless thermal excitations above the critical ground state damp the power law correlations and we get exponential decay of correlations; for the equal time correlations we have

$$G_{R,L}(x,\tau = 0) \sim \exp\left(-\frac{\pi k_B T}{\hbar v_F}|x|\right), \tag{2.54}$$

which gives us a correlation length $\xi = \hbar v_F/(\pi k_B T)$. Notice that ξ (which scales like a length) is proportional to $1/T$ (which scales as a time), and this is precisely what is expected from the scaling transformations (2.44) of the $\mu > 0$ ground state. In general, we expect the T dependent correlation length above a critical ground state to scale as $T^{-1/z}$.

Before concluding this subsection, we note an important relationship between (2.52) and (2.53). Observe that it is possible to obtain from the $T = 0$ correlator (2.53), the finite T correlator (2.52) by the mapping

$$v_F\tau \pm ix \rightarrow \frac{\hbar v_F}{\pi k_B T}\sin\left(\frac{\pi k_B T}{\hbar v_F}(v_F\tau \pm ix)\right) \tag{2.55}$$

This is actually an example of a very general and extremely powerful result. It can be shown that the mapping (2.55) in fact relates the $T = 0$ and $T > 0$ values of *any* two-point correlator of $\mathcal{L}_{FL}$; this result is a consequence of conformal symmetry of $\mathcal{L}_{FL}$, which was mentioned in passing earlier. Indeed, the mapping (2.55) relates two-point $T = 0$ and $T > 0$ correlators of any conformally invariant theory whose excitations move with the velocity v_F [8], and we will use it again on several occasions.

(c) High T limit, $k_B T \gg |\mu|$

This is the last, and in many ways the most interesting, region of Fig 3. Now $k_B T$ is the most important energy scale controlling the deviation from the $\mu = 0$, $T = 0$ quantum critical point. It should be emphasized that while the value of $k_B T$ is significantly larger than $|\mu|$, it cannot be so large that it exceeds the limits of applicability for the continuum action $\mathcal{L}_0$. The continuum theory fails above an upper energy scale, Λ, at which the quadratic dispersion $\hbar^2 k^2 / 2m$ is no longer valid, and we must have $k_B T \ll \Lambda$ (see Fig 3). For the simple tight-binding model (2.9) of the XX chain, this means that $\Lambda \sim J$.

We discuss first the behavior of the of the fermion density. In the high T limit of the continuum theory $\mathcal{L}_0$, $|\mu| \ll k_B T \ll \Lambda$ we have from (2.36,2.37) the universal result

$$
\begin{aligned}
\langle \Psi^\dagger \Psi \rangle &= -\left(\frac{2mk_B T}{\hbar^2}\right)^{d/2} \Phi_G(0, 0^-, 0) \\
&= \left(\frac{2mk_B T}{\hbar^2}\right)^{d/2} \int \frac{d^d w}{(2\pi)^d} \frac{1}{e^{w^2} + 1} \\
&= \left(\frac{2mk_B T}{\hbar^2}\right)^{d/2} \zeta(d/2) \frac{(1 - 2^{d/2})}{(4\pi)^{d/2}}
\end{aligned}
\tag{2.56}
$$

This density implies an interparticle spacing which is of order the de Brogie wavelength $= (\hbar^2/2mk_B T)^{1/2}$: thermal and quantum effects are therefore expected to be equally important, and neither dominate.

Let us also consider the fermion density for $T \gg \Lambda$ (the region above the hatched marks in Fig 3), to illustrate the limitations on the continuum description discussed above. Now the result depends upon the details of the non-universal fermion dispersion; on a hypercubic lattice with dispersion $\epsilon_k - \mu$, we obtain

$$
\begin{aligned}
\langle \Psi^\dagger \Psi \rangle &= \int_{-\pi/a}^{\pi/a} \frac{d^d k}{(2\pi)^d} \frac{1}{e^{(\epsilon_k - \mu)/T} + 1} \\
&= \frac{1}{2a^d} - \frac{1}{4T} \int_{-\pi/a}^{\pi/a} \frac{d^d k}{(2\pi)^d} (\epsilon_k - \mu) + \mathcal{O}(1/T^2).
\end{aligned}
\tag{2.57}
$$

The limits on the integration, which extend from $-\pi/a$ to π/a for each momentum component, had previously been sent to infinity in the continuum limit $a \to 0$. In the presence of lattice cutoff, we are able to make a naive expansion of the integrand in powers of $1/T$, and the result therefore only contains negative integer powers of T. Contrast this with the universal continuum result (2.56) where we had non-integer powers of T dependent upon the scaling dimension of Ψ.

We now return to the universal high T region, $|\mu| \ll k_B T \ll \Lambda$, and describe the behavior as a function of temporal and spatial scales. The reader will find it useful to compare the following discussion with the corresponding discussion in the Fermi liquid region discussed in Section II B 1(b). As before, we will consider the behavior of the Greens function $G(x, \tau)$; its leading behavior in the present high T region is obtained simply by setting $y = 0$ in the scaling function in (2.37). The results are schematically sketched in Fig 4. At the shortest scales we again have the free quantum particle behavior of the $\mu = 0$, $T = 0$ critical point

$$G(x, \tau) \text{ is given by (2.22) for } |x| \ll \left(\frac{\hbar^2}{2mk_BT}\right)^{1/2} \quad |\tau| \ll \frac{\hbar}{k_BT}. \tag{2.58}$$

Notice that the limits on x and τ in (2.58) are different from those in (2.41), in that they are determined by $k_B T$ and not μ. The results (2.41) and (2.58) illustrate two very important general principles of quantum critical points:

(i) At short distance and time scales the correlators are those of the quantum critical point

(ii) At larger scales, the relevant perturbations away from the critical point become apparent, but it is the largest energy scale which gets noticed first.

So here thermal effects modify the system *before* it has a chance to feel the effects of μ. In particular, even the sign of μ has a minor effect on the properties of the high T region: *the system does not look like either $\mu > 0$ the Fermi liquid ground state, or the $\mu < 0$ dilute classical gas, at any scale.* Instead for $|x| \gg \left(\hbar^2/2mk_BT\right)^{1/2}$ or $|\tau| \gg \hbar/k_BT$ it crosses over to a novel behavior characteristic of the high T region. We illustrate this by looking at the large x asymptotics of the equal time G in $d = 1$ (other d are quite similar)

$$G(x, 0) = \int \frac{dk}{2\pi} \frac{e^{ikx}}{1 + e^{-\hbar^2 k^2/2mk_BT}} \tag{2.59}$$

For large x this can be evaluated by a contour integration which picks up contributions from the poles at which the denominator vanishes in the complex k plane. The dominant contributions come from the poles closest to the real axis, and gives the leading result

$$G(|x| \to \infty, 0) = -\left(\frac{\pi\hbar^2}{2mk_BT}\right)^{1/2} \exp\left(-(1-i)\left(\frac{m\pi k_BT}{\hbar^2}\right)^{1/2} x\right) \tag{2.60}$$

Thermal effects therefore lead to an exponential decay of equal-time correlations, with a correlation length $\xi = \left(\hbar^2/m\pi k_BT\right)^{1/2}$. Notice that the T dependence is precisely that expected from the exponent $z = 2$ associated with the $\mu = 0$ quantum critical point and the general scaling relation $\xi \sim T^{-1/z}$. The additional oscillatory term in (2.60) is a reminder that quantum effects are still present at the scale ξ, which is clearly of order the de Broglie wavelength of the particles. We will discuss other properties of the high T region of quantum critical points in more detail in other contexts later in this review; in particular, the dynamic correlations are particularly interesting but it appears useful to postpone their analysis to a case where the critical theory has interactions between its degrees of freedom.

We summarize the results of Section II B 1 by drawing the readers attention again to Figs 3 and 4, which show the crossovers as a function of frequency scale in the different regimes of the T, μ plane.

III. THE ISING CHAIN IN A TRANSVERSE FIELD

This section will study another simple solvable model of a quantum phase transition in one dimension. Like the XX model considered in Section II, its solvability is ensured by a mapping to free spinless fermion system via the Jordan Wigner transformation. However unlike the XX model, the present model does possess a quantum phase with true long-range order at $T = 0$, and this leads to important differences in the physics at low temperatures.

An important property of the Hamiltonian (2.1) of the XX model is that it possesses a global, continuous $U(1)$ symmetry: all of the physics is invariant under rotations by an arbitrary angle θ in the XX plane: $\sigma_i^x + i\sigma^y + i \rightarrow e^{i\theta}(\sigma_i^x + i\sigma_i^y)$. Strong low energy fluctuations associated with the orientation of the local ordering in the XY plane prevent the appearance of true long-range order at $T = 0$, $\mu > 0$; instead, correlations of the order parameter decays as $1/\sqrt{x}$ for large x. In this section we will break the continuous $U(1)$ symmetry down to a discrete Z_2 symmetry, and find phases with true long-range order.

The simplest route to accomplishing this is to introduce an XY anisotropy in the exchange couplings; this amounts to replacing the exchange terms in (2.1) by $-\sum_i (J_X \sigma_i^x \sigma_{i+1}^x + J_Y \sigma_i^y \sigma_{i+1}^y)$, while the field coupling to σ_i^z remains unchanged. Such a model with $J_X \neq J_Y$ is also completely solvable, and its properties have been discussed recently [9], following an earlier analysis [10]. This analysis shows that the properties of the special case $J_X \neq 0$, $J_Y = 0$ are in the same universality class as the general $J_X \neq J_Y$. As the computations are somewhat simpler for this special case, we will restrict our attention to it here, and consider

$$H_I' = -J \sum_i \left(g\sigma_i^z + \sigma_i^x \sigma_{i+1}^x \right) \tag{3.1}$$

where we have written it in a form in which $J > 0$ is an overall energy scale, $g > 0$ is a dimensionless coupling constant. As most of us are more accustomed to working with eigenstates of σ^z, rather than σ^x, we will therefore map

$$\sigma^z \rightarrow \sigma^x \quad , \quad \sigma^x \rightarrow -\sigma^z \tag{3.2}$$

(this preserves the commutation relations), and work instead with the completely equivalent

$$H_I = -J \sum_i \left(g\sigma_i^x + \sigma_i^z \sigma_{i+1}^z \right) \tag{3.3}$$

This Hamiltonian H_I is often referred to as the quantum Ising model, or the Ising chain in a transverse field. It is invariant under a global unitary transformation, performed by the operator $\prod_i \sigma_i^x$ under which

$$\sigma_i^z \to -\sigma_i^z \qquad \sigma_i^x \to \sigma_i^x \qquad (3.4)$$

Notice that this global Ising Z_2 symmetry is present in the presence of the transverse field. A longitudinal field, coupling to σ^z would break the Z_2 symmetry.

A. Limiting cases

We begin by examining the spectrum of H_I under strong $(g \gg 1)$ and weak $(g \ll 1)$ coupling [11]. The analysis is relatively straightforward in these limits, and two very different physical pictures emerge. The exact solution, to be discussed later, shows that there is a critical point exactly at $g = 1$, but that the qualitative properties of the ground states for $g > 1$ $(g < 1)$ are very similar to those for $g \gg 1$ $(g \ll 1)$. One of the two limiting descriptions is therefore always appropriate, and only the critical point $g = 1$ has genuinely different properties.

1. Strong coupling $g \gg 1$

Let $|\uparrow\rangle_i$ and $|\downarrow\rangle_i$ denote the eigenstates of σ_i^z. Then $|\pm\rangle_i = (|\uparrow\rangle_i \pm |\downarrow\rangle_i)/\sqrt{2}$ are the eigenstates of σ_i^x. Then at $g = \infty$ the ground state of H_I is clearly determined by the transverse field term to be

$$|0\rangle = \prod_i |+\rangle_i \qquad (3.5)$$

The values of σ_i^z on different sites are totally uncorrelated in this state, and so $\langle 0|\sigma_i^z \sigma_j^z|0\rangle = \delta_{ij}$. Perturbative corrections in $1/g$ will build in correlations in σ^z which increase in range at each order in $1/g$; for g large enough these correlations are expected to remain short-ranged, and the σ^z correlator to decay exponentially with separation. There is thus no magnetic long-range order and this state is a "quantum paramagnet". Notice that this state is invariant under the Z_2 symmetry described above.

What about the excited states ? For $g = \infty$ these can also be listed exactly. The lowest excited states are

$$|i\rangle = |-\rangle_i \prod_{j \neq i} |+\rangle_j, \qquad (3.6)$$

obtained by flipping the state on site i to the other eigenstate of σ^x. All such states are degenerate, and we will refer to them as the "single-particle" states. Similarly, the next degenerate manifold of states are the two-particle states $|i, j\rangle$, obtained by flipping the states at sites i and j, and so on to the general n-particle states. To leading order in $1/g$, we can neglect the mixing between states between different particle number, and just study how the degeneracy within each manifold is lifted. For the one-particle states, the exchange term in H_I leads only to the off-diagonal matrix element

$$\langle i|H_I|i+1\rangle = -J \tag{3.7}$$

which hops the 'particle' between nearest neighbor sites. As in the tight-binding models of solid state physics, the Hamiltonian is therefore diagonalized by going to the momentum space basis

$$|k\rangle = \frac{1}{\sqrt{N}} \sum_i e^{ikx_i}|i\rangle \tag{3.8}$$

where N is the number of sites. This eigenstate has energy (we have choosen an overall constant in H_I to make the energy of the ground state zero)

$$\epsilon_k = Jg\left(2 - (2/g)\cos ka + \mathcal{O}(1/g^2)\right) \tag{3.9}$$

where a is the lattice spacing. The lowest energy one-particle state is therefore at $\epsilon_0 = 2g - 2J$

Now consider the two-particle states. As long as the two particles are well separated from each other, the eigenstate is formed simply by taking the tensor product of two single particle eigenstates. However these particles will collide, and will then have a non-trivial S matrix which mixes the states in the single particle basis. If we exclude the possibility of two-particle bound states (which do not occur here), the total energy of the state is determined by the configuration where the particles are well separated, and is simply the sum of the single particle energies. Thus the energy of a two-particle state with total momentum p is given by $E_p = \epsilon_{p_1} + \epsilon_{p_2}$ where $p = p_1 + p_2$. Notice that for a fixed p, there is still an arbitrariness in the single particle momenta $p_{1,2}$ and so the total energy E_p can take a range of values. There is thus no definite energy momentum relation, and we have instead a 'two particle continuum'. It should be clear, however, that the lowest energy two-particle state in the infinite system (its "threshold") is at $2\epsilon_0$. Similar considerations apply to the n-particle continua, which have thresholds at $n\epsilon_0$.

At next order in $1/g$ we have to account for the mixing between states with differing numbers of particles. Non-zero matrix elements like

$$\langle 0|H_I|i,i+1\rangle = -J \tag{3.10}$$

lead to a coupling between n and $n+2$ particle states. It is clear that these will renormalize the one-particle energies ϵ_p. However qualitative features of the spectrum will not change, and we will still have renormalized one-particle states with a definite energy-momentum relationship, and renormalized $n \geq 2$ particle continua with thresholds at $n\epsilon_0$.

The spectrum described above has simple, but important, consequences for the dynamic spin susceptibility $\chi(p,\omega_n)$. This is defined in imaginary time as the Fourier transform into momentum and frequency of the σ^z correlator:

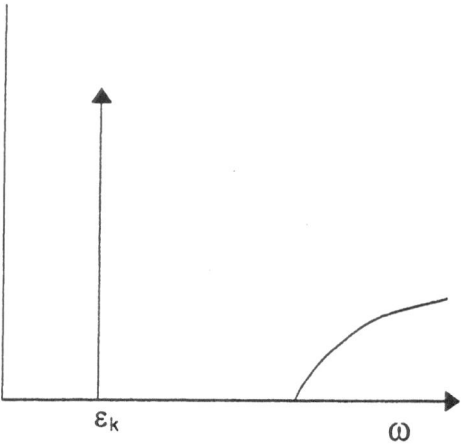

FIG. 5. Schematic of the spectral density $\chi''(k,\omega)$ of H_I as a function of ω at $T = 0$ and a small k. There is a quasiparticle delta function at $\omega = \epsilon_k$, and a three-particle continuum at higher frequencies.

$$\chi(k, i\omega_n) = \int_0^{1/T} d\tau \int dx \, \langle \sigma^z(x, \tau)\sigma^z(0,0)\rangle \, e^{i(kx - \omega_n\tau)} \tag{3.11}$$

Its spectral density $\chi''(p, \omega)$ is the imaginary part of the real frequency $\chi(k, \omega)$, and it given by

$$\chi''(k, \omega) = \pi \sum_a |\langle 0|\sigma^z(k)|a\rangle|^2 \delta(\omega - E_a) \tag{3.12}$$

where the sum over a extends over all the eigenstates of H_I with energy E_a. The eigenstates and energies described above allow us to simply deduce the qualitative form of $\chi''(p, \omega)$ which is sketched in Fig 5. The operator σ^z flips the state at a single site, and so the matrix element in (3.12) is non-zero for the single particle states: only the state with momentum p will contribute, and so there is an infinitely sharp delta function contribution to $\chi''(k, \omega) \sim \delta(\omega - \epsilon_k)$. This delta function is the "quasiparticle peak" and its co-efficient is the quasiparticle amplitude. At $g = \infty$ this quasiparticle peak is the entire spectral density, but for smaller g the quasiparticle amplitude decreases and the multiparticle states also contribute to the spectral density. The mixing between the one and three particle states discussed above, means that the next contribution to $\chi''(p, \omega)$ occurs above the 3 particle threshold $\omega > 3\epsilon_0$; because there are a continuum of such states, their contribution is no longer a delta function, but a smooth function of omega (apart from a threshold singularity), as shown in Fig 5. Similarly there are continua above higher odd number particle thresholds; only states with odd numbers of particles contribute because the matrix element in (3.12) vanishes for even numbers of particles.

2. Weak coupling $g \ll 1$

Now the energy is dominated by the exchange term. There are two degenerate ground states at $g = 0$ with the spins either all up or down (in eigenstates of σ^z):

$$|\uparrow\rangle = \prod_i |\uparrow\rangle_i \qquad |\downarrow\rangle = \prod_i |\downarrow\rangle_i \qquad (3.13)$$

Turning on a small g will mix in a small fraction of spins of the opposite orientation, but the degeneracy will survive as the two states are related to each other by the global Z_2 symmetry noted above (3.4). A thermodynamic system will always choose one or the other of the states as its ground states (which may be preferred by some infinitesimal external perturbation), and hence the Z_2 symmetry will be spontaneously broken. The correlations of the magnetization σ^z have an infinite range in either state as

$$\lim_{|x|\to\infty} \langle \sigma^z(x,0)\sigma^z(0,0)\rangle = N_0^2 \neq 0 \qquad (3.14)$$

The quantity N_0 is the spontaneous magnetization, and equals $\langle \sigma^z \rangle$ in either of the two ground states. All of the statements made in this paragraph clearly hold for $g = 0$, and will hold for some $g > 0$ provided the perturbation theory in g has a non-zero radius of convergence. The exact solution of the model to be discussed later will verify that this is indeed the case.

The excited states can be described in terms of an elementary domain wall (or kink) excitation. For instance the state

$$\cdots |\uparrow\rangle_i |\uparrow\rangle_{i+1} |\downarrow\rangle_{i+2} |\downarrow\rangle_{i+3} |\downarrow\rangle_{i+4} |\uparrow\rangle_{i+5} |\uparrow\rangle_{i+6} \cdots$$

has domain walls, or nearest neighbor pairs of antiparallel spins, between sites $i + 1$, $i + 2$ and sites $i + 4$, $i + 5$. At $g = 0$ the energy of such a state is clearly $2J\times$number of domain walls. The consequences of a small non-zero g are now very similar to those due to $1/g$ corrections in the complementary large g limit: the domain walls become "particles" which can hop and form momentum eigenstates with excitation energy

$$\epsilon_k = J \left(2 - 2g \cos(ka) + \mathcal{O}(g^2)\right). \tag{3.15}$$

The spectrum can be interpreted in terms n-particle scattering states, although it must be emphasized that the interpretation of the particle is now very different from that in the large g limit. Again, the perturbation theory in g only mixes states which differ by even numbers of particles, although now the matrix element in (3.12) is non-zero only for states a with an *even* number of particles; these assertions can easily be checked to hold in a perturbation theory in g. So $\chi''(p, \omega)$ will now have a pole at $p = 0$, $\omega = 0^+$, from the term in (3.12) where $a = $ one of the ground states, indicating the presence of long-range order. Further, there is now no single particle contribution, and the first finite ω spectral density is the continuum above the two particle threshold. The absence of a single particle delta function in this case is a very special feature of the $d = 1$ quantum Ising model, and is not expected to hold for Ising models in higher d.

B. Exact spectrum

The qualitative considerations of the previous section are quite useful in developing an intuitive physical picture. We will now take a different route, and set up a formalism that will eventually lead to an exact determination of many physical correlators; these results will vindicate the approximate methods for $g > 1$, $g < 1$, and also provide an understanding of the novel physics at $g = 1$.

The central idea, as in Section II, is the application of the Jordan-Wigner transformation [4]. We apply (2.5,2.6), while recalling the mapping (3.2), to obtain H_I in the form

$$H_I = -J \sum_i \left(c_i^\dagger c_{i+1} + c_{i+1}^\dagger c_i + c_i^\dagger c_{i+1}^\dagger + c_{i+1} c_i - 2g c_i^\dagger c_i - g\right) \tag{3.16}$$

This fermionic Hamiltonian differs from the one for the XX model (Eqn (2.9)) by terms like $c^\dagger c^\dagger$ which violate the fermion conservation number. So the eigenstates of H_I will not have a definite fermion number. Nevertheless, the new terms are still quadratic in the fermion operators, and H_I can be diagonalized by elementary means. First, use the momentum eigenstates (2.10) to get

$$H_I = J \sum_k \left(2(g - \cos(ka)) c_k^\dagger c_k - i \sin(ka)(c_{-k}^\dagger c_k^\dagger + c_{-k} c_k) - g\right) \tag{3.17}$$

Next, use the Bogoluibov transformation to map into a new set of fermionic operators (γ_k) whose number is conserved. These new operators are defined by a unitary transformation on the pair $c_k, c_{-k}^\dagger$

$$\gamma_k = u_k c_k - i v_k c_{-k}^\dagger, \tag{3.18}$$

where u_k, v_k are real numbers satisfying $u_k^2 + v_k^2 = 1$, $u_{-k} = u_k$, and $v_{-k} = -v_k$. It can now be checked that canonical fermion anticommutation relations for the c_k imply that the same relations are also satisfied by the γ_k *i.e.*

$$\left\{\gamma_k, \gamma_{k'}^\dagger\right\} = \delta_{k,k'} \quad \left\{\gamma_k^\dagger, \gamma_{k'}^\dagger\right\} = \{\gamma_k, \gamma_{k'}\} = 0 \tag{3.19}$$

We also note the inverse of (3.18)

$$c_k = u_k \gamma_k + i v_k \gamma_{-k}^\dagger \tag{3.20}$$

We now insert (3.20) into (3.17), and demand that H_I not contain any terms like $\gamma^\dagger \gamma^\dagger$ which violate conservation of the γ fermions. The as yet undefined constants, u_k, v_k can always be chosen to ensure this: we define $u_k = \cos(\theta_k/2)$, $v_k = \sin(\theta_k/2)$, and a simple calculation then shows that the choice

$$\tan \theta_k = \frac{\sin(ka)}{\cos(ka) - g} \tag{3.21}$$

satisfies our requirements. The final form of H_I is now

$$H_I = \sum_k \epsilon_k (\gamma_k^\dagger \gamma_k - 1/2) \tag{3.22}$$

where

$$\epsilon_k = 2J \left(1 + g^2 - 2g \cos k\right)^{1/2} \tag{3.23}$$

is the single particle energy. As $\epsilon_k \geq 0$, the ground state, $|0\rangle$, of H_I has has no γ fermions and therefore satisfies $\gamma_k |0\rangle = 0$ for all k. The excited states are created by occupying the single particle states; they can clearly be classified by the total number of occupied states, and a n-particle state has the from $\gamma_{k_1}^\dagger \gamma_{k_2}^\dagger \cdots \gamma_{k_n}^\dagger |0\rangle$, with all the k_i distinct.

The above structure of the spectrum confirms the approximate considerations of Section III A. We have now found that the particles are in fact free fermions, and two fermions will not scatter even when they are close to each other; alternatively they can be considered as hard core bosons which have an S matrix of -1 for two-particle scattering; this remarkable feature was not anticipated in the simple approach of Section III A, where we constructed general features of the spectrum by only considering well-separated particles. It is also reassuring to see that the exact single-particle excitation energy (3.23) agrees with (3.9) in the limit $g \gg 1$, and with (3.15) in the limit $g \ll 1$.

C. Continuum theory

The excitation energy ϵ_k in (3.23) is non-zero and positive for all k provided $g \neq 1$. The energy gap, or the minimum excitation energy is always at $k = 0$, and equals $2J|1 - g|$. This gap vanishes at $g = 1$, and it is natural to expect that $g = 1$ is the phase boundary between the two qualitatively different phases discussed in Section III A. Precisely at $g = 1$, fermions with low momenta can carry arbitrarily low energy, and therefore must dominate the low temperature properties. These properties suggest that the state at $g = 1$ is critical, and there is a universal continuum quantum field theory which decribes the critical properties in its vicinity.

We shall now obtain this critical theory using a strategy very similar to that followed in Section II B. As the important excitations are near $k = 0$, we define the continuum Fermi field

$$\Psi(x_i) = \frac{1}{\sqrt{a}} c_i \tag{3.24}$$

We express H_I in terms of Ψ, and expand in spatial gradients. We present the final answer in terms of the Lagrangean $\mathcal{L}_I$, which will enter a Grassman path integral like (2.18)

$$\mathcal{L}_I = -\Psi^\dagger \frac{\partial \Psi}{\partial \tau}^\dagger + \frac{c}{2} \left(\Psi^\dagger \frac{\partial \Psi}{\partial x} - \Psi \frac{\partial \Psi}{\partial x} \right) + \Delta \Psi^\dagger \Psi \tag{3.25}$$

The field Ψ is now implicitly assumed to be a function of space and imaginary time (τ). The coupling constants in (3.25) are given for H_I by

$$\Delta = 2J(1 - g) \qquad c = 2Ja. \tag{3.26}$$

However, the relations (3.26) are very specific to the solvable model H_I. For more complicated, non-solvable, Ising models which have a similar naive continuum limit (e.g. models with second neighbor exchange), the values of Δ and c appearing in the continuum quantum field theory cannot be determined exactly. Indeed Δ and c are parameters which depend upon details of the microscopic Hamiltonian, i.e. they have a non-universal dependence upon a coupling constant like g, as can be checked by a simple estimate of perturbative fluctuation corrections due to irrelevant operators. This is to be contrasted by the behavior of the XX model, where the values of physical parameters like m and μ were given exactly by their naive continuum limit values for a arbitrary microscopic Hamiltonians; recall that this was called "no scale factor" universality, which is absent here.

In general then, determination of Δ and c requires relating them to a physically measurable observable. The continuum theory $\mathcal{L}_I$ can be diagonalized much like the lattice model H_I, and the excitation energy now takes a "relativistic" form

$$\epsilon_k = \left(\Delta^2 + c^2 k^2 \right)^{1/2} \tag{3.27}$$

which shows that $|\Delta|$ is the $T = 0$ energy gap (we will choose the sign of Δ to be different on the two sides of the critical value of g), and c is the velocity of the excitations, both measurable quantities. The form of ϵ_k correctly suggests that $\mathcal{L}_I$ is invariant under Lorentz transformations. This can be made explicit by writing the complex Grassman field Ψ in terms of two real Grassman fields, when the action becomes what is known as the field theory of Majorana fermions of mass Δ/c^2 [12]; we will not explicitly display this here.

It is now easy to see that $\mathcal{L}_I$ is scale-invariant at $\Delta = 0$. The rescaling transformation

$$x' = xe^{-\ell}$$
$$\tau' = \tau e^{-\ell}$$
$$\Psi' = \Psi e^{\ell/2} \tag{3.28}$$

leaves $\mathcal{L}_I$ unchanged in form. The velocity c is taken to be invariant under rescaling, and its role is merely to specify the relative units of space and time; this role is similar to that of m at the $\mu = 0$ critical point of dilute spinless Fermi gas in Section II B. The transformation (3.28) specifies the scaling dimensions

$$z = 1$$
$$\dim[\Psi] = 1/2. \tag{3.29}$$

The coupling Δ is a *relevant* perturbation as it transforms like $\Delta' = e^{\ell}\Delta$, or

$$\dim[\Delta] = 1 \tag{3.30}$$

As shown below, Δ is the only relevant perturbation, and therefore the exponent ν (defined below (2.25) is given by

$$\nu = 1 \tag{3.31}$$

We now show that no other relevant perturbations to $\mathcal{L}_I$ which respect the symmetry (3.4) of the Ising model, and the simplicity of the argument is another illustration of the power of the scaling analysis. There are two different types of perturbations to $\mathcal{L}_I$ that are possible. The first type arises higher spatial gradients in the mapping from the particular Hamiltonian H_I, and the simplest of these is

$$\lambda_1 \Psi^\dagger \frac{\partial^2 \Psi}{\partial x^2}. \tag{3.32}$$

The second type comes from additional terms we could add to H_I, like $\sigma_i^x \sigma_{i+1}^x$, which respect the symmetry (3.4) and are therefore not expected to modify qualitative features of the transition; after the Jordan-Wigner transformation, and expansion in spatial gradients, such a term induces in the continuum

$$\lambda_2 \Psi^\dagger \frac{\partial \Psi^\dagger}{\partial x} \frac{\partial \Psi}{\partial x} \Psi. \tag{3.33}$$

A simple computation now shows that

$$\dim[\lambda_1] = -1 \qquad \dim[\lambda_2] = -2 \tag{3.34}$$

and hence both are irrelevent.

The absence of other relevant perturbations at $\Delta = 0$ implies that $\mathcal{L}_I$ is the universal continuum quantum field theory describing crossovers near the $\Delta = 0$, $T = 0$ quantum-critical point. It is fortunate that this universal theory happens to be expressible as a free fermion model. Although our original motivation for examining H_I was its solvability, the arguments of this section have shown that this choice also happily co-incides with that required for obtaining a universal critical theory.

Let us now compute finite temperature correlators of the free fermion field Ψ. These correlators are not related to any local observable of the Ising chain, and therefore cannot be measured experimentally. Our main purpose in discussing them is to present scaling ideas in a simple context; for simplicity we will restrict ourselves to the critical point $\Delta = 0$, although the results for $\Delta \neq 0$ are easy to obtain. The two-point Ψ correlators can be computed by performing the analog of the lattice Bogoluibov transformation on the continuum theory. We found for imaginary time $\tau > 0$

$$
\begin{aligned}
\langle \Psi(x, \tau)\Psi^{\dagger}(0,0) \rangle &= \frac{1}{2} \int_{-\infty}^{\infty} \frac{dk}{2\pi} \frac{e^{ikx}}{e^{c|k|/T} + 1} \left(e^{c|k|(1/T - \tau)} + e^{c|k|\tau} \right) \\
&= \left(\frac{T}{4c} \right) \left(\frac{1}{\sin(\pi T(\tau - ix/c))} + \frac{1}{\sin(\pi T(\tau + ix/c))} \right).
\end{aligned} \tag{3.35}
$$

We are now using units in which $\hbar = k_B = 1$, and will continue to do so in the remainder of the paper. In a similar manner, we can find

$$
\langle \Psi(x, \tau)\Psi(0,0) \rangle = \left(\frac{iT}{4c} \right) \left(\frac{1}{\sin(\pi T(\tau - ix/c))} - \frac{1}{\sin(\pi T(\tau + ix/c))} \right). \tag{3.36}
$$

The results (3.35,3.38) have precisely the scaling forms that would have been expected under the scaling dimensions in (3.29). At $T = 0$, (3.35) simplifies to

$$
\langle \Psi(x, \tau)\Psi^{\dagger}(0,0) \rangle = \frac{1}{4\pi} \left(\frac{1}{c\tau - ix} + \frac{1}{c\tau + ix} \right), \tag{3.37}
$$

when we notice that the analog of the transformation (2.55)

$$
c\tau \pm ix \rightarrow \frac{c}{\pi T} \sin\left(\frac{\pi T}{c}(c\tau \pm ix) \right) \tag{3.38}
$$

connects the $T = 0$ and $T > 0$ results. This is again due to the conformal invariance of $\mathcal{L}_I$. We will use (3.38) in an important way later in this section.

Correlators of σ^x can be constructed out of those of simple bilinears of the fermion operators, and we will not display them explicitly. More interesting, however, are the correlators of the order parameter σ^z. Computing just the equal time two-point correlator, or even simply the value of $\dim[\sigma^x]$ at the $\Delta = 0$ critical point, involves a rather lengthy and involved computation, which will not be discussed here. Rather, in the next section, we will quote some recently obtained technical results and then proceed to a physical discussion of the crossovers at finite temperature.

D. Finite temperature crossovers

A useful way to begin our of study finite T crossovers in the correlations of the order parameter σ^z is to examine the equal time correlators. These were studied recently in Ref [13], where some old results of McCoy [14] were used to obtain their exact long-distance behavior in the continuum limit. As the computations are somewhat technical, we will not go over them here; we simply quote the final result:

$$\langle \sigma^z(x, \tau = 0)\sigma^z(0,0)\rangle = Z T^{1/4} G_I(\Delta/T)$$

$$\exp\left(-\frac{T|x|}{c} F_I(\Delta/T)\right) \text{ as } |x| \to \infty \tag{3.39}$$

where Z is non-universal constant, and $F_I(s)$ and $G_I(s)$ are universal scaling functions.

The result (3.39) implies that the correlation length ξ obeys

$$\xi^{-1} = \frac{T}{c} F_I\left(\frac{\Delta}{T}\right) \tag{3.40}$$

The form of this result is precisely what could have been expected on general scaling grounds with $\dim[T] = z = 1$, and the length (as for all lengths) $\dim[\xi] = -1$. The velocity c has scaling dimension 0, but is present to ensure the correct engineering dimensions. The function $F_I(s)$ was determined exactly and is given by

$$F_I(s) = \frac{1}{\pi}\int_0^\infty dy \ln \coth \frac{(y^2 + s^2)^{1/2}}{2} + |s|\theta(-s) \tag{3.41}$$

Despite appearances, the function $F_I(s)$ is smooth as a function of s for all real s, and is analytic at $s = 0$. The analyticity at $s = 0$ is required by the absence of any thermodynamic singularity at finite T for $\Delta = 0$. This is a key property, which was in fact used to obtain the answer in (3.41). The exact expression for the function $G_I(s)$ is also known

$$\ln G_I(s) = \int_s^1 \frac{dy}{y}\left[\left(\frac{dF_I(y)}{dy}\right)^2 - \frac{1}{4}\right] + \int_1^\infty \frac{dy}{y}\left(\frac{dF_I(y)}{dy}\right)^2, \tag{3.42}$$

and its analyticity at $s = 0$ follows from that of F_I. We show a plot of F_I and G_I in Fig 6.

The prefactor Z in (3.39) is a non-universal constant, as noted earlier. For the solvable H_I, we chose the overall normalization of G_I such that $Z = J^{-1/4}$. In general, the value of Z is set by relating it to an observable, as we will show below. Also note that Z has no dependence on Δ, and is therefore non-singular at the quantum critical point.

The power of T is front of (3.39), and the value $z = 1$, tell us

$$\dim[\sigma^z] = 1/8. \tag{3.43}$$

Armed with the knowledge of the scaling dimensions, we can write down the full scaling form for the time-dependent σ^z correlator, which applies to the lattice model in the limits $\Lambda \sim\to \infty$, $a \to 0$ at fixed Δ, c and T

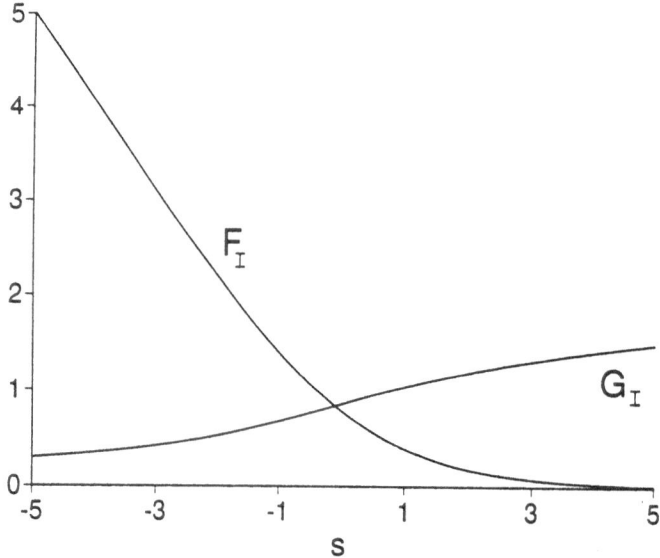

FIG. 6. The crossover functions for the correlation length (F_I) and the amplitude (G_I) as a function of $s = \Delta/T$.

$$\langle \sigma^z(x,\tau)\sigma^z(0,0)\rangle = ZT^{1/4}\Phi_I\left(\frac{Tx}{c}, T\tau, \frac{\Delta}{T}\right) \tag{3.44}$$

where Φ_I is a universal function which is analytic as a function of its third argument $s = \Delta/T$ on the real s axis. The result (3.39) obviously specifies Φ_I for large Tx/c and $\tau = 0$.

We will now describe the physical properties of Φ_I in different regions of the phase diagram as a function of g and T, sketched in Fig 7.

1. Low T on the ordered side, $\Delta > 0$, $T \ll \Delta$

This is the "renormalized classical" [15] region of Fig 7, and the reasons for this name will become clear below.

Assuming that it is valid to interchange the limits $T \to 0$ and $x \to \infty$ in (3.39), we can use the limiting values $F_I(\infty) = 0$, $G_I(s \to \infty) = s^{1/4}$ to deduce that (recall (3.14):

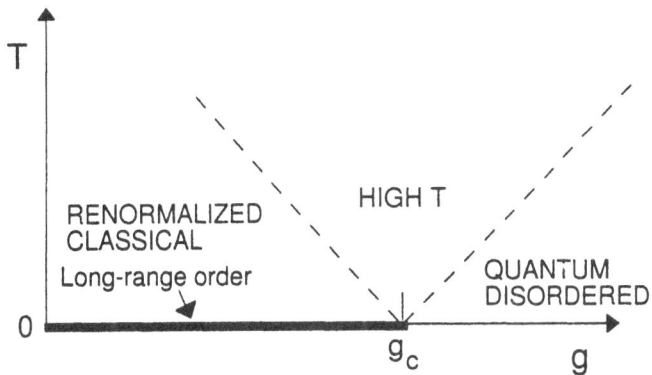

FIG. 7. Finite T phase diagram of the $d = 1$ quantum Ising model, H_I, as a function of the coupling g and temperature T. Long-range order ($N_0 = \langle \sigma_z \rangle \neq 0$) is present only for $T = 0$ and $g < g_c$. The energy scale $\Delta \sim g_c - g$, and dashed lines are crossovers at $|\Delta| \sim T$

$$N_0^2 \equiv \lim_{|x| \to \infty} \langle \sigma^z(x, 0) \sigma^z(0, 0) \rangle = Z \Delta^{1/4} \qquad \text{at } T = 0. \tag{3.45}$$

Thus, as claimed earlier, there is long-range order in the $g < 1$ ground state of H_I, with the order parameter $N_0 \langle \sigma^z \rangle = Z^{1/2} \Delta^{1/8}$ (this relates the value of Z to a physical observable). For small $T \ll \Delta$, we obtain from the large s behavior of $F_I(s)$ (see (3.41)) that

$$\langle \sigma^z(x, 0) \sigma^z(0, 0) \rangle = N_0^2 e^{-|x|/\xi_c} \qquad \text{large } |x|, \tag{3.46}$$

where the correlation length

$$\xi_c^{-1} = \left(\frac{2\Delta T}{\pi c^2} \right)^{1/2} e^{-\Delta/T}. \tag{3.47}$$

We have put a subscript c on the correlation length to emphasize that the system is expected to behave *classically* in this low temperature region [13]. The excitations above the ground states consists of particles (the kinks and anti-kinks of Section III A) whose mean separation ($\sim \xi_c$) is much larger than their de Broglie wavelengths ($\sim (c^2/\Delta T)^{1/2}$, as the mass of these particles $= \Delta/c^2$), which is precisely the canonical condition for the applicability of classical physics. More generally, we expect classical models to apply near any phase with true long-range order. It is also reassuring to note that (3.46) is precisely the form of equal-time correlations in the classical Ising model at low T. The prefactor N_0^2 is the true ground state magnetization including the effects of quantum fluctuations, and this is the reason for the adjective "renormalized" in the name for this region.

We now conjecture that the mathematical structure of the emergence of classical physics in the scaling form (3.44), is analogous to the appearance of Fermi liquid behavior in the dilute spinless Fermi gas, as discussed in Section II B(b). This means that the function Φ_I should collapse into a *reduced scaling function*, Φ_{Ic}, in the limit $\Delta \ll T$, and further that Φ_{Ic} should be obtainable completely from an effective classical model. Using the form (3.46), it is easy to see that

$$\langle \sigma^z(x,t)\sigma^z(0,0)\rangle = N_0^2 \Phi_{Ic}\left(\frac{x}{\xi_c}, \gamma\frac{t}{\xi_c^{z_c}}\right), \tag{3.48}$$

where t now denotes *real* physical time. The dynamics of the classical Ising model determines the dynamic exponent z_c; there is no fundamental scaling constraint which states that z_c must equal the dynamic exponent $z = 1$ of the $\Delta = 0$ quantum critical point, just as the dynamic exponent of the Fermi liquid scaling ($z = 1$, Eqn (2.45)) was different from that the $\mu = 0$ critical point of the dilute spinless Fermi gas ($z = 2$, Eqn (2.21)) in Section II B. The function Φ_{Ic} is contained wholly within the universal function Φ_I, and so the prefactor γ must also be universal; on purely dimensional grounds we expect

$$\gamma = c\left(\frac{c}{\Delta}\right)^{z_c-1}\left(\frac{\Delta}{T}\right)^\rho \mathcal{R} \tag{3.49}$$

where $\mathcal{R}$ and ρ are universal numbers.

This is as far as general scaling ideas will take us; further progress requires specification of the classical dynamics of the effective model and its solution. This has very recently been done by A.P. Young and the author, who obtained the full space-time dependent function Φ_{Ic} and the values of $\mathcal{R}$ and ρ exactly. The reader is referred to Ref [16] for more details. The effective classical dynamics of the spins is *relaxational*, but an earlier conjecture by the author [13] that the relaxational dynamics is described by the Glauber model [17] is incorrect.

2. Low T on the disordered side, $\Delta < 0$, $T \ll |\Delta|$

This is the "quantum disordered" region of Fig 7.

Now we need to take the $s \to -\infty$ limit of the functions $F_I(s)$, $G_I(s)$; from these limits we find

$$\langle \sigma^z(x,0)\sigma^z(0,0)\rangle = \frac{ZT}{|\Delta|^{3/4}}e^{-|r|/\xi} \quad |x| \to \infty \text{ at fixed } 0 < T \ll |\Delta|, \tag{3.50}$$

with the correlation length ξ given by

$$\xi^{-1} = \frac{|\Delta|}{c} + \left(\frac{2|\Delta|T}{\pi c^2}\right)^{1/2} e^{-|\Delta|/T} \tag{3.51}$$

So correlations decay exponentially on a scale $\sim c/|\Delta|$, and there is no long-range order. We have a similar result for equal time correlations at $T = 0$, although the limits $T \to 0$ and $|x| \to \infty$ do not commute for the prefactor of the exponential decay. The $T = 0$ result is determined by the form-factor expansion technique [18], which we will not discuss here; it yields [19,20]

$$\langle \sigma^z(x,0)\sigma^z(0,0)\rangle = Z|\Delta|^{1/4}\left(\frac{c}{2\pi|\Delta||x|}\right)^{1/2} e^{-\Delta|r|/c} \quad |x| \to \infty \text{ at } T = 0. \quad (3.52)$$

The form factor expansion also yields the $T = 0$ dynamic susceptibility (defined in (3.11)). The leading term is in fact precisely the quasiparticle pole at energy $\epsilon_k = (c^2k^2 + \Delta^2)^{1/2}$ that was argued to exist in this phase in Section III A. We have

$$\chi(k,\omega) = \frac{2Z\Delta^{1/4}}{c^2k^2 + \Delta - (\omega + i\epsilon)^2} + \ldots \quad T = 0 \quad (3.53)$$

where ϵ is a positive infinitesimal. The imaginary part of this gives the delta function sketched in Fig 5; the continuum of excitations above the three particle threshold come from higher order terms in the form factor expansion, and are respresented by the ellipses in (3.53). It can now be checked that the Fourier transform of (3.53) yields the leading term (3.52) in the equal time correlation function.

The result (3.53) shows that the *quasiparticle residue* is $Z\Delta^{1/4}$ (this is another relation between Z and a physical measureable, and along with (3.45), it implies a relationship between the values of the residue and N_0 as we approach the critical point from either side). The residue vanishes at the critical point $\Delta = 0$, where the quasiparticle picture breaks down, and we will have a completely different structure of excitations.

The above is an essentially complete description of the correlations and excitations of the quantum paramagnetic ground state. At finite T, there will be a small density of quasi-particle excitations which will behave classical for the same reasons as in Section III D 1: their mean spacing is much larger than their de Broglie wavelength. The motion of these thermally excited particles leads to a small dissipative broadening for the quasi-particle delta function at a spatial scales of order the mean spacing between the particles. The details of the classical relaxational dynamics leading to this broadening have recently been worked out by A.P. Young and the author, and the reader is referred to Ref [16] for further details.

3. High T, $T \gg |\Delta|$

Right at the critical point, $\Delta = 0$, this regime extends all the way down to $T = 0$. We begin by writing the $T = 0$ equal-time correlator of the continuum theory; from the scaling dimension of σ^z this must have the form

$$\langle \sigma^z(x,0)\sigma^z(0,0)\rangle \sim \frac{1}{(|x|/c)^{1/4}} \quad \text{at } T = 0, \Delta = 0, \quad (3.54)$$

We will now fix the prefactor in (3.54) using our earlier results. The key ingredient is our knowledge that the underlying continuum model $\mathcal{L}_I$ is is relativistically and conformally invariant at the $T = 0$ critical point. Further we assume that the mapping (3.38) between correlators at $T \neq 0$ from those at $T = 0$ holds also for the two point correlator. of σ^z Then from (3.54) we must have at $T \neq 0$

$$\langle \sigma^z(x, \tau) \sigma^z(0, 0) \rangle$$

$$\sim T^{1/4} \frac{1}{[\sin(\pi T(\tau - ix/c)) \sin(\pi T(\tau + ix/c))]^{1/8}} \qquad \text{at } \Delta = 0. \qquad (3.55)$$

Let us now use this result in the equal-time case in the regime $xT/c \gg 1$. Precise results for this regime where quoted earlier in (3.39), where using the values $F_I(0) = \pi/4$ (from evaluation of (3.41)) and $G_I(0) = 0.858714569\ldots$ we have

$$\langle \sigma^z(x, \tau = 0) \sigma^z(0, 0) \rangle =$$

$$Z T^{1/4} G_I(0) \exp\left(-\frac{\pi T |x|}{4c}\right) \qquad \text{as } |x| \to \infty \text{ at } \Delta = 0 \qquad (3.56)$$

Finally, comparing with (3.55) we obtain

$$\langle \sigma^z(x, \tau) \sigma^z(0, 0) \rangle =$$

$$Z T^{1/4} \frac{2^{-1/8} G_I(0)}{[\sin(\pi T(\tau - ix/c)) \sin(\pi T(\tau + ix/c))]^{1/8}} \qquad \text{at } \Delta = 0. \qquad (3.57)$$

As expected, this result is of the scaling form (3.44), and indeed completely determines the function Φ_I for the case where its last argument is zero.

Now let us turn to a physical interpretation of the main result (3.57). Consider first the case $T = 0$. By a Fourier transformation of the $T = 0$ limit of (3.57) we obtain the dynamic susceptibility

$$\chi(k, \omega) = Z (4\pi)^{3/4} G_I(0) \frac{\Gamma(7/8)}{\Gamma(1/8)} \frac{c}{(c^2 k^2 - (\omega + i\epsilon)^2)^{7/8}} \qquad T = 0, \Delta = 0 \quad (3.58)$$

We plot $\text{Im}\chi(k, \omega)/\omega$ in Fig 2. Notice that there are no delta functions in the spectral density like there were in the quantum disordered phase (Fig 5), indicating the absence of any well-defined quasiparticles. Instead, we have a critical continuum of excitations.

Now let us turn to non-zero T. We Fourier transform (3.57) to obtain $\chi(k, i\omega_n)$ at the Matsubara frequencies ω_n and then analytically continue to real frequencies (there are some interesting subtleties in the Fourier transform to $\chi(k, i\omega_n)$ and its analytic structure in the complex ω plane, which are discussed elsewhere [6]). This gives us the leading result for $\chi(k, \omega)$ in the high T region

$$\chi(k, \omega) = \frac{Zc}{T^{7/4}} \frac{G_I(0)}{4\pi} \frac{\Gamma(7/8)}{\Gamma(1/8)}$$

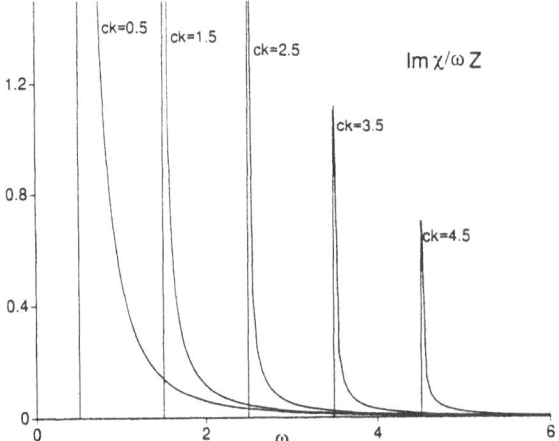

FIG. 8. Spectral density, $\mathrm{Im}\chi(k,\omega)/\omega Z$, of H_I at its critical point $g = 1$ ($\Delta = 0$) at $T = 0$, as a function of frequency ω, for a set of values of k.

$$\frac{\Gamma\left(\frac{1}{16} + i\frac{\omega + ck}{4\pi T}\right)\Gamma\left(\frac{1}{16} + i\frac{\omega - ck}{4\pi T}\right)}{\Gamma\left(\frac{15}{16} + i\frac{\omega + ck}{4\pi T}\right)\Gamma\left(\frac{15}{16} + i\frac{\omega - ck}{4\pi T}\right)} \qquad \Delta = 0. \tag{3.59}$$

We show a plot of $\mathrm{Im}\chi/\omega$ in Fig 9. This result is the finite T version of Fig 8. Notice that the sharp features of Fig 8 have been smoothed out on the scale T, and there is non-zero absorption at all frequencies. For $\omega, k \gg T$ there is a well-defined peak in $\mathrm{Im}\chi/\omega$ (Fig 9) rather like the $T = 0$ critical behavior of Fig 8. However, for $\omega, k \ll T$ we cross-over to the **quantum relaxational** regime [21] and the spectral density $\mathrm{Im}\chi/\omega$ is similar to (but not identical) a Lorentzian around $\omega = 0$. This relaxational behavior can be characterized by a relaxation rate Γ_R defined as

$$\Gamma_R^{-1} = -i\left.\frac{\partial \ln \chi(0,\omega)}{\partial \omega}\right|_{\omega=0} \tag{3.60}$$

(this is motivated by the phenomenological relaxational form $\chi(0,\omega) = \chi_0/(1 - i\omega/\Gamma_R + \mathcal{O}(\omega^2))$). From (3.59) we determine:

$$\Gamma_R = \left(2\tan\frac{\pi}{16}\right)\frac{k_B T}{\hbar}, \tag{3.61}$$

where we have returned to physical units. At the scale of the characteristic rate Γ_R, the dynamics of the system involves intrinsic quantum effects (responsible

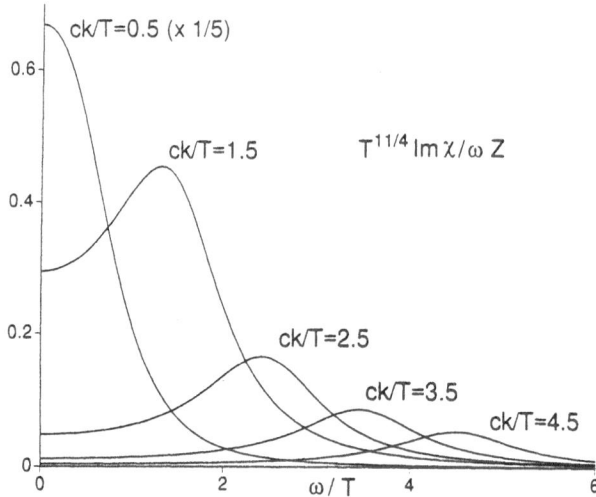

FIG. 9. The same observable as in Fig 8, $T^{11/4}\mathrm{Im}\chi(k,\omega)/\omega Z$, but for $T \neq 0$. This is the leading result for $\mathrm{Im}\chi$ for $T \gg |\Delta|$ *i.e.* in the high T region of Fig 7. All quantities are scaled appropriately with powers of T, and the absolute numerical values of both axes are meaningful.

for the non-Lorentzian lineshape) which cannot be neglected; description by an effective classical model (as was appropriate in both the renormalized classical and quantum disordered regions of Fig 7) would require that $\Gamma_R \ll k_B T/\hbar$, which is thus not satisfied in the high T region of Fig 7. The ease with which (3.61) was obtained belies its remarkable nature. Notice that we are working in a closed Hamiltonian system, evolving unitarily in time with the operator $e^{-iH_I t/\hbar}$, from an initial density matrix given by the Gibbs ensemble at a temperature T. Yet, we have obtained relaxational behavior at low frequencies, and determined an exact value for a dissipation constant. Such behavior is more typically obtained in phenomenological models which couple the system to an external heat bath and postulate an equation of motion of the Langevin type.

4. Summary

The main features of the finite temperature physics of the quantum Ising model are summarized in Figs 7 and Fig 10. At short enough times or distances in all three regions of Fig 7, the systems displays critical fluctuations characterized by the

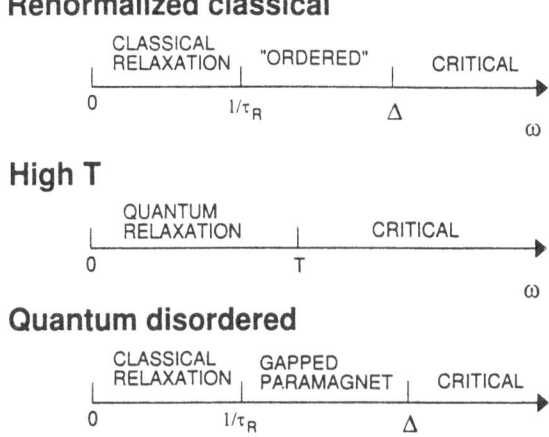

FIG. 10. Crossovers as a function of frequency for the Ising model in the different regimes of Fig 7. The correlations in the two classical relaxational regimes are quite different from each other. The "ordered" regime is in quotes, because there is no long-range order, and the system only appears ordered between spatial scales c/Δ and ξ_c. The reader is referred to Ref [16] for a precise specification of the frequency scales $\tau_R \sim e^{|\Delta|/T}$.

dynamic susceptibility (3.58). The regions are distinguished by their behaviors at the low frequencies and momenta. In both the low T regimes of Fig 7 (renormalized classical and quantum disordered), the long time dynamics is relaxational and is described by effective classical models. In contrast, the dynamics in the high T region is also relaxational, but involves quantum effects in an essential way, as was described above.

IV. QUANTUM ROTORS BETWEEN ONE AND THREE DIMENSIONS

This section is our first foray into the examination of quantum critical points in greater than one dimension. Much of the technology and the physical ideas introduced earlier for $d = 1$ will generalize rather straightforwardly to higher dimensions, although we will no longer be able to obtain exact results for crossover functions. The characterization of the physics in terms of three regions separated by smooth crossovers, the high T and the two low T regions on either side of the quantum critical point, will continue to be extremely useful in higher dimensions, and will again be the basis of our discussion. We will however, meet several genuinely new features. It is now possible for systems to have a thermodynamic phase transition at a non-zero temperature. We shall be particularly interested in the interplay between the critical singularties of the finite temperature transition and those of the quantum critical point. The concept of a 'reduced scaling function', introduced earlier in the discussion of the free Fermi gas, and used again to describe renormalized classical behavior in Section III D 1, will turn out to be precisely what is needed to deal with this intricate phenomenon.

The $O(N)$ quantum rotor model is defined by the following Hamiltonian on the sites i of a regular d dimensional lattice:

$$H_R = \frac{Jg}{2} \sum_i \overleftrightarrow{L}^2{}_i - J \sum_{\langle ij \rangle} \vec{n}_i \cdot \vec{n}_j, \tag{4.1}$$

where the sum $\langle ij \rangle$ is over nearest neighbors, $J > 0$ is an overall energy scale, and $g > 0$ is a dimensionless coupling constant. The N component vectors The N-component vectors $\vec{n}_i$, with $N \geq 2$, are of unit length, $\vec{n}_i^2 = 1$, and represent the orientation of the rotors on the surface of a sphere in N-dimensional rotor space. The operators $L_{i\mu\nu}$ ($\mu < \nu$, $\mu, \nu = 1 \ldots N$) are the $N(N-1)/2$ components of the angular momentum $\overleftrightarrow{L}_i$ of the rotor: the first term in H_R is the kinetic energy of the rotor with $1/g$ the moment of inertia. The different components of $\vec{n}_i$ constitute a complete set of commuting observables and the state of the system can be described by a wavefunction $\Psi(\vec{n}_i)$. The action of $\overleftrightarrow{L}_i$ on Ψ is given by the usual differential form of the angular momentum

$$L_{i\mu\nu} = -i \left(n_{i\mu} \frac{\partial}{\partial n_{i\nu}} - n_{i\nu} \frac{\partial}{\partial n_{i\mu}} \right). \tag{4.2}$$

The commutation relations among the $\overleftrightarrow{L}_i$ and $\vec{n}_i$ can now be easily deduced. We emphasize the difference of the rotors from Heisenberg-Dirac quantum spins: the components of the latter at the same site do not commute, whereas the components of the $\vec{n}_i$ do.

There is a strong analogy between the rotor Hamiltonian H_R in (4.1) and the Ising Hamiltonian H_I in (3.3). We will be looking at the transition between a magnetically ordered state with $\langle \vec{n} \rangle \neq 0$ and $O(N)$ symmetry broken, and a quantum paramagnet in which correlations of $\vec{n}$ are short ranged. As in the Ising model, it is the exchage term, proportional to J, that favors the ordered state, while the 'kinetic energy', proportional to Jg leads to fluctuations in the orientation of the order parameter and eventually to loss of long-range order. The similarity between the two models will also be apparent in the strong (large g) and weak coupling (small g) analyses in the following section.

A. Limiting cases

The pictures which emerge in the following two perturbative analyses are expected to hold on either side of a quantum critical point at $g = g_c$.

1. Strong coupling $g \gg 1$

At $g = \infty$, the exchange term in H_R can be neglected, and the Hamiltonian decouples into independent sites, and can be diagonalized exactly. The eigenstates on each site are the eigenstates of $\overleftrightarrow{L}^2$; for $N = 3$ these are the states

$$|\ell, m\rangle_i \quad \ell = 0, 1, 2, \ldots, \quad -\ell \leq m \leq \ell \tag{4.3}$$

and have eigenenergy $Jg\ell(\ell + 1)/2$. Compare this single site spectrum with that of a pair of Heisenberg-Dirac spin S quantum spins with an antiferromagnetic exchange K; for a suitably chosen K we get the same sequence of levels and energies but with a maximum allowed value of $\ell = 2S$. Assuming this upper cutoff is not crucial for the low energy physics, we can use a single quantum rotor is an effective model for a *pair* of spins.

The ground state of H_R in the large g limit consists of the quantum paramagnetic state with $\ell = 0$ on every site:

$$|0\rangle = \prod_i |\ell = 0, m = 0\rangle_i \tag{4.4}$$

Compare this with strong coupling ground state (3.5) of the Ising model. Indeed, the remainder of the strong coupling analysis of Section III A can borrowed here for the rotor model, and we can therefore be quite brief. The lowest excited state is a 'particle' in which a single site has $\ell = 1$, and this excitation hops from site to site. An important difference from the Ising model is that this particle is three-fold degenerate, corresponding to the three allowed values m. The dynamic susceptibility has a quasiparticle pole at the energy of this particle, and odd particle continua above the three particle threshold.

2. Weak coupling, $g \ll 1$

At $g = 0$, the ground state breaks $O(N)$ symmetry, and all the $\vec{n}_i$ vectors orient themselves in a common, but arbitrary direction. Excitations above this state consist of 'spin waves' which can have an arbitrarily low energy. This is a crucial difference from the Ising model, in which there was an energy gap above the ground state. The presence of gapless spin excitations is a direct consequence of the continuous $O(N)$ symmetry of H_R: we can make very slow deformations in the orientation of $\langle \vec{n} \rangle$, and get an orthogonal state whose energy is arbitrarily close to that of the ground state. Explicitly, for $N = 3$, and a ground state polarized along $(1, 0, 0)$ we parametrize

$$\vec{n}(x, t) = (1, \pi_1(x, t), \pi_2(x, t)) \tag{4.5}$$

where $|\pi_1|, |\pi_2| \ll 1$, and look at the linearized equations of motion for π_1, π_2. A standard calculation then gives harmonic spin waves with energy $\omega = ck$ (c is the spin wave velocity); their wavefunctions can then be contructed using harmonic oscillator states.

B. Continuum theory and large N limit

To obtain the path integral representation of the quantum mechanics of H_R, we interpret the $\vec{n}_i$ as the co-ordinates of particles constrained to move on the surface of a sphere in N dimensions: we can then simple use the standard Feynman path integral representation of single-particle quantum mechanics. After taking the continuum limit, such a procedure gives

$$Z = \int \mathcal{D}\vec{n}(x,\tau)\delta(\vec{n}^2(x,\tau) - 1) \exp\left(-\int_0^{1/T} d\tau \int d^dx \mathcal{L}\right)$$

$$\mathcal{L} = \frac{N}{2c\tilde{g}}\left[c^2\left(\frac{\partial\vec{n}}{\partial x_i}\right)^2 + \left(\frac{\partial\vec{n}}{\partial\tau}\right)^2\right] \tag{4.6}$$

Here $c \sim Ja$ is a velocity which will turn out to be the spin wave velocity, and we have set $\hbar = k_B = 1$. The prefactor of N is for future convenience. The coupling constant $\tilde{g} \sim ga^{d-1}$ has the dimensions of $(length)^{d-1}$; we will not use the original g in H_R further in this discussion, and we will drop the tilde in $\tilde{g}$ from now. The above action is valid only at long distances and times, so there is an implicit cutoff above momenta of order $\Lambda \sim 1/a$ and frequencies of order $c\Lambda$. Our main interest here shall be the universal physics at scales much smaller than Λ.

This section shall present a study of the model (4.6) in the limit of a large number of components of $\vec{n}$ $i.e.$ at $N = \infty$ [21]. The model is exactly soluble in this limit, and displays an interesting quantum phase transition. Most features of the finite temperature crossovers turn out not to be artifacts of the $N = \infty$ point, and thus provide a useful introduction to quantum critical points in dimensions $d > 1$. Some dynamic properties are however not correctly captured at $N = \infty$, and require loop corrections which are discussed in the literature [21].

The framework of the $N = \infty$ solution is quite easy to set up, at least in the phase without long range order in the order parameter $\vec{n}$; we will consider the case with long range order later in this chapter. We rescale the $\vec{n}$ field to

$$\vec{\tilde{n}} = \sqrt{N}\vec{n}, \tag{4.7}$$

and impose the $\vec{\tilde{n}}^2 = N$ constraint by a Lagrange multiplier, λ. The action is then quadratic in the $\vec{\tilde{n}}$ field, which can then be integrated out to yield

$$Z = \int \mathcal{D}\lambda(x,\tau)$$

$$\exp\left[-\frac{N}{2}\left(\text{Tr}\ln(-c^2\partial_i^2 - \partial_\tau^2 + i\lambda) - \frac{i}{cg}\int_0^{1/T} d\tau \int d^dx\lambda\right)\right] \tag{4.8}$$

The action now has a prefactor of N, and the $N = \infty$ limit of the functional integral is therefore given exactly by its saddle point value. We assume that the saddle-point value of λ is space and time independent, and given by $i\lambda = \sigma^2$. The saddle-point equation determining the value of the parameter σ^2 is

$$\int^\Lambda \frac{d^d k}{(2\pi)^d} T \sum_{\omega_n} \frac{1}{c^2 k^2 + \omega_n^2 + \sigma^2} = \frac{1}{cg}, \tag{4.9}$$

where the sum over ω_n extends over the Matsubara frequencies $\omega_n = 2n\pi T$, n integer. It is also not difficult to see that the retarded dynamic susceptibility $\chi(k,\omega)$, which is the two-point correlator of the n field, is given by

$$\chi(k,\omega) = \frac{cg/N}{c^2 k^2 - (\omega + i\epsilon)^2 + \sigma^2}, \tag{4.10}$$

at $N = \infty$. The Eqns (4.9,4.10) are the central results of the $N = \infty$ theory, and most of the remainder of this chapter will be spent on analyzing their consequences. In spite of their extremely simple structure, these equations contain a great deal of information, and it takes a rather subtle and careful analysis to extract the universal information contained in them [21]. We will display all the details, as similar ideas can be used in a variety of contexts. We will begin by characterizing the $T = 0$ ground states, and then turn to the finite temperature crossovers.

C. Zero temperature

At $T = 0$, we can make use of the relativistic invariance of the action (4.6) to simplify our analysis. The summation over Masubara frequencies in (4.9) turns into an integral, and after introducing spacetime momentum $p \equiv (k, \omega/c)$, the constraint equation (4.9) becomes

$$\int^\Lambda \frac{d^{d+1} p}{(2\pi)^d} \frac{1}{p^2 + (\sigma/c)^2} = \frac{1}{g} \tag{4.11}$$

The integral on the left hand side increases monotonically with decreasing σ, and has its maximum finite value at $\sigma = 0$. It is then clear that there is no solution to (4.11) for $g < g_c$ where

$$\int^\Lambda \frac{d^{d+1} p}{(2\pi)^d} \frac{1}{p^2} = \frac{1}{g_c} \tag{4.12}$$

For $g \geq g_c$ there is a unique solution of the saddle-point equation (4.11), which describes a quantum paramagnetic ground state: we will study its properties in the following subsections for $g > g_c$ and $g = g_c$, and find that they are quite similar to those of the Ising chain. Determination of the ground state for $g \leq g_c$ requires a reanalysis of the derivation of the large N saddle equation. This will be done in Section IV C 3, where we find a state with magnetic long-range order and spontaneous breakdown of the $O(N)$ symmetry.

1. Quantum paramagnet, $g > g_c$

Subtract (4.11) from (4.12), and obtain

$$\frac{1}{g_c} - \frac{1}{g} = \int \frac{d^{d+1}p}{(2\pi)^d} \left(\frac{1}{p^2} - \frac{1}{p^2 + (\sigma/c)^2} \right)$$
$$= X_{d+1}(\sigma/c)^{d-1}, \tag{4.13}$$

where the constant $X_d \equiv 2\Gamma((4-d)/2)(4\pi)^{-d/2}/(d-2)$. In the last equation, we have used the fact that the integral was convergent at large momenta to send the cut-off Λ to infinity. This result gives us the value of σ for all $g \geq g_c$.

A key step in the analysis of any ground state of a continuum theory, is the determination of an energy scales which characterizes it. In this case, the quantum paramagnet has a gap, Δ_+, given by

$$\Delta_+ = \sigma \qquad T = 0. \tag{4.14}$$

We emphasize that, by definition, the gap Δ_+ is a temperature independent quantity, and equals the temperature dependent value of σ only at $T = 0$. The presence of a gap is apparent in the structure of the spectral density $\chi''(k, \omega)$, which from (4.10) is given by

$$\chi''(k, \omega) = \mathcal{A} \frac{\pi}{2\sqrt{c^2 k^2 + \Delta_+^2}}$$

$$\left(\delta(\omega - \sqrt{c^2 k^2 + \Delta_+^2}) - \delta(\omega + \sqrt{c^2 k^2 + \Delta_+^2}) \right) \tag{4.15}$$

which has weight only at frequencies greater than Δ. The spectral weight appears entirely in the form of delta functions which indicate the presence of magnon quasiparticles; the quantity $\mathcal{A} = cg/N$ is the quasi-particle residue. This magnon is obviously the same as the three-fold degenerate particle that appeared earlier in the strong-coupling analysis of the $O(3)$ model. The n-particle continua ($n \geq 3$, odd) are absent here in the $N = \infty$ theory, but appear at higher orders in $1/N$.

Also justifying our identification of this phase as a quantum paramagnet, is that equal-time n correlations decay exponentially in space

$$\frac{1}{N} \langle \vec{n}(x, 0) \cdot \vec{n}(0, 0) \rangle = \mathcal{A} \int \frac{d^{d+1}p}{(2\pi)^{d+1}} \frac{e^{ip \cdot x}}{p^2 + (\Delta_+/c)^2}$$
$$= \frac{\mathcal{A}}{2(2\pi)^{d/2}(\Delta_+/c)^{(2-d)/2}} \frac{e^{-x\Delta/c}}{x^{d/2}} \tag{4.16}$$

which identifies Δ/c as the inverse correlation length. Notice the close similarity of these results to those in Section III D 2 on the Ising model, where $2Z\Delta^{1/4}$ played the role of the quasiparticle residue, $\mathcal{A}$.

2. Critical point, $g = g_c$

As g approaches g_c from above, the energy gap, Δ_+, vanishes as

$$\Delta_+ \sim (g - g_c)^{1/(d-1)} \tag{4.17}$$

The critical state at $g = g_c$ turns out to be scale-invariant at scales much longer than Λ^{-1}, as expected by analogy with the Ising model. The coupling g is the parameter which tunes the system away from this scale-invariant point, and as Δ is an energy (inverse time) scale, the result (4.17) identifies the exponent

$$z\nu = \frac{1}{d-1} \tag{4.18}$$

The equal-time correlations now decay as

$$\langle \vec{n}(x,0) \cdot \vec{n}(0,0) \rangle \sim \int \frac{d^{d+1}p}{(2\pi)^{d+1}} \frac{e^{ikx}}{p^2}$$

$$\sim \frac{1}{x^{d-1}} \tag{4.19}$$

which is a power-law, as expected for a scale-invariant theory; the decay as a function of time has the same exponent, and so

$$z = 1, \tag{4.20}$$

as must be the case for a Lorentz-invariant theory. The application of the scaling transformation on (4.19), also tells us that

$$\dim[\vec{n}] = \frac{d-1}{2} \tag{4.21}$$

The value of the exponent z is exact, as it is fixed by Lorentz invariance of the critical theory, but the values of ν and $\dim[\vec{n}]$ will have corrections at order $1/N$. In general, it is conventional to parametrize

$$\dim[\vec{n}] = \frac{d+z-2+\eta}{2}, \tag{4.22}$$

with η the "anomalous dimension" of the field. Comparing with (4.21) we see that $\eta = 0$ in the $N = \infty$ theory. The exact solution of the Ising chain had $\eta = 1/4$, as that gives $\dim[\sigma^z] = 1/8$. A non-zero, positive, value of η will appear upon consideration of fluctuation corrections, and has important physical consequences. In particular, notice that in the present $N = \infty$ theory, the quasiparticle residue $\mathcal{A}$ was non-zero all the way upto $g = g_c$. A standard scaling argument, which demands the consistency of an expression like (4.15) with the scaling dimension of $\vec{n}$, tells us that we must have

$$\mathcal{A} \sim (g - g_c)^{\eta\nu}, \tag{4.23}$$

i.e. the quasiparticle residue vanishes the system approaches the critical point. Again this scaling is consistent with the Ising model in which $\mathcal{A} = 2Z\Delta^{1/4} \sim (g - g_c)^{1/4}$.

If there are no quasiparticles, what do the excitations look like ? As in the Ising chain, there is a critical continuum of excitations, whose spectral density is determined by η. The dynamic susceptibility has the form

$$\chi(k, \omega) \sim \frac{1}{(c^2 k^2 - \omega^2)^{1-\eta/2}} \tag{4.24}$$

and its imaginary part looks much like Fig 8. The $\eta = 0$ case is of course special, in that the spectral density has a single delta function at $\omega = ck$, and the critical excitations have a particle-like nature: this is clearly an artifact of the $N = \infty$ theory, and is its major failing.

3. Magnetically ordered ground state, $g < g_c$

Our analysis so far has shown no meaningful solution of the saddle-point equations in the large N limit for $g < g_c$. The culprit for this shortcoming lies in the step before (4.8), where we indiscriminately integrated out all N components of the $\vec{n}$ field [22]. As we expect a magnetically ordered phase to appear for $g < g_c$, it seems sensible to allow for the possibility that fluctuations of $\vec{n}$ along the direction of the ordered ground state will be different from those orthogonal to it. So we write

$$\vec{n} = (\sqrt{N} r_0, \pi_1, \pi_2 \ldots \pi_{N-1}), \tag{4.25}$$

where it is assumed that the order parameter is polarized along the 1 direction. Inserting this and (4.7) into (4.6), imposing the constraint with a Lagrange multiplier λ, and integrating out *only* the $\pi_{1...N-1}$ fields, we find

$$Z = \int \mathcal{D}\lambda \mathcal{D}r_0$$

$$\exp\left[-\frac{N-1}{2} \operatorname{Tr} \ln(-c^2 \partial_i^2 - \partial_\tau^2 + i\lambda) + \frac{iN}{cg} \int_0^{1/T} d\tau \int d^d x \lambda(1 - r_0^2) \right] \tag{4.26}$$

In the large N limit, we can ignore the difference between $N-1$ and N, and obtain the saddle point equations with respect to variations in λ *and* r_0. As before, σ^2 is taken to be the saddle-point value of $i\lambda$. The mean value of r_0 will be the spontaneous magnetization at $N = \infty$, which we denote by N_0; so

$$N_0 = \langle n_1 \rangle. \tag{4.27}$$

The saddle point equations are

$$N_0^2 + g \int^\Lambda \frac{d^{d+1}p}{(2\pi)^{d+1}} \frac{1}{p^2 + (\sigma/c)^2} = 1$$
$$\sigma^2 N_0 = 0 \qquad (4.28)$$

where we have set $T = 0$. One solution of the second equation is $N_0 = 0$, but then the first equation for σ becomes identical to the one considered earlier, and is known to fail for $g < g_c$. So we choose the other solution, where

$$\sigma = 0$$
$$N_0^2 = 1 - g \int^\Lambda \frac{d^{d+1}p}{(2\pi)^{d+1}} \frac{1}{p^2}$$
$$= 1 - \frac{g}{g_c} \qquad (4.29)$$

It is satisfying to find that N_0 is non-zero precisely for $g < g_c$, reinforcing our belief in the correctness of our procedure in finding the saddle point. Notice that N_0 vanishes as $(g_c - g)^{1/2}$ as g approaches g_c. It is conventional to define the critical exponent β by the dependence $N_0 \sim (g_c - g)^\beta$, and we therefore have $\beta = 1/2$ in the present $N = \infty$ theory. More generally, the scaling dimension of N_0 must be the same as the scaling dimension of $\vec{n}$, and we therefore have from (4.22) that

$$2\beta = (d + z - 2 + \eta)\nu, \qquad (4.30)$$

an exponent relation that is satisfied by the $N = \infty$ theory.

The above approach also determines the two-point correlator of spin components orthogonal to the axis of the spontaneous magnetization. We denote the corresponding susceptibility by $\chi_\perp(k, \omega)$, and it is the Fourier transform of the n_2, n_2 correlator (say); we have at $N = \infty$

$$\chi_\perp(k, \omega) = \frac{cg/N}{c^2 k^2 - (\omega + i\epsilon)^2} \qquad (4.31)$$

Notice that there is a quasiparticle pole at $\omega = ck$, and the energy of this excitation vanishes as $k \to 0$. These are the spin-wave excitations discussed earlier in the weak-coupling analysis. These spin waves survive fluctuation corrections as $k \to 0$, although the nature of the spectral density becomes different at larger k, as we will discuss shortly.

As was the case on the disordered side, we need an energy scale to characterize the ordered ground state, and its distance from the critical point. A convenient choice is to build an energy out of the **spin stiffness**, ρ_s. This quantity is a measure of how easy it is to make smooth changes in the order parameter orientation. Imagine, if instead of the uniform condensate $\langle \vec{n} \rangle = N_0(1, 0, 0, 0, \ldots)$ the system would choose on its own, we constrain the magnetization to precess smoothly in the $1 - 2$ plane (say)

$$\langle \vec{n} \rangle = N_0(\cos \varphi(x), \sin \varphi(x), 0, 0, \ldots) \qquad (4.32)$$

where $\varphi(x)$ is a very slowly varying function of x. A constant $\varphi(x)$ cannot change the ground state energy of the constrained system, so the change in energy can depend only $\nabla\varphi(x)$. By inversion symmetry in x, the change cannot be linear in $\nabla\varphi$, and so the lowest order term in the change in energy has to be of the form

$$\delta E = \frac{\rho_s}{2} \int d^d x (\nabla\varphi)^2 \qquad (4.33)$$

The coefficient appearing in the expression above is defined to be the spin stiffness ρ_s. We emphasize that the this stiffness is defined by changes in the ground state energy, and will always be assumed to be a $T = 0$ quantity, unless otherwise stated. The dimension of the stiffness under the scaling transformation of the $g = g_c$ point can now be easily deduced. The angle φ is a variable with period 2π, and must therefore necessary be dimensionless. By definition we have $\dim[\delta E] = z$, and therefore

$$\dim[\rho_s] = d + z - 2 \qquad (4.34)$$

We can now construct the energy scale, which we denote Δ_-, which characterizes ground state for $g < g_c$. The requirement is that Δ_- should have scaling dimension z, and physical units of $(\text{time})^{-1}$. Such an object has to made out of powers of ρ_s, whose scaling dimension is above and whose physical units are $(\text{length})^{2-d}(\text{time})^{-1}$, and the velocity c, whose scaling dimension is 0 and physical units $(\text{length})(\text{time})^{-1}$; the unique combination is

$$\Delta_- \equiv (\rho_s/N)^{1/(d-1)} c^{(d-2)/(d-1)}. \qquad (4.35)$$

The factor of N has been chosen for future convenience.

Knowledge of the spin stiffness allows us to make an exact statement on the form of the static transverse susceptibility $\chi_\perp(k, 0)$ in the limit $k \to 0$. This susceptibility is the response of the system to a very slowly varying static field, $h(x)$, which couples linearly to the 2 component (say). The system will respond to such an external field by a slowly varying shift in the angular orientation of the order parameter, and the net energy cost with then be

$$\delta E = \int d^d x \left[\frac{\rho_s}{2} (\nabla\varphi)^2 - hN_0 \sin\varphi \right] \qquad (4.36)$$

Minimizing the energy cost with respect to variations in φ we get in Fourier space

$$\langle n_2(k) \rangle \approx N_0 \varphi(k)$$
$$= \frac{N_0^2}{\rho_s k^2} h(k). \qquad (4.37)$$

This gives us the exact result

$$\lim_{k \to 0} \chi_\perp(k, 0) = \frac{N_0^2}{\rho_s k^2} \qquad (4.38)$$

Combining the $N = \infty$ results (4.29,4.31) with (4.38), we have

$$\rho_s = cN \left(\frac{1}{g} - \frac{1}{g_c} \right) \tag{4.39}$$

In general, from (4.34), ρ_s is expected to vanish $(g_c - g)^{(d-1)\nu}$, and the result (4.39) is consistent with the $N = \infty$ values of the exponents.

We conclude this subsection by remarking on two features of the response functions of the ordered ground state which depend upon having a non-zero η, and are therefore absent in the $N = \infty$ theory: these features can be verified by explicit computation in a $1/N$ expansion. First, from (4.38), we deduce that the residue at the spin-wave pole (for $k \to 0$) is N_0^2/ρ_s; as g approaches g_c, this vanishes as $(g_c - g)^{\eta\nu}$, unlike the result (4.31) in which the spin-wave residue remains non-zero all the way upto g_c. Second, with energy scale Δ_- in hand, we can also define a corresponding length scale ξ_J

$$\xi_J = \frac{c}{\Delta_-}. \tag{4.40}$$

This is known as the Josephson length. The forms (4.38) and (4.31), which are characteristic long-wavelength transverse responses of a phase with spontanesouly broken continuous symmetry, remain valid at length scales larger than ξ_J, and times longer than Δ_-^{-1}. At shorter scales, the responses crossover to the isotropic response of the critical points like in (4.24).

D. Nonzero temperatures

The structure of the $N = \infty$ theory for $T > 0$ is especially simple, as the dynamic susceptibility retains the form (4.10) for all T, with the parameter σ determined by the solution of (4.9). Indeed, by Fourier transforming (4.10), we see that σ/c is the correlation length, which is a useful physical interpretation to keep in mind. The imaginary part of (4.10) also implies that there is a gap in the spectrum equal to σ. This feature is an artifact of the $N = \infty$ limit: the response of any interacting system has a non-zero spectral density at all frequencies (in certain cases, the response could vanish above some large ultraviolet cutoff $\sim c\Lambda$), as there are essentially no restrictions on the set of frequencies at which all the possible thermally excited states can absorb energy. The proper physical interpretation of the energy scale σ varies at different points in the g, T plane, and will be discussed further below.

As σ/c is a correlation length, it is quite easy to deduce its expected scaling form near the $g = g_c$ quantum critical point for $T > 0$. The scaling dimension of σ is $z = 1$, and from this, or by analogy with the result for the correlation length of the Ising chain in (3.40) we deduce [21]

$$\sigma = TF_\pm \left(\frac{\Delta_\pm}{T} \right) \tag{4.41}$$

We have written two separate scaling functions, F_+, F_-, which hold for $g \geq g_c$, $g \leq g_c$ respectively. We were forced into this somewhat awkward construction because the energy scales $\Delta_\pm \sim |g - g_c|^{z\nu}$, which characterize the ground states on either side of the $g = g_c$ quantum critical point, have quite different physical interpretations, and, in general, a rather different dependence on the bare coupling $g - g_c$. In the case of the Ising chain, we were able to use a single function because of some special features of that model: there was a gap in both ground states, the exponent $z\nu = 1$ implied that the gap $\sim |g - g_c|$, and we were able to define an energy scale $\Delta \sim g_c - g$ such that the gap $\Delta_+ = -\Delta$ for $g > g_c$, and the gap $\Delta_- = \Delta$ for $g < g_c$. In the present case, there is a gap only for $g > g_c$ and $z\nu \neq 1$. Nevertheless, the analyticity requirements on $T > 0$ physical properties, as a function of the bare coupling g, at the quantum critical coupling $g = g_c$ still hold. This means that given (say) F_+, and the dependencies of $\Delta_\pm$ on $g - g_c$, it is in principle possible to determine F_- by analytic continuation in g. It is perhaps worthwhile to reiterate that the analyticity holds *not* as a function of the physical, renormalized, energy scales $\Delta_\pm$, but as function of the bare coupling g at $g = g_c$. A new feature of models in $d > 1$, as we will see below, that for $T > 0$, there *can* be non-analyticities at couplings away from $g = g_c$. These non-analyticities correspond to finite T, thermal phase transitions.

We now determine the universal functions $F_\pm$, and will subsequently turn to a description of the physics in the various regions. The method used here introduces a number of tricks which repeatedly appear in the extraction of universal, cut-off independent crossover functions in other contexts [23].

We present first the calculation on the disordered side $g \geq g_c$. The first step is to subtract from (4.9) the corresponding equation (4.11) at the same coupling constants at $T = 0$; this gives us

$$\int^\Lambda \frac{d^d k}{(2\pi)^d} T \sum_{\omega_n} \frac{1}{c^2 k^2 + \omega_n^2 + \sigma^2} - \frac{1}{c} \int^\Lambda \frac{d^{d+1} p}{(2\pi)^d} \frac{1}{p^2 + (\Delta_+/c)^2} = 0 \qquad (4.42)$$

where Δ_+ is the gap at the current value of g. Next, subtract from the summation over frequencies of any quantity, the integration over frequences of precisely the same function; so we rewrite (4.42) as

$$\int^\Lambda \frac{d^d k}{(2\pi)^d} \left(T \sum_{\omega_n} \frac{1}{c^2 k^2 + \omega_n^2 + \sigma^2} - \int \frac{d\omega}{2\pi} \frac{1}{c^2 k^2 + \omega^2 + \sigma^2} \right)$$
$$+ \frac{1}{c} \int^\Lambda \frac{d^{d+1} p}{(2\pi)^d} \left(\frac{1}{p^2 + (\sigma/c)^2} - \frac{1}{p^2 + (\Delta_+/c)^2} \right) = 0 \qquad (4.43)$$

Now we use the general relation

$$T \sum_{\omega_n} \frac{1}{\omega_n^2 + a^2} - \int \frac{d\omega}{2\pi} \frac{1}{\omega^2 + a^2} = \frac{1}{a} \frac{1}{e^{a/T} - 1} \qquad (4.44)$$

valid for any positive a. Notice that the right-hand side falls off exponentially as a becomes large. This is a key property, and was the reason for considering the combination in (4.44). Applying this identity to (4.43), we see that the first integration over k has an integrand which is exponentially small for large k, and hence is quite insensitive to Λ which can safely be sent to infinity. The integration over p in the second term is also ultraviolet convergent, again allowing Λ to be set to infinity. The resulting expression is then cutoff independent, and hence universal; we obtain

$$\int \frac{d^d k}{(2\pi)^d} \frac{1}{\sqrt{c^2 k^2 + \sigma^2}} \frac{1}{e^{\sqrt{c^2 k^2 + \sigma^2}/T} - 1} - \frac{X_{d+1}}{c^d} \left(\sigma^{d-1} - \Delta_+^{d-1} \right) = 0 \qquad (4.45)$$

The solution of this equation is clearly of the form (4.41); after rescaling momenta by c/T in (4.45), we find that the function $F_+(s)$ is determined implicitly by solution of the equation

$$\int \frac{d^d k}{(2\pi)^d} \frac{1}{\sqrt{k^2 + F_+^2}} \frac{1}{e^{\sqrt{k^2 + F_+^2}} - 1} - X_{d+1} \left(F_+^{d-1} - s^{d-1} \right) = 0 \qquad (4.46)$$

We will discuss asymptotic features of the solution of this equation in the subsections below. We note here that precisely in $d = 2$, the equation (4.46) has a simple, explicit solution [21]

$$F_+(s) = 2 \sinh^{-1} \left(\frac{e^{s/2}}{2} \right) \qquad d = 2 \qquad (4.47)$$

Now we turn to the ordered side, $g \le g_c$. We assume that T is large enough that the magnetization is zero; the case of the magnetized state with $T \ne 0$ can be treated similarly, and will be referred to below. Subtract from (4.9), the value of ρ_s/N in (4.39), and insert the value of $1/g_c$ in (4.12). Evaluating the frequency summation as above we find

$$\int \frac{d^d k}{(2\pi)^d} \frac{1}{\sqrt{c^2 k^2 + \sigma^2}} \frac{1}{e^{\sqrt{c^2 k^2 + (\sigma/c)^2}/T} - 1}$$
$$+ \frac{1}{c} \int \frac{d^{d+1} p}{(2\pi)^d} \left(\frac{1}{p^2 + \sigma^2} - \frac{1}{p^2} \right) = \frac{\rho_s}{N c^2} \qquad (4.48)$$

The solution of this is also in the form (4.41), and the function $F_-(s)$ is given by

$$\int \frac{d^d k}{(2\pi)^d} \frac{1}{\sqrt{k^2 + F_-^2}} \frac{1}{e^{\sqrt{k^2 + F_-^2}} - 1} - X_{d+1} F_-^{d-1} - s^{d-1} = 0 \qquad (4.49)$$

Again, there is a simple explicit solution in $d = 2$ [21]

$$F_-(s) = 2 \sinh^{-1} \left(\frac{e^{-2\pi s}}{2} \right) \qquad d = 2 \qquad (4.50)$$

With expressions for the crossover functions $F_\pm$ in hand, let us discuss the physical properties of the system in different regimes of the g, T, plane

1. Low T on the disordered side, $g > g_c$, $T \ll \Delta_+$

Properties in this phase are essentially identical to those of the Ising model in Section III D 2, and so we will be quite brief. For the parameter σ we have

$$\sigma = \Delta_+ + \mathcal{O}(e^{-\Delta_+/T}) \tag{4.51}$$

There is a finite correlation length c/σ which has exponentially small corrections from its $T = 0$ value c/Δ_+. The gap in the spectrum is filled in at non-zero temperatures by classical relaxation in the dilute gas of thermally excited quasi-particles: this process is however not apparent in the $N = \infty$ susceptibility which has a gap even at non-zero T.

2. High T, $T \gg \Delta_+, \Delta_-$

Again properties are the same those of the Ising chain as discussed in Section III D 3. Now we have

$$\sigma = TF_+(0) = TF_-(0) \tag{4.52}$$

where $F_+(0)$, $F_-(0)$ are pure numbers. This represents a correlation length $\sim c/T$. Spin correlations decay through a 'quantum relaxation' process [21] at a rate $\sim T$, and the $N = \infty$ gap $\sim T$ is an artifact of the approximation.

3. Low T on the ordered side, $g < g_c$

Now there are some differences from the Ising chain for $2 < d < 3$.

Let us assume first that T is large enough so that $\langle \vec{n} \rangle = 0$ and so (4.49) can be used to determine F_-. For $1 < d \leq 2$, one finds that there is a solution of (4.49) for all T ($s \to \infty$). We find

$$\sigma = \begin{cases} \dfrac{T^{1/(2-d)}}{\Delta_-^{(d-1)/(2-d)}|X_d|^{1/(2-d)}} & 1 < d < 2 \\ T \exp(-2\pi\Delta_-/T) & d = 2 \end{cases} \tag{4.53}$$

So the correlation length $\sim c/\sigma$ diverges as $T \to 0$, but remains finite for all non-zero T. This was exactly the situation as in the Ising model, and the phase diagram for this model is therefore identical to Fig 7, except that the crossover boundaries do not come in linearly as $z\nu \neq 1$. There is a reduced scaling function which describes the classical physics of the system as $T \to 0$, as in the Ising model.

Now let us consider the case $2 < d < 3$. Although there is no physical dimension in this region, the results obtained below will apply in $d = 3$ with cutoff-dependent logarithmic corrections we do not want to discuss here. Further, the physics of the quantum Ising model in $d = 2$ is expected to be similar to that of the large N solution with $2 < d < 3$. The key observation in this case is that there is no solution of (4.49) for F_- below a critical value of $s = s_c$ given by

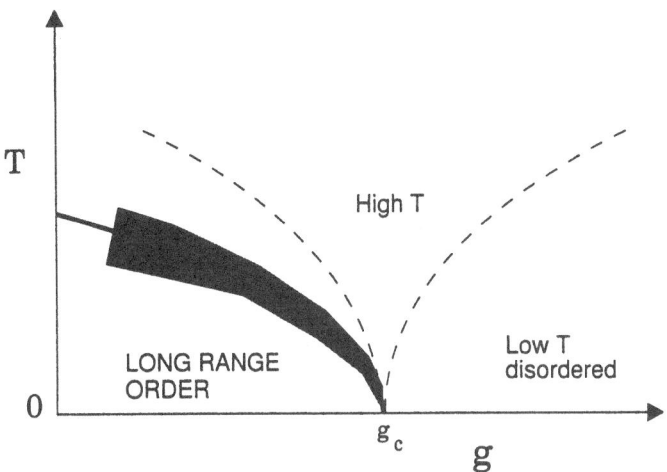

FIG. 11. Phase diagram for the $O(N)$ rotor model with $2 < d < 3$ for $N > 2$, or $2 \leq d < 3$ for $1 \leq N \leq 2$. The dashed lines are crossovers, while the full line is the locus of finite temperature phase transitions with T_c given by (4.55). The shaded region is where the reduced classical scaling functions apply.

$$s_c^{d-1} = \int \frac{d^d k}{(2\pi)^d} \frac{1}{k} \frac{1}{e^k - 1}$$
$$= \frac{2\Gamma(d-1)\zeta(d-1)}{\Gamma(d/2)(4\pi)^{d/2}} \tag{4.54}$$

This defines a critical temperature T_c given precisely by

$$T_c \equiv \Delta_-/s_c \tag{4.55}$$

such that the system is in the paramagnetic phase only for $T > T_c$: the resulting phase diagram is shown in Fig 11 There is a finite temperature phase transition at $T = T_c$, and a magnetically ordered phase for $T < T_c$. As T approaches T_c, the conventional classical phase transition theory becomes applicable in the region $|T - T_c| \ll T_c$. The classical scaling functions of this transition emerge as reduced scaling functions of the quantum functions, in a manner very similar to the other cases that have been discussed above. One consequence of this behavior is that all the scale factors of the classical scaling functions, which are usually considered non-universal, are universally determined by the parameters Δ_-, c, and N_0 of the quantum crossover functions. We have already seen an example of this in (4.55), where T_c was universally determined by Δ_- [23].

Let us explicitly observe the collapse of the scaling function (4.41) in this classical region. As the primary quantum crossover function has only one argument, the reduced function would have no arguments i.e. it is a pure power law. Indeed, solution of (4.49) for s close to but above s_c gives us

$$\sigma = T_c \left[\left(\frac{T - T_c}{T_c} \right) \frac{(d-1)s_c^{d-1}}{X_d} \right]^{1/(d-2)} \tag{4.56}$$

The correlation length c/σ diverges with the classical exponent $\nu_c = 1/(d-2)$ with an amplitude that is universal. A similar collapse will occur in all other static and dynamic observables.

V. CONCLUSION

This review has described the physics of the vicinity of second-order quantum phase transition using three simple examples which had the virtue of capturing some essential physical phenomena. Among the physical issues we have not discussed are:

(i) Crossovers near quantum-critical points above their upper critical dimension: a simple understanding of this can be obtained from the large N solution of the $O(N)$ quantum rotor model with $d > 3$, or by a perturbative analysis [23].

(ii) Quantum phase transition in the presence of quenched randomness: this is a complicated subject still in its infancy, with only a few firm results.

ACKNOWLEDGMENTS

I thank Profs. German Sierra and Miguel Martin-Delgado for the kind invitation to lecture at this summer school. This article is adapted from a forthcoming book by the author on quantum critical phenomena, to be published by Cambridge University Press; I thank Cambridge University Press for permission to reproduce this material here. I am grateful to K. Damle, T. Senthil and S. Majumdar for numerous useful discussions and helpful remarks on the manuscript. The title of this article was inspired by A. Georges. This research was supported by the National Science Foundation under Grant DMR-96-23181.

---

[1] E. Brezin, J.C. Le Guillou and J. Zinn-Justin in *Phase Transitions and Critical Phenomena* , vol. 6, C. Domb and M.S. Green eds., Academic Press, London (1976).

[2] *Quantum Field Theory and Critical Phenomena* by J. Zinn-Justin, Oxford University Press, Oxford (1993).

[3] S. Sachdev in the *Proceedings of the 19th IUPAP International Conference on Statistical Physics, Xiamen, China, July 31 - August 4 1995*, edited by B.-L. Hao, World Scientific, Singapore (1996); Report No.cond-mat/9508080.

[4] E. Lieb, T. Schultz, and D. Mattis, Ann. of Phys. **16**, 406 (1961).

[5] J.A. Hertz, Phys. Rev. B **14**, 525 (1976).

[6] S. Sachdev, T. Senthil, and R. Shankar, Phys. Rev. B **50**, 258 (1994).

[7] R. Shankar, Rev. Mod. Phys. **66**, 129 (1994).

[8] J.L. Cardy, J. Phys. A **17**, L385 (1984).

[9] K. Damle and S. Sachdev, Phys. Rev. Lett. **76**, 4412 (1996).

[10] E. Barouch and B.M. McCoy, Phys. Rev. A **3**, 786 (1971).

[11] J.B. Kogut, Rev. Mod. Phys. **51**, 659 (1979).

[12] *Statistical Field Theory* by C. Itzykson and J.-M. Drouffe, Cambridge University Press, Cambridge (1989).

[13] S. Sachdev, Nucl. Phys. B **464**, 576 (1996)

[14] B.M. McCoy, Phys. Rev. **173**, 531 (1968).

[15] S. Chakravarty, B.I. Halperin, and D.R. Nelson, Phys. Rev. B **39**, 2344 (1989).

[16] S. Sachdev and A.P. Young, cond-mat/9609185.

[17] R.J. Glauber, J. Math. Phys. **4**, 294 (1963).

[18] J.L. Cardy and G. Mussardo, Nucl. Phys. B **340**, 387 (1990); V.P. Yurov and Al.B. Zamalodchikov, Int. J. Mod. Phys. A **6**, 3419 (1991).

[19] O. Babelon and D. Bernard, Physics Letters. B **288**, 113 (1992).

[20] A. Leclair, F. Lesage, H. Saleur, and S. Sachdev, Nucl. Phys. B in press; cond-mat/9606104

[21] S. Sachdev and J. Ye, Phys. Rev. Lett. **69**, 2411 (1992); A.V. Chubukov and S. Sachdev, Phys. Rev. Lett. **71**, 169 (1993); A.V. Chubukov, S. Sachdev and J. Ye, Phys. Rev. B **49**, 11919 (1994).

[22] E. Brezin and J. Zinn-Justin, Phys. Rev. B **14**, 3110 (1976).

[23] S. Sachdev, Phys. Rev. B **55**, January 1, 1997; cond-mat/9606083.

Notes on the Density Matrix Renormalization Group; Applications to Ladder Systems

Steven R. White

Department of Physics and Astronomy, University of California, Irvine, CA 92697-4575
(May 13, 1997)

In these notes we review both the formulation and some applications of the density matrix renormalization group (DMRG). We illustrate the use of DMRG with applications to ladder systems. We first consider Heisenberg ladders, which exhibit an interesting odd-even alternation of behavior as the number of chains increases. We then consider the behavior upon doping, using the *t-J* model. A simple theoretical framework is developed to explain why holes bind in pairs in two-dimensional antiferromagnets.

I. INTRODUCTION

In these notes we review both the formulation and some applications of the density matrix renormalization group (DMRG). In Sect. II (see Ref. [1]) we develop the basic formulation of DMRG, starting with a discussion of Wilson's original numerical renormalization group approach. The key new idea of DMRG is that rather than keep the lowest-lying eigenstates of the Hamiltonian in forming a new effective Hamiltonian of a block of sites, one should keep the most significant eigenstates of the block density matrix, obtained from diagonalizing the Hamiltonian of a larger section of the lattice which includes the block. This approach is much more accurate than Wilson's original numerical renormalization group approach. DMRG can be applied to almost any one-dimensional quantum lattice system with local interactions, and can provide a wide variety of static properties. Recent progress has been made in applying DMRG to two-dimensional systems.

In Sect. III (see Ref. [2]) we illustrate the use of DMRG with applications to ladder systems. We first consider Heisenberg ladders, which exhibit an interesting odd-even alternation of behavior as the number of chains increases. We discuss a resonating valence bond variational ansatz which provides an intuitive understanding of this behavior.

In Sect. IV (see Ref. [3]) we consider the behavior upon doping, using the *t-J* model. A simple theoretical framework is developed to explain why holes bind in pairs in two-dimensional antiferromagnets. For intermediate interaction strengths in the *t-J* model, the hole pairs reside predominantly on a 2×2 core plaquette with the probability that the holes are on diagonal sites greater than nearest-neighbor sites. There is a strong singlet bond connecting the spins on the two remaining sites of the plaquette. We show that a general characteristic of dynamic holes in an antiferromagnet is the presence of frustrating antiferromagnetic bonds connecting next-nearest-neighbor sites across the holes. Pairs of holes bind in order to share the frustrating bonds.

II. INTRODUCTION TO THE DENSITY MATRIX
RENORMALIZATION GROUP

Shortly after Wilson developed his numerical renormalization group (RG) procedure to solve the Kondo problem [4], there was considerable interest in applying
closely related techniques to a variety of problems. In particular, it seemed that a
number of quantum lattice models (such as the Hubbard and Heisenberg models),
particularly in one dimension (1D), could be treated with a real-space blocking
version of this technique. It was clear from the beginning that one could not hope
to achieve the accuracy Wilson obtained for the Kondo problem in these other systems, but it was hoped that the method would yield qualitatively reliable results.
Unfortunately, the approach proved to be rather unreliable, particularly in comparison with other numerical approaches, such as Monte Carlo, which were being
developed at the same time. Until recently, the method was only used occasionally.

Recent developments in renormalization group algorithms have changed this
picture completely. The first significant advance came in the understanding of the
effect of boundary conditions on the basic renormalization group procedure [5].
The standard approach of neglecting all connections to neighboring blocks during
the diagonalization of the block Hamiltonian introduces large errors which cannot
be corrected by any reasonable increase in the number of states kept. However, by
varying the boundary conditions on a block, and keeping states from several diagonalizations with different boundary conditions, one can eliminate these errors, at
least for single-particle problems. The second advance came in the development
of a technique suitable for many-particle systems, using a formulation in terms
of density matrices, the density matrix renormalization group [6]. For systems
such as 1D Heisenberg spin chains, the density matrix approach makes the numerical renormalization group approach not just qualitatively reliable, it makes it
substantially more accurate and powerful for calculating many zero temperature
properties than current quantum Monte Carlo approaches.

We first describe in detail the standard RG approach in the simplest possible
context, a real space blocking approach for a 1D lattice system. The notation and
many of the central ideas will be very similar in the density matrix approach described later. The Hamiltonian considered here could describe a spin system, such
as the Heisenberg model, or an interacting electron system, such as the Hubbard
model. The approach is relevant for zero temperature, and one obtains the ground
state and some low-lying excited states.

One begins by breaking the 1D chain into finite identical blocks. It is usually
convenient to start at the first iteration with blocks consisting of just one site. We
will label the blocks B and the block Hamiltonian H_B. H_B contains all terms of
H involving only sites contained in B. For example, for the Hubbard model at the
first iteration, where B consists of one site, $H_B = U n_{i\uparrow} n_{i\downarrow} - \mu(n_{i\uparrow} + n_{i\downarrow})$. For the
Heisenberg model at the first iteration, $H_B = 0$.

Rather than describe B and H_B in the usual way by listing the sites of B and
using second-quantized operator expressions for H_B, we describe B by a list of the
many-body states on the block, and by quantum numbers and matrix elements
between these states. We store the number of states m, and for each state we list
all quantum numbers which are to be used, such as S_z and S for a spin system,

or $N_\uparrow, N_\downarrow$, and S for an electron system. H_B is represented as an $m \times m$ matrix. In order to reconstruct H, additional information is needed besides H_B. The additional information describes the interactions between blocks. For a Heisenberg system with interaction

$$\vec{S}_i \cdot \vec{S}_{i+1} = S_i^z S_{i+1}^z + \frac{1}{2}(S_i^+ S_{i+1}^- + S_i^- S_{i+1}^+) \qquad (1)$$

one needs to store $m \times m$ matrix representations of S_i^z, S_i^+, and S_i^- for i equal to both the left and right end sites of B. (In practice, one need not store S_i^-, since it can be obtained by taking the Hermitian conjugate of S_i^+). For a Hubbard model one would have to store matrices for $c_{i\sigma}^\dagger$ and $c_{i\sigma}$, with $\sigma =\uparrow$ and $\downarrow$, in order to reconstruct the hopping term $\sum_\sigma (c_{i+1\sigma}^\dagger c_{i\sigma} + c_{i\sigma}^\dagger c_{i+1\sigma})$.

The standard procedure is summarized in Table I. At the beginning of an iteration one forms the Hamiltonian for two blocks joined together, H_{BB}. BB has m^2 states. The states are labeled by two indices, $i_1 i_2$. For a Heisenberg system with $J = 1$ the $m^2 \times m^2$ matrix for H_{BB} is given by

$$[H_{BB}]_{i_1 i_2; i_1' i_2'} = [H_B]_{i_1 i_1'} \delta_{i_2 i_2'} + [H_B]_{i_2 i_2'} \delta_{i_1 i_1'} + [S_r^z]_{i_1 i_1'} [S_\ell^z]_{i_2 i_2'} +$$
$$\frac{1}{2} [S_r^+]_{i_1 i_1'} [S_\ell^-]_{i_2 i_2'} + \frac{1}{2} [S_r^-]_{i_1 i_1'} [S_\ell^+]_{i_2 i_2'} \qquad (2)$$

where r represents the rightmost site of the left block, and ℓ the leftmost site of the right block.

TABLE I. Standard numerical renormalization group algorithm for a 1D quantum system.

1. Isolate two blocks BB, and form H_{BB}.

2. Diagonalize H_{BB}, obtaining the m lowest eigenvectors u^α.

3. Form matrix representations of S_ℓ^z, etc., for BB from the corresponding matrices for B.

4. Change basis to the u^α, keeping only the lowest m states, using $H_{B'} = OH_{BB}O^\dagger$, etc., with $O(\alpha; i_1, i_2) = u_{i_1, i_2}^\alpha$, $\alpha = 1, \ldots, m$.

5. Replace B with B'.

6. Go to step 1.

In diagonalizing H_{BB} it is useful to separate the basis states by quantum numbers, since H_{BB} is block diagonal. It is very simple to use S_z or $N_\uparrow$ and $N_\downarrow$ in this way. Utilizing the total spin S is more tedious (especially when one puts four blocks together, as we do below), and we have not used S to further reduce the dimension of H_{BB}. (The value of S for a state can easily be inferred by degeneracies for different values of S_z.)

The lowest lying eigenstates $u_{i_1 i_2}^\alpha$, $\alpha = 1, \ldots m$, of H_{BB} are the states used to describe B' ($BB \to B'$). The new block Hamiltonian matrix $H_{B'}$ is diagonal. However, in the more general case where the states kept, the u^α, are not eigenstates of H_{BB} we can write

$$H_{B'} = O H_{BB} O^\dagger \tag{3}$$

where the $m \times m^2$ matrix $O_{i;i_1 i_2} = u_{i_1 i_2}^i$, i.e. the rows of O are the states kept. If O were square, this would be a unitary transformation. Since O is not square, the transformation truncates away (integrates out) the high energy states.

In order to obtain new matrices for S_ℓ^z, S_r^z, etc., it is necessary to use O again. First, one must construct the operators for S_ℓ^z, S_r^z, etc. for BB, which we denote by $\tilde{S}_\ell^z, \tilde{S}_r^z$, etc. For example

$$\left[\tilde{S}_\ell^z\right]_{i_1 i_2; i_1' i_2'} = [S_\ell^z]_{i_1 i_1'} \, \delta_{i_2 i_2'} \tag{4}$$

$$\left[\tilde{S}_r^z\right]_{i_1 i_2; i_1' i_2'} = [S_r^z]_{i_2 i_2'} \, \delta_{i_1 i_1'} \tag{5}$$

Then the new matrices for B' are given by

$$S_\ell^z = O \tilde{S}_\ell^z O^\dagger \tag{6}$$

etc.

After these new operator matrices are formed, we can replace B by B' and start the next iteration. The iteration is continued until the system is large enough to represent properties of the infinite system. As our main concern here is the iterative diagonalization procedure discussed above, we will not discuss the analysis, using fixed points, relevant and irrelevant operators, etc., of the effective Hamiltonians obtained with the procedure.

Wilson's approach to the Kondo problem is closely related to the method described here, despite some important differences. One difference is that rather than joining two identical blocks, the degrees of freedom associated with a single interval (an "onion-layer") were added to the system at each iteration. The analogous procedure for a 1D system would be to add a single site to a block at each iteration. From a computational point of view, this has a distinct advantage in that many more states can be kept (m can be made larger) since at each iteration a system with nm states, as opposed to m^2, must be diagonalized, where n is the number of states on a single site ($n = 4$ for Hubbard models, $n = 2S + 1$ for spin models).

The most important difference between the Kondo system and a 1D system is that the couplings between adjacent layers or "sites" decreases exponentially in the Kondo system, whereas it remains constant for a 1D system. This exponential decrease is the key to the success of the method for the Kondo system and related impurity systems. More discussion concerning how the detailed form of the Hamiltonian makes the numerical approach accurate are given by Wilson [4].

When applied to other systems, such as 1D spin systems, where the couplings do not decrease exponentially, this standard numerical RG approach generally performs poorly.

The fundamental difficulty in the standard approach discussed above lies in choosing the eigenstates of H_{BB} to be the states kept. Since H_{BB} contains no connections to the rest of the lattice, its eigenstates have inappropriate features at the block ends. This is clearly illustrated in the work of White and Noack [5], who suggested two alternatives to the standard approach. These methods shared a common feature: the states that were kept were *not* the eigenstates of H_{BB}. They differed in how the states to be kept were chosen. In the first method, the combination of boundary conditions (CBC) approach, the lowest lying eigenstates of several different block Hamiltonians were kept. The several block Hamiltonian differed only in the boundary condition applied to a block, e.g. one Hamiltonian might have periodic boundary conditions applied and another antiperiodic. The rationale for this was that quantum fluctuations in the rest of the system effectively apply a variety of boundary conditions to the block. States from any single boundary condition cannot respond properly to these fluctuations. By applying a representative set of boundary conditions, which is in some sense "complete" enough for the problem at hand, one obtains a set of states which are able to respond to these fluctuations. This approach proved very effective for the simple single-particle problems studied by White and Noack, as well as for Anderson localization models [7].

The CBC approach appears to be ill-suited to interacting systems. It is useful to consider a noninteracting many-particle system, such as the Hubbard model with $U = 0$. An arbitrary state of this system can be described in terms of the single particle wavefunctions of each of its particles. Some of these single-particle wavefunctions may have nodes at the ends of a block, and some may have antinodes.

It is easy to choose boundary conditions with generate block states where every particle on the block has a node or every particle has an antinode, but it is difficult to get different boundary behavior for different particles. In order to properly represent the block, the states kept not only need to allow for different end behavior for different particles, they must represent a complete range of boundary behavior.

This general line of reasoning is supported by numerical tests on Heisenberg chains. We have tried to find a simple set of boundary conditions which can be used to treat the $S = \frac{1}{2}$ Heisenberg chain. We tried combinations of periodic and antiperiodic couplings between the ends of the block, as well as varying the magnitude of the coupling between the ends of a block. We were unable to find any set of boundary conditions which was at all satisfactory.

The other approach suggested by White and Noack, the superblock method, forms the basis for the density matrix approach. In the superblock method, one diagonalizes a larger system (the "superblock"; the name is analogous to "supercell", as used in electronic structure calculations) composed of three or more blocks which includes the two blocks BB which are used to form B'. The wavefunctions for the superblock are projected onto BB, and these projected states of BB are kept. For a single particle wavefunction, this projection is single-valued and trivial. The superblock method works quite well in the single-particle model, with the accuracy increasing rapidly with the number of extra blocks used. However, for a many-particle wavefunction, the "projection" of a wavefunction onto BB is many-valued, and, in fact, a single many-particle state for the entire lattice generally "projects" onto a complete set of block states. However, some of these states are more important than others; the density matrix tells us which states are the most important. (The reader is urged to review Feynman's introduction to density matrices [8] before proceeding further.)

It is very natural to use the density matrix to choose the states which we wish to keep. Consider first the following argument by analogy. For an *isolated* block at finite temperature, the probability that the block is in an eigenstate α of the block Hamiltonian is proportional to its Boltzmann weight $\exp(-\beta E_\alpha)$. The Boltzmann weight is an eigenvalue of the density matrix $\exp(-\beta H_B)$, and an eigenstate of the Hamiltonian is also an eigenstate of the density matrix. Since lowest energy corresponds to highest probability in the Boltzmann weight, we can view the standard RG approach as choosing the m most probable eigenstates to represent the block given the *assumption* that the block is isolated. (Alternatively, we can view the rest of the lattice as a heat bath at an effective inverse temperature β, to which the system is very weakly coupled.) However, in reality the block is not isolated, the density matrix is not $\exp(-\beta H_B)$, and eigenstates of the block Hamiltonian are not eigenstates of the block's density matrix. For a system which is stronglye coupled to the outside universe, it is much more appropriate to use the eigenstates of the density matrix to describe the system rather than the eigenstates of the system's Hamiltonian. Thus a natural generalization of the standard approach is to choose to keep the m most probable eigenstates of the block density matrix.

This conclusion–that the optimal states to keep are the most probable eigenstates of the block density matrix–can be justified precisely [1].

Incorporating this result in a numerical renormalization group algorithm involves a fundamental change in the way the calculation is carried out. In the standard RG approach, to find the states to be kept, one diagonalizes only the system BB, which becomes B'. In the density matrix approach, in order to obtain any reasonable approximation to the density matrix, it is necessary to diagonalize the Hamiltonian of a larger system which includes BB, namely some sort of superblock, and then use the eigenstates of the superblock to determine the density matrix. The density matrix is then diagonalized, and its most significant eigenstates are the states kept. The number of eigenstates of the Hamiltonian of the superblock used to produce the density matrix can be as small as one; this single state produces a density matrix for BB which has many eigenstates to be used as block states to be kept.

A density matrix algorithm is defined mainly by the form of the superblock and the manner in which the blocks are enlarged (such as by doubling the block, $B' = BB$, or by adding a single site, $B' = B + $ site), and by the choice of superblock eigenstates used in constructing the density matrix (e.g., the two lowest-lying $S_z = 0$ states). An eigenstate of the superblock Hamiltonian is called a *target state* if it is used in forming the block density matrix. The most efficient algorithms use only a single target state (usually the ground state) in constructing the density matrix. By targeting only one state, the block states are more specialized for representing that state, and fewer are needed for a given accuracy. Probably the most important characteristic of a density matrix algorithm is the rate at which the accuracy increases with the number of states m. We have found that the accuracy of the representation of the target states increases roughly exponentially with m, at least for open boundary conditions. The coefficient governing the increase of accuracy with m is largest with a single target state.

Several considerations enter in the construction of the superblock to be used in an algorithm. Generally it is more efficient to enlarge the block by adding a single site, rather than doubling a block. The superblock configuration used here is represented symbolically as $B_\ell \bullet \bullet B_{\ell'}^R$, where B_ℓ represents a block composed of ℓ sites, B_ℓ^R is a reflected block (right interchanged with left) of length ℓ', $\bullet$ represents a single site, and the total length of the superblock is $L = \ell + \ell' + 2$. Here $B' = B_{\ell+1}$ is formed from the left block plus a single site, i.e. $B_{\ell+1} = B_\ell \bullet$. Open boundary conditions are used. The right-hand block and site $\bullet B_{\ell'}^R$ are only used to help form the density matrix for $B_{\ell+1}$; in the construction of the density matrix, the states of $\bullet B_{\ell'}^R$ are traced over. This configuration can be used in two different ways: in an infinite chain method, in which the chain size increases by two at each step, and in a finite chain method, in which the chain size is fixed.

A. The infinite system method

In the first step of the infinite system method, we start with a four site chain and diagonalize the Hamiltonian of the superblock configuration $B_1 \bullet \bullet B_1^R$, where B_1 and B_1^R both represent a single site. We use the Davidson algorithm [9] for the sparse matrix diagonalization, but one could also use the more well known Lanczos method. Using the target states calculated with this configuration, we calculate a density matrix and form an effective Hamiltonian for $B_2 = B_1 \bullet$. In the second step we diagonalize $B_2 \bullet \bullet B_2^R$, where we have formed B_2^R by reflecting B_2. We continue in this manner, diagonalizing the configuration $B_\ell \bullet \bullet B_\ell^R$, and setting $B_{\ell+1} = B_\ell \bullet$, and using $B_{\ell+1}$ and its reflection in the next step of the iteration. At each step, both blocks increase in length by one site, and the total length of the chain increases by two at each step of the iteration. The infinite chain method is usually used when one is interested in ground state properties of the infinite chain. Each step of the iteration pushes the ends of the chain farther from the two sites in the center. After many steps, each block approximately represents one half of an infinite chain. In order to represent one half of an infinite chain, B must not only contain many sites itself, its effective Hamiltonian must be formed from a system in which the rest of the chain has many sites. The effective Hamiltonian formed from the left-hand side $B_\ell \bullet$ depends strongly on the right-hand side $\bullet B_\ell^R$. The infinite chain algorithm converges in two senses simultaneously: in the length of B_ℓ going to infinity and in the sense that B_ℓ is adapted to respond to an infinite chain connected to it on the right.

The infinite system algorithm is summarized in Table II. The representation of the blocks is identical to that of the standard algorithm: we describe a block by listing how many states it has and the quantum numbers for each state, and by storing matrices for H_B, S_i^z, etc. Once the matrix O is constructed using the most significant eigenvectors of the density matrix, the change of basis procedure is also identical to that of the standard algorithm. For the purposes of organizing the algorithm, it is easiest to think of the two sites in the middle as blocks which can be treated similarly to the two outer blocks, although they contain only a few states.

TABLE II. Infinite system density-matrix algorithm for a 1D system.

1. Make four initial blocks, each consisting of a single site, representing the initial four site system. Set up matrices representing the block Hamiltonian and other operators.

2. Form the Hamiltonian matrix (in sparse form) for the superblock.

3. Using the Davidson or Lanczos method, diagonalize the superblock Hamiltonian to find the target state $\psi(i_1, i_2, i_3, i_4)$. ψ is usually the ground state. Expectation values of various operators can be measured at this point using ψ.

4. Form the reduced density matrix for the two-block system 1-2, using
$\rho(i_1, i_2; i_1', i_2') = \sum_{i_3, i_4} \psi(i_1, i_2, i_3, i_4) \psi(i_1', i_2', i_3, i_4)$.

5. Diagonalize ρ to find a set of eigenvalues w_α and eigenvectors u_{i_1, i_2}^α. Discard all but the largest m eigenvalues and associated eigenvectors.

6. Form matrix representations of operators (such as H) for the two-block system 1-2 from operators for each separate block.

7. Form a new block 1 by changing basis to the u^α and truncating to m states using $H^{1'} = O H^{12} O^\dagger$, etc. If blocks 1 and 2 have m_1 and m_2 states, then O is an $m \times m_1 m_2$ matrix, with matrix elements $O(\alpha; i_1, i_2) = u_{i_1, i_2}^\alpha$, $\alpha = 1, \ldots, m$.

8. Replace old block 1 with new block 1.

9. Replace old block 4 with the reflection of new block 1.

10. Go to step 2.

B. The finite system method

The finite system algorithm is designed to calculate accurately the properties of a finite system of size L, which we will assume for simplicity to be even. It is summarized in Table III. It begins with the use of the infinite system algorithm for $L/2 - 1$ steps, so that the final superblock used is of size L. In the infinite system method, there is no need to store B_ℓ once we have $B_{\ell+1}$; we need only store the latest block. In the finite system method, we need to store $L - 3$ blocks, B_1 to B_{L-3}, and the infinite system method is used to get initial, approximate versions of B_1 to $B_{L/2}$. After the system $B_{L/2-1} \bullet \bullet B_{L/2-1}^R$ is used to form $B_{L/2}$, the next step is to use the configuration $B_{L/2} \bullet \bullet B^R_{L/2-2}$ to form $B_{L/2+1}$. This system, and all the other superblocks to follow, contain L sites. We continue to form the other blocks up to size $L - 3$, using the superblock $B_\ell \bullet \bullet B^R_{L-\ell-2}$ to form $B_{\ell+1}$. This sequence of steps is the first iteration of the finite system algorithm.

The second and subsequent iterations use the blocks obtained from the previous iteration as the right-hand reflected blocks in each superblock. The first step starts by diagonalizing the superblock $B_1 \bullet \bullet B^R_{L-3}$, where B_1 is a single site and is always known exactly, and B^R_{L-3} is obtained from the last step of the previous iteration. Once a new B_ℓ is formed, it replaces the old B_ℓ, so that only one set of blocks need be stored. Consequently, for the second half of the iteration, starting with the superblock $B_{L/2-1} \bullet \bullet B^R_{L/2-1}$, we use a block formed in the current iteration, rather than the last iteration, as the right-hand block. On the very last iteration, we usually stop after the diagonalization of $B_{L/2-1} \bullet \bullet B^R_{L/2-1}$, and then use this wavefunction of the L-site system to measure various properties, such as the local magnetization or correlation functions.

After a few iterations each B_ℓ accurately represents an ℓ-site block which is the left-hand ℓ sites of an L-site chain. For many systems, such as most 1D spin chains, the method converges by the middle of the second iteration, although sometimes three iterations are necessary.

TABLE III. Finite system density-matrix algorithm for a 1D system consisting of L sites. A calculation consists of several iterations, indexed by I, with each iteration consisting of $L - 3$ steps, indexed by ℓ, where ℓ is the size of the first block.

1. (First half of $I = 1$.) Use the infinite system algorithm for $L/2 - 1$ steps to build up the lattice to L sites. At each iteration store the block Hamiltonian and end operator matrices for block 1. Label the blocks by their size, B_ℓ, $\ell = 1, \ldots, L/2$.

2. (Start of second half of $I = 1$) Set $\ell = L/2$. Use B_ℓ as block 1, and the reflection of $B_{L-\ell-2}$ as block 4.

3. Steps 2–8 of Table II.

4. Store the new block 1 as $B_{\ell+1}$, replacing the old $B_{\ell+1}$.

5. Replace block 4 with the reflection of $B_{L-\ell-2}$, obtained from the first half of this iteration.

6. If $\ell < L - 3$, set $\ell = \ell + 1$ and go to step 3.

7. (Start of iteration I, $I \geq 2$). Make four initial blocks, the first three consisting of a single site, and the fourth consisting of the reflection of B_{L-3} from the previous iteration. Set $\ell = 1$.

8. Steps 2–8 of Table II.

9. Store the new block 1 as $B_{\ell+1}$, replacing the old $B_{\ell+1}$.

10. Replace block 4 with the reflection of $B_{L-\ell-2}$, obtained from the previous iteration (if $\ell \leq L/2 - 1$) or the first half of this iteration (if $\ell > L/2 - 1$)

11. If $\ell < L - 3$, set $\ell = \ell + 1$ and go to step 8. If $\ell = L - 3$, start a new iteration by going to step 7. (Stop after 2 or 3 iterations.)

III. HEISENBERG LADDERS

In this section we summarize results of a density matrix renormalization group (DMRG) study [2] of isotropic Heisenberg coupled-chain systems with $n_c = 1$, 2, 3, and 4. We find that the $n_c = 2$ and $n_c = 4$ systems have a spin gap, while the $n_c = 1$ and $n_c = 3$ systems are gapless. Based on these results, we discuss how a variational RVB wave function, originally introduced by Liang et. al. [10] to describe the 2D antiferromagnetic Heisenberg system, provides an intuitive picture for understanding the results and suggests behavior for larger n_c than can be studied numerically. We conclude that, indeed, all even n_c systems are spin liquids, and we give a simple expanation for the difference between even and odd n_c systems based on confinement of topological defects.

We consider the Heisenberg Hamiltonian

$$H = J \sum_{\langle i,j \rangle} \mathbf{S}_i \cdot \mathbf{S}_j \tag{7}$$

defined on an $L \times n_c$ lattice with $S = \frac{1}{2}$. We will also consider the anisotropic system where the exchange along the chains is J and between the chains is J', but unless otherwise noted, $J' = J$. We begin by calculating the spin gap Δ defined by

$$\Delta(L) = E_0(L, 1) - E_0(L, 0). \tag{8}$$

Here $E_0(L, S_z)$ is the ground state energy for an $L \times n_c$ lattice with open boundary conditions and z-component of total spin S_z. For a single chain we expect that the finite size corrections scale as L^{-1}, and we have plotted $\Delta(L)$ versus L^{-1} in Fig. 1 for $n_c = 1$–4.

The solid curves are fits to the data of the form

$$\Delta(L) = \Delta + a_1 L^{-1} + a_2 L^{-2} + \dots \tag{9}$$

For both $n_c = 1$ and $n_c = 3$, an accurate fit is obtained with $\Delta = 0$. For $n_c = 2$ and $n_c = 4$, the data are fit very well with $a_1 = 0$, which is the expected form for an $S = 1$ Heisenberg chain [11]. For $n_c = 2$, we find $\Delta = 0.504$, and for $n_c = 4$, we find $\Delta = 0.190$. The spin-spin correlation functions $\langle \mathbf{S}_i \cdot \mathbf{S}_j \rangle$ are shown in Fig. 2. Here, because of the open boundary conditions, we have chosen i and j so that they are as symmetrically located about the center of the lattice as possible. The semilog plot in the inset of Fig. 2(a) shows the exponential decay of the spin correlations for the two and four chain systems. The correlation length for $n_c = 2$ is $\xi = 3.19(1)$, and for $n_c = 4$, $\xi \sim 5 - 6$. The spin-spin correlations for $n_c = 3$

decay as a power law, similar to those for a single ($n_c = 1$) chain, as shown in the inset of Fig. 2(b).

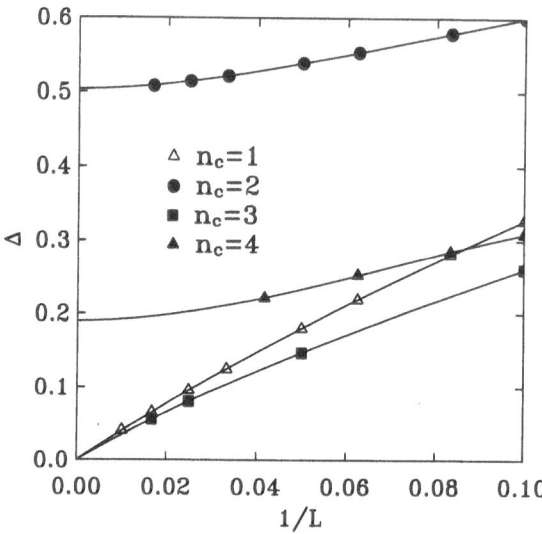

FIG. 1. Spin gaps as a function of system size L for open $L \times n_c$ coupled chain Heisenberg systems.

To obtain an intuitive picture of this odd-even behavior, we examine an RVB variational wavefunction [12,13]. We consider both short-range and long-range RVB states and conclude that a short-range RVB picture applies for even n_c, whereas a long-range RVB picture describes systems with odd n_c.

The RVB states we consider are specific to bipartite lattices, and contain only bonds connecting one sublattice (A) to the other (B). We consider wave functions of the form [10]

$$|\psi\rangle = \sum_{\substack{i_\alpha \in A \\ j_\alpha \in B}} h(i_1 - j_1) \ldots h(i_n - j_n)(i_1 j_1) \ldots (i_n j_n), \qquad (10)$$

Here (ij) represents a singlet bond between sites i and j, and the non-negative bond amplitude h can be chosen variationally. We consider a short-range RVB wavefunction to be one with a bond amplitude $h(l)$ which decays exponentially in l or faster, while a long-range RVB wave function will typically have a power-law decay, $h(l) \sim l^{-p}$. The state with the shortest possible range is the dimer RVB state, for which $h(i - j) = 1$ for i and j nearest neighbors, and is zero otherwise.

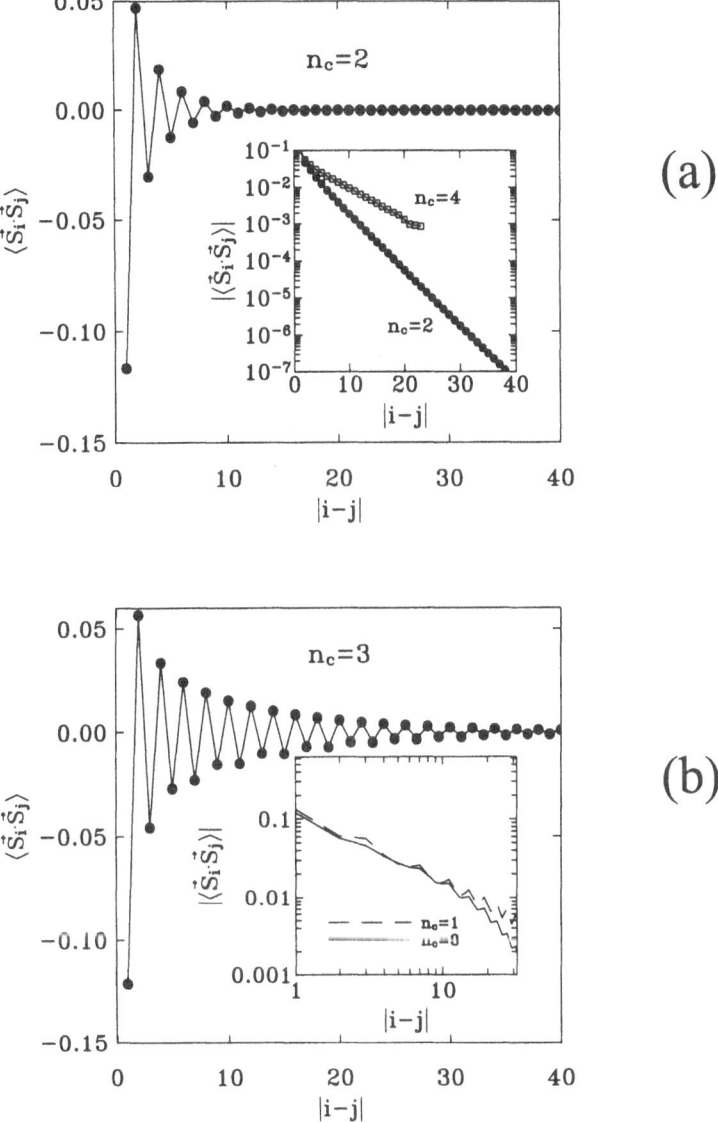

FIG. 2. Spin-spin correlations $\langle \mathbf{S}_i \cdot \mathbf{S}_j \rangle$ versus $|i-j|$ with i and j located on the top chain for (a) n_c even. The semilog plot in the inset shows the exponential decay of the correlations. (b) n_c odd. The log-log plot in the inset shows that the correlations for $n_c = 3$ and $n_c = 1$ decay with similar power-laws. The deviation from pure power-law behavior visible for the largest values of $|i-j|$ is due to finite-size effects from the open boundaries.

We first consider the dimer wavefunction for the two-chain system. A valence bond configuration for this state is formed by drawing dimer bonds connecting pairs of adjacent sites, with every site part of one bond. The resonance between different valence bond configurations leads to a substantial lowering of the energy. The simplest and perhaps most important type of resonance consists of a square of four adjacent sites fluctuating between two adjacent vertical bonds and two adjacent horizontal bonds [13]. Consider the possible resonances for a ladder system. The two types of bond configurations, "resonating" and "staggered", are shown in Fig. 3(a) and 3(b), respectively. The staggered type of configuration is incapable of resonance, and thus has higher energy. It is possible to form a local region of staggered bond order only by placing soliton spin defects at the edges of the region, as shown in Fig. 3(c). Hence to lowest order the staggered bond configurations can be ignored. One assumes that all resonating configurations are equally likely, and the ground state, within this variational estimate, is taken as the sum of all such configurations. The ground state energy within this variational state has been calculated analytically [14,15]. One obtains an average energy per site of -0.556029. Compared with the essentially exact result from the DMRG calculations of -0.578043140, the simple dimer RVB energy differs by less than 4%. While the variational energy is reasonable, the spin-spin correlation length $\xi = 0.238012$ calculated with this dimer RVB state is more than an order of magnitude smaller than our DMRG result of $\xi = 3.19$. This implies that $h(l)$ has a larger range. However, as discussed for the 2D lattice in Ref. [10], as long as $h(l)$ falls off exponentially one finds an exponential decay of spin correlations and a spin gap.

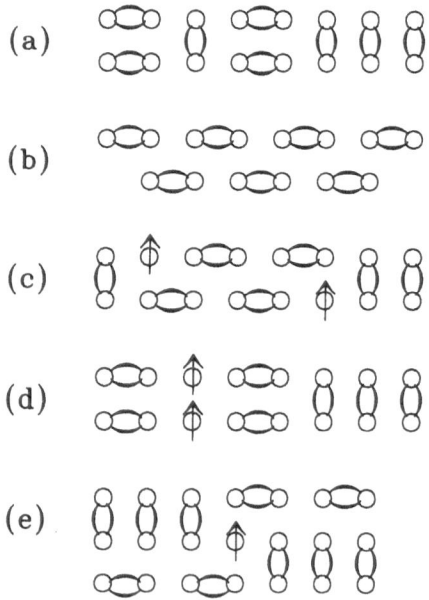

FIG. 3. Various dimer valence bond configurations, with and without topological spin defects present.

Although, as we have seen, the correlation length is poorly determined with the dimer RVB ansatz, a variety of qualitative feautures predicted by the ansatz are indeed present. For example, the variational state has a greater bond strength for interchain nearest-neighbor bonds compared to nearest-neighbor intrachain bonds. Most importantly, within the short-range RVB picture one expects to find that pairs of topological spin defects are bound. We see from Fig. 3(c) that two spin defects produce a region of staggered bond order between them if they are separated. Furthermore, one expects from this picture that the pair of defects should reside predominantly on a single rung, as in Fig. 3(d), rather than on adjacent sites on a single chain, in order to maximize resonance. Each of these predictions is supported by the DMRG calculations.

Now, it is possible to represent *any* singlet state as an RVB state [10], provided long-range singlet bonds are allowed. The crucial point in considering such an RVB representation is whether the amplitude for long-range bonds decays exponentially or algebraically, and if algebraically, with what exponent. Our DMRG results indicate that for the $n_c = 2$ and $n_c = 4$ systems the universality class is that of the short-range RVB, i.e. exponential decay of bond amplitudes. For quantitative results from the variational state, we expect that we must include some longer bonds and optimize over the bond amplitudes, but for qualitative results, the dimer state is adequate. The universality class for odd n_c is the long-range RVB state.

What is the behavior for even $n_c > 4$, and why is there different behavior for odd and even n_c? We believe the answer to this can be understood in terms of the confinement of topological defects present within a dimer RVB state with even n_c. The confinement for $n_c = 2$ is represented in Fig. 3(c), and the lack of confinement for $n_c = 3$ is shown in Fig. 3(e). In general, for even n_c, the presence of a single defect puts the system into a generalized form of staggered order characterized by an odd number of bonds crossing any vertical line separating rungs. We expect that this staggered order, although still capable of resonance for $n_c \geq 4$, is higher in energy than the "resonating" type of order. Thus defects are confined for an even number of chains, just as for the $n_c - 2$ case illustrated in Fig. 3(c) [16]. For odd n_c, there is only one type of order, characterized by an alternation as one moves along the chains of an odd number and an even number of bonds crossing a vertical line. A defect shifts the alternation by one lattice spacing, but with no cost in energy away from the defect.

The confinement of defects relates to the presence of long-range bonds in the ground state because a long-range bond can be considered to be a pair of separated topological defects. Thus considering a single long bond in a background of dimer bonds, we expect "confinement" of the long bond for even n_c; in other words, we expect it to be supressed exponentially with the separation, since the energy difference grows linearly with the size of the staggered region. (In making this argument, we are allowing the region between the two sites connected by the long bond to resonate between different valence bond configurations, while holding the long bond fixed. The same conclusion is obtained if we instead consider the *number* of valence bond configurations which have such a long bond.) Note also that the presence of non-dimer, but still short-ranged bonds does not heal the staggered order induced by the long-range bond. Such a short-range bond only heals the staggered order within the region of the bond. The presence of these short-range non-dimer bonds can be considered as "dressing" the dimer state, lowering the energies of regions with resonant bond order and with staggered order, but not changing the result that the staggered-order region is higher in energy. If a sufficiently high density of non-dimer bonds were present, the confinement picture might not be valid, but variational calculations for the 2D Heisenberg model show that even in long-range, low-energy RVB states, dimer bonds are much more probable than any other type of bond [10]. Thus it appears that this confinement mechanism is very effective at suppressing long-range bonds.

IV. DOPED LADDERS

In this section we discuss a theoretical framework for understanding hole motion in the t-J model, and show a few DMRG calculations for t-J clusters. The Hamiltonian is [17]

$$\mathcal{H} = \mathcal{H}_S + \mathcal{H}_K = J \sum_{\langle ij \rangle} (\vec{S}_i \cdot \vec{S}_j - \tfrac{1}{4} n_i n_j) - t \sum_{\langle ij \rangle, s} P_G(c_{i,s}^\dagger c_{j,s} + c_{j,s}^\dagger c_{i,s}) P_G \quad (11)$$

where $\langle ij \rangle$ denotes nearest-neighbor sites, s is a spin index, $\vec{S}_i$ and $c_{i,s}^\dagger$ are electron spin and creation operators, $n_i = c_{i,\uparrow}^\dagger c_{i,\uparrow} + c_{i,\downarrow}^\dagger c_{i,\downarrow}$, and the Gutzwiller projector P_G excludes configurations with doubly occupied sites. In the calculations shown here, we set the hopping $t = 1$ and the exchange $J = 0.5$.

Let $|\psi\rangle$ be the ground state of a particular t-J system with N sites and $N - m$ electrons. Define a hole projection operator for site i as $P_h(i) = c_{i,\downarrow} c_{i,\downarrow}^\dagger c_{i,\uparrow} c_{i,\uparrow}^\dagger$. $P_h(i)$ projects out the part of a wavefunction in which site i is vacant. Although we call this vacant site a "hole", there is not necessarily any spin associated with the vacancy. We define an operator $P_h(h)$, which projects out a particular configuration of m holes, as $P_h(h) = P_h(h_1) \ldots P_h(h_m)$, where $h = (h_1, \ldots, h_m)$, and $h_1 < \ldots < h_m$. We can then separate $|\psi\rangle$ into parts with specified hole locations as

$$|\psi\rangle = \sum_h P_h(h)|\psi\rangle = \sum_h a_h |\psi_h\rangle, \quad (12)$$

where $|\psi_h\rangle$ is a normalized wavefunction with holes at the specified sites, and $a_h > 0$. The ground state energy is given by

$$E = \sum_h a_h^2 \langle \psi_h | \mathcal{H}_S | \psi_h \rangle + \sum_h \sum_{h'} a_h a_{h'} \langle \psi_h | \mathcal{H}_K | \psi_{h'} \rangle. \quad (13)$$

The first term we refer to as the exchange energy, denoted by E_S. The second term in Eq. (13), the hopping energy or kinetic energy, can be written as

$$E_K = -t \sum_{\langle ij \rangle, s} \sum_h a_h a_{h'} \langle \psi_h | c_{i,s}^\dagger c_{j,s} | \psi_{h'} \rangle, \quad (14)$$

where the hole configurations h and h' are the same, except that h has a hole at site j and h' has one at site i. In general, we consider two hole configurations h and h' *adjacent* if they differ by a near-neighbor hop of a single hole. Define the hopping overlap between h and h' as

$$O_{h,h'} = \langle \psi_h | \sum_{\langle ij \rangle, s} (c_{i,s}^{\dagger} c_{j,s} + c_{j,s}^{\dagger} c_{i,s}) | \psi_{h'} \rangle. \tag{15}$$

Clearly a necessary condition for $O_{h,h'}$ to be nonzero is that h and h' are adjacent, in which case only one pair of sites i, j appears in the sum. If h and h' differ only in the position of hole m, $h_m \neq h'_m$, then

$$O_{h,h'} = \langle \psi_h | \sum_{s} c_{h'_m,s}^{\dagger} c_{h_m,s}) | \psi_{h'} \rangle. \tag{16}$$

It is easy to see that $|O_{h,h'}| \leq 1$. The kinetic energy can be written as

$$E_K = -t \sum_{h,h'} a_h a_{h'} O_{h,h'} \tag{17}$$

We see that we can view the ground state as the result of a set of coupled variational calculations, where the exchange energy of each wavefunction $|\psi_h\rangle$ is minimized, subject to having as much overlap as possible with adjacent hole configurations. For $t > J$, the interplay between the kinetic and exchange terms is interesting. In the low doping regime, since there are more exchange terms which come into play, the bulk spin behavior is dominated by exchange. Close to any holes, however, since $t > J$, substantial modifications of the local spin arrangements can occur. At higher doping, the bulk spin behavior can be changed substantially as well.

Using DMRG, we can study $|\psi_h\rangle$ directly: we calculate $|\psi\rangle$, and then measure operators of the form $AP_h(h)$ (or $P_h(h)AP_h(h)$), normalizing by $\langle \psi | P_h(h) | \psi \rangle$. It is useful to use $A = \vec{S}_i \cdot \vec{S}_j$, where i and j are near a hole or pair of holes. This measurement gives us a "snapshot" of the spin configuration around a dynamic hole. If this expectation value is close to -0.75 for two sites i and j, we say that there is a "singlet bond" connecting i and j, even if there is no term in the Hamiltonian directly coupling i and j. We use the terms "antiferromagnetic bond", "valence bond", or just "bond" simply to indicate that $\langle \vec{S}_i \cdot \vec{S}_j \rangle < 0$. Of course, Néel order makes weak "bonds" connecting widely separated sites on opposite sublattices, but we will be particularly concerned here with bonds connecting nearby sites on the *same* sublattice.

We can also take a snapshot of the spin configuration using $A = S_i^z$, for a single hole on an even number of sites. In that case, the ground state is degenerate with $S^z = \pm 1/2$, so that the expectation value of S_i^z in one of the ground states is finite. One can also project out some of the holes, and use $A = n_{i,s} = c_{i,s}^{\dagger} c_{i,s}$, to find out where the unprojected holes are, or $A = K_{ij} = -t \sum_s (c_{i,s}^{\dagger} c_{j,s} + c_{j,s}^{\dagger} c_{i,s})$, to study their motion.

A. Single chain

As a warmup exercise, consider the 1D t-J model, with one hole. One might consider as a variational ansatz for $|\psi_h\rangle$ a Néel arrangement of the electron spins, with one electron removed. In this ansatz we have made a "quasiparticle", since an extra spin 1/2 is associated with the hole. However, this is a very poor ansatz: $|\psi_h\rangle$ has no overlap with $|\psi_{h\pm 1}\rangle$. Alternatively, one can arrange the spins as shown in Fig. 4(a), with shifted Néel arrangements separated by the hole [18]. There are two spin wavefunctions $|\psi_h\rangle$, plus translations: for h odd (even), an up spin is to the left of the hole, while for h even (odd), a down spin is to the left of the hole. In this case there is complete overlap, and the kinetic energy associated with the hole takes on the maximal (in magnitude) value $-2t$. This is a simple intuitive argument for spin–charge separation in a 1D t-J model. Since a single hole moves freely, it also suggests that there is no kinematic reason for the binding of pairs of holes, although for unphysically large J/t the diagonal term in Eq.(13) can cause binding.

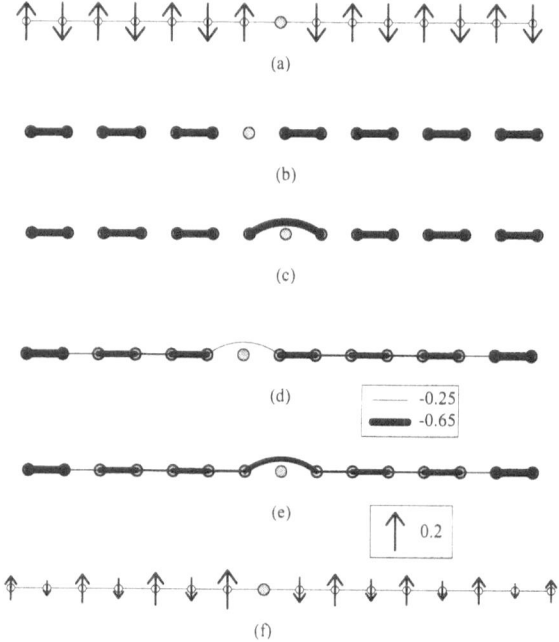

FIG. 4. Spin structure near a single hole (the gray circle) on a 1D t-J lattice. (a) Néel spin configuration, shifted by one spacing to the right of the hole. (b,c) Valence bond configurations with a hole. (d,e) Results of a DMRG calculation for the ground state of a 15 site t-J system, with $J/t = 0.5$, and open boundary conditions. The thickness of the lines is proportional to the bond strengths, $\langle \psi | \vec{S}_i \cdot \vec{S}_j P_h(h) | \psi \rangle / \langle \psi | P_h(h) | \psi \rangle$, according to the scale shown. In (d), $h = 7$, and in (e), $h = 8$. (f) Results of a DMRG calculation for the ground state of a 16 site system, with $J/t = 0.5$, and open boundary conditions. The length of the arrow is proportional to $\langle S^z P_h(h) \rangle / \langle \psi | P_h(h) | \psi \rangle$.

A justification for considering these Néel configurations for the 1D t-J model is the existence of power-law decaying antiferromagnetic correlations in the 1D Heisenberg model. Bond-bond correlations $\langle \vec{S}_i \cdot \vec{S}_{i+1} \vec{S}_j \cdot \vec{S}_{j+1} \rangle$ also decay as a power law, suggesting a valence bond configuration as a complementary ansatz: valence bonds occupy odd (even) links to the left of the hole, and even (odd) links to the right, as shown in Fig. 4(b). If one takes this configuration, and applies $\sum_s c_{i,s}^\dagger c_{j,s}$ to move the hole to a neighboring site, one obtains the configuration in Fig. 4(c), with a valence bond straddling the hole. Consequently, if we let the valence bond configuration of Fig. 4(b) define $|\psi_h\rangle$ for all odd sites h, and let the configuration of Fig. 4(c) define $|\psi_h\rangle$ for all even sites, then the hole moves freely, with the kinetic energy taking on its maximal value $-2t$.

In Fig. 4(d)-(e), we show DMRG results for the bond strength $A = \vec{S}_i \cdot \vec{S}_j$ for a single hole in a 15 site 1D chain, with open boundary condtions. The width of the line corresponding to each bond has been made proportional to the bond strength, as indicated by the scale in the box. The maximum possible bond strength is $-3/4$. The boundaries induce dimerization in the system, and the results are quite similar to the valence bond configurations shown in Fig. 4(b)-(c). It is also possible to obtain results which look like Fig. 4(a). In Fig. 4(f), we show results for the $S^z = 1/2$ ground state of a system with an even number of sites and one hole. The excess spin $1/2$ is spread out over the lattice.

Particularly interesting is the strength of the bond across the hole in Fig. 4(e). In order to maximize the hopping overlap with adjacent hole configurations, in addition to having antiferromagnetic correlations on nearest-neighbor links, we expect such correlations between *next-nearest-neighbor* sites i and j if there is a hole at site k which is a nearest-neighbor to both i and j. Such a valence bond becomes a nearest-neighbor link after one hop of the hole to either site, since moving the hole also moves the bond. For example, suppose the hole configuration h has a hole at site k, with i and j nearest-neighbor sites to k. Let h' be the hole configuration after the hole hops from k to i. Since j and k are nearest-neighbor sites, we expect a strong antiferromagnetic bond between these sites in $|\psi_{h'}\rangle$. In order to maximize the hopping overlap $O_{h,h'}$, there will also be a strong antiferromagnetic bond between sites i and j in $|\psi_h\rangle$. This tendency applies to two dimensions as well as one, and appears as an essential ingredient for pair binding in ladders and two dimensions.

B. Results: An 8×6 cluster

In Fig. 5 we show DMRG results for two holes on an 8×6 cluster. We kept 600 states per block, the truncation error was about 2×10^{-4}. This was sufficient to determine the structure of the pair with reasonable accuracy and to determine the pair binding energy. In Fig. 5(a) we show the expectation value of the kinetic energy of a hole, when the other hole has been projected onto a central site. The two holes are clearly bound. The most likely configuration for the holes is at diagonal next-nearest-neighbor sites. These configurations are more likely than nearest-neighbor configurations because they connect with more hole configurations h', giving them more weight in the kinetic part of the energy.

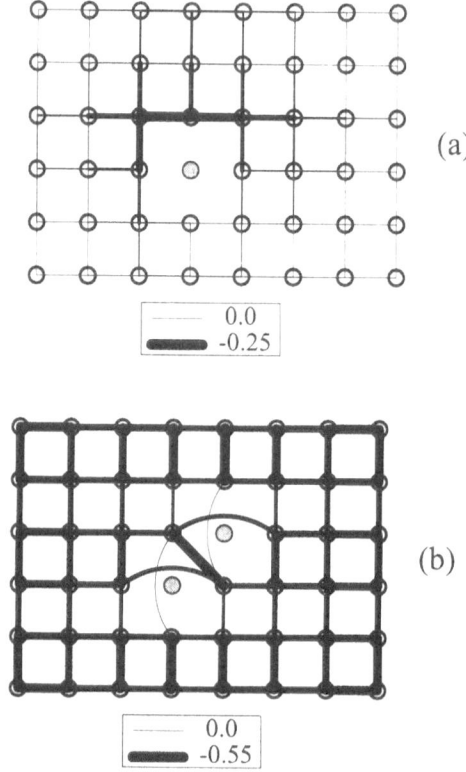

FIG. 5. Two dynamic holes on a 8 × 6 system, with open boundary conditions. (a) The hopping energy for each link when one hole is projected onto a particular site. (b) The bond strengths about the pair of holes.

In Fig. 5(b) we show the bond strengths surrounding several likely configurations of the pair. A frustrating diagonal singlet crossing the pair is present in Fig. 5(b): this is the clearest "signature" of a bound pair of holes, and is present in all the systems in which we have found pair binding. For four of the six hops available to (d), this bond becomes a nearest-neighbor bond, and in each of those neighboring configurations the bond is quite strong. Therefore, the kinetic term strongly favors a singlet bond connecting these sites. In addition, additional frustrating bonds crossing the holes are present in both directions. Vertical dimerization is present above and below the holes, where it is expected, and to the left and right, where we might have expected horizontal dimerization. Even on a system of width 6, the boundaries are still substantially affecting the spin structure surrounding the pair, and it is not clear which type of dimerization would appear in a large system. The most probable configuration of the pair is not shown: (3,4)-(4,3), with probability 0.018. Configuration (b) has a probability of 0.014.

The kinetic energy of a pair of holes on an 8 × 6 cluster is -5.36t. Twice the kinetic energy of a single hole is -5.38t. The increase in exchange energy caused by a pair of holes is 2.71t, compared with 2.96t for two separate holes. The increase in kinetic energy from binding a pair of holes is very tiny, and is more than made up for by the decrease in exchange energy. The pair binding energy is $E_b = 0.24(2)t$.

Frustrating next-nearest-neighbor bonds forming across holes are a universal feature in all of the clusters we have studied. These bonds are necessary for hole motion. Holes bind in pairs in order to share their frustration. This mechanism for pairing is quite different from simple "broken-bond" counting, which predicts nearest-neighbor pairing for *static* holes: for two static holes, a nearest neighbor configuration eliminates seven bonds, while anything else eliminates eight. For physical values of J/t, such as $J/t = 0.5$, the "broken-bond" effect enhances pair-binding somewhat, but is not dominant. Consider once again the 8×6 cluster, with two holes. Results for the hole-hole correlation function indicate that the pair resides on nearest-neighbor sites only 22% of the time. Even if a broken bond results in an extra exchange energy of $J = 0.5$, the effect on pair binding is only $0.22J = 0.11t$, while the actual pair binding energy is $0.24(2)t$. A more accurate estimate of the effect of broken bonds comes from considering two *static* holes on an 8×6 cluster: the difference in energy between nearest-neighbor static holes and widely separated holes is $0.62J$, rather than J, suggesting that the broken-bond energy for dynamic holes is about 0.07t. The majority of the pair binding energy comes from the reduction in frustration, which is induced by hole motion. for bound pairs of holes.

ACKNOWLEDGEMENTS

We acknowledge support from the from the NSF under Grant Nos. DMR-9509945. Some of the calculations were performed at the San Diego Supercomputer Center.

[1] S.R. White, Phys. Rev. B **48**, 10345 (1993).

[2] S.R. White, R.M. Noack, and D.J. Scalapino, Phys. Rev. Lett. **73**, 886 (1994).

[3] S.R. White and D.J. Scalapino, preprint, cond-mat/9605143.

[4] K.G. Wilson, *Rev. Mod. Phys.* 47, 773 (1975).

[5] S.R. White and R.M. Noack, *Phys. Rev. Lett.* **68** , 3487 (1992).

[6] S.R. White, *Phys. Rev. Lett.* **69** , 2863 (1992).

[7] R.M. Noack and S.R. White, to appear in *Phys. Rev.* , (manuscript number BW4753).

[8] R.P. Feynman, *Statistical Mechanics: A Set of Lectures*, (Benjamin, Reading, MA, 1972).

[9] E.R. Davidson, *J. Comp. Phys.* **17**, 87 (1975).

[10] S. Liang, B. Doucot, and P.W. Anderson, Phys. Rev. Lett. **61**, 365 (1988).

[11] Erik S. Sørensen and Ian Affleck, Phys. Rev. Lett. **71**, 1633 (1993).

[12] P.W. Anderson, Science **235**, 1196 (1987).

[13] S.A. Kivelson, D.S. Rokhsar, and J.P. Sethna, Phys. Rev. B **35**, 8865 (1987).

[14] Y. Fan and M. Ma, Phys. Rev. B **37**, 1820 (1988).

[15] Thomas Blum and Yonathan Shapir, *J. Stat. Phys.* **59**, 333 (1990).

[16] If it turns out that the staggered order is lower in energy, then the ground state has staggered order and there is still binding of defects because of the induced resonating order between the defects.

[17] F.C. Zhang, T.M. Rice, Phys. Rev. B **37**, 3759 (1988).

[18] M. Ogata and P.W. Anderson, Phys. Rev. Lett. **70**, 3087 (1993).

An Introduction to Quantum Monte Carlo Methods

Anders W. Sandvik

National High Magnetic Field Laboratory,
Florida State University, 1800 E. Paul Dirac Dr.,
Tallahassee, Florida 32306

This is an overview of some numerical many-body methods in the quantum Monte Carlo (QMC) simulation family. First, the classical Monte Carlo technique (the Metropolis algorithm) is briefly reviewed. Then, the problem of rewriting a quantum mechanical expectation value as a form accessible to similar stochastic methods is considered. The transformations leading to three different finite-temperature lattice QMC algorithms — the worldline method, the fermion determinant method and a power series expansion method ("stochastic series expansion") — are discussed in some detail.

I. INTRODUCTION

Quantum Monte Carlo (QMC) simulation methods are very efficient calculational tools for some classes of interacting many-particle models, in particular those for which the so called "sign problem" can be avoided (i.e., a positive definite weight function can be constructed). This is the case, e.g., for one-dimensional fermion systems and non-frustrated quantum spin systems in any number of dimensions. Even in cases where simulations are hampered by the sign problem, non-perturbative results can often be obtained that are not accessible in any other way.

These notes provide a brief introduction to some of the many QMC methods used today. The techniques considered are suitable for studies of lattice Hamiltonians, such as the Heisenberg and Hubbard models, at finite temperature. The main goal is to make the reader acquainted with a few of the different ways in which a quantum mechanical expectation value can be transformed into a form accessible to stochastic methods analogous to the powerful Monte Carlo methods used in classical statistical mechanics. For readers not familiar with classical Monte Carlo, the Metropolis algorithm is first briefly reviewed. The transformations forming the basis of three different QMC algorithms are then discussed. Two of these, the worldline [1] and the fermion determinant [2] methods, are based on the so called Trotter approximation [3], whereas the third instead employs a power series expansion ("stochastic series expansion") [4–7]. While the formalisms of the transformations are discussed in some detail, the computations involved in actual QMC simulations are only briefly touched upon. Such technical aspects are of course important if one wants to write a program to study a particular model, but are not critical to obtaining an understanding of the basic principles. More details can be found in the articles cited. The notes conclude with a brief discussion of some problems where QMC methods such as those described here have been successfully applied.

II. CLASSICAL MONTE CARLO

The purpose of a classical Monte Carlo simulation is to obtain estimates of thermal expectation values of physical observables for some model described by a classical Hamiltonian H, by sampling a representative set of configurations of its degrees of freedom [8]. The perhaps simplest model in the context of which to discuss the principles of the technique is the Ising model, defined by

$$H = -J \sum_{\langle i,j \rangle} \sigma_i \sigma_j, \tag{1}$$

where $\langle i,j \rangle$ is a pair of nearest-neighbor lattice sites, and there is a "spin" degree of freedom $\sigma_i = \pm 1$ (up and down) at each site i. The coupling $-J$ ($J > 0$) favors a ferromagnetic alignment of the spins. Without going too deep into the physics of the Ising model, it should be mentioned that it can be solved exactly in one and two dimensions. A one-dimensional Ising chain has long-range order only at $T = 0$, whereas on a two-dimensional square lattice there is a second order phase transition at $T/J \approx 2.17$. In higher dimensions the model cannot be solved exactly. Many results, including critical exponents, have been obtained using Monte Carlo methods such as the one described below.

The ensemble average of a quantity A at inverse temperature $\beta = 1/T$ is given by

$$\langle A \rangle = \frac{\sum_i A(C_i) e^{-\beta E(C_i)}}{\sum_i e^{-\beta E(C_i)}} = \frac{\sum_i A(C_i) W(C_i)}{\sum_i W(C_i)}, \tag{2}$$

where the sum runs over all the configurations $\{C_i\} = \{\sigma_1, \sigma_2, \ldots, \sigma_N\}$ of the spins, $E(C_i)$ is the energy of configuration C_i and $A(C_i)$ is the corresponding value of the observable A. A typical quantity of interest would be the spin correlation function $\langle \sigma_i \sigma_j \rangle$, which decays exponentially to zero as the distance $|\mathbf{r}_i - \mathbf{r}_j| \to \infty$ if the system is not ordered, and approaches a finite value if the system has long range order. Thermodynamic quantities such as the susceptibility and the heat capacity are of course also of interest.

A trivial, and very inefficient, stochastic method for estimating an expectation value of the form (2) is to randomly generate a number M of configurations, $C_{R(1)}, C_{R(2)}, \ldots, C_{R(M)}$, and simply use Eq. (2) with this subset. That this indeed is an inefficient procedure is clear at low temperatures, where only configurations with almost all spins pointing in the same direction contribute significantly. Most randomly generated configurations, on the other hand, have approximately equal numbers of up and down spins. While one would in principle obtain the correct expectation value with M very large, a practically feasible sampling would typically lead to few, if any, important configurations. This would be reflected by large statistical fluctuations in calculated expectation values.

An efficient way to select configurations is *importance sampling*, and this is generally what the term "Monte Carlo simulation" refers to in statistical mechanics. Importance sampling amounts to generating M configurations $C_{I(1)}, C_{I(2)}, \ldots, C_{I(M)}$, which are statistically distributed according to the *weight function* $W(C_i)$ in Eq. (2). With the weighting taken into account already in the selection of configurations, the estimate of the expectation value $\langle A \rangle$ is simply given by the arithmetic average of the function $A(C_{I(i)})$ over the generated subset:

$$\langle A \rangle = \frac{1}{M} \sum_{j=1}^{M} A(C_{I(j)}). \tag{3}$$

The accuracy of this estimate scales as $1/\sqrt{M}$, with the prefactor depending on the way the configurations are generated (i.e., the presence of correlations among successive configurations).

In order to achieve the correct distribution, a certain "random walk" in the configuration space is performed. Starting from an arbitrary configuration $C_I(0)$, a series of configurations $C_{I(1)}, C_{I(2)}, \ldots$ is generated by making random changes ("updates"), that are accepted or rejected according to rules which guarantee that an equilibrium will be reached for which the desired probability distribution is obeyed. This is the case if the *detailed balance condition*,

$$\frac{P(C_i \rightarrow C_j)}{P(C_j \rightarrow C_i)} = \min\left[\frac{W(C_j)}{W(C_i)}, 1\right], \tag{4}$$

is satisfied. Here $P(C_i \rightarrow C_j)$ denotes the probability of making a change in the current configuration $C_{I(n)} = C_i$, such that the next configuration $C_{I(n+1)} = C_j$, and $\min(x, y)$ is the smaller of the arguments x, y. A process satisfying detailed balance eventually leads to configurations distributed according to W, provided that the process is *ergodic*, i.e., that any configuration can (in principle) be reached by a series of updates. The perhaps most widely used simulation method is the Metropolis algorithm [9], which involves the following simple steps:

(1) Start with an arbitrary configuration $C_{I(0)}$ ($n = 0$).

(2) Make a change in $C_{I(n)}$: $C_{I(n)} \rightarrow C'_{I(0)}$.

(3) Evaluate the weight ratio $R = W(C'_{I(n)})/W(C_{I(n)}) = \exp(\beta[E(C_{I(n)}) - E(C'_{I(n)})])$.

(4) If $R \geq 1$ [$E(C'_{I(n)}) < E(C_{I(n)})$]: Accept the new configuration, i.e., let $C_{I(n+1)} = C'_{I(n)}$. If $R < 1$: Accept only with a probability $P = R$. Reject (i.e., keep the old configuration; $C_{I(n+1)} = C_{I(n)}$) with a probability $P = 1 - R$.

(5) Repeat from (2), with $n \rightarrow n + 1$.

For the Ising model, updates involving "flips" of single spins, $\sigma_i \rightarrow -\sigma_i$, suffice for making the process ergodic. One can either go through every spin $i = 1, \ldots, N$, in some given order, or select a spin at random at every step. The calculation of the weight ratio involves only the nearest-neighbors of the flipped spin, and can be rapidly evaluated. The acceptance with a probability $P = R$ in the case $R < 1$ in (4) is of course achieved by generating a random number between 0 and 1, and comparing this number with R.

Hence, at each iteration ("time"), the system can either evolve to a new configuration, or remain in the previous configuration. The average probability of evolving is called the *acceptance rate*. The acceptance rate of course depends on the types of updates used. Attempting to flip more than one spin at a time typically leads to a lower acceptance rate, but each accepted update represents a larger evolution of the system, and therefore may still be more efficient.

The dynamics of a simulation is characterized by an *autocorrelation time*. The *autocorrelation function* for a quantity A is defined as the normalized time-averaged correlation between measurements separated by t updating steps:

$$C_A(t) = \frac{\langle A(n+t)A(n) \rangle - \langle A(n) \rangle^2}{\langle A(n)^2 \rangle - \langle A(n) \rangle^2}. \tag{5}$$

Asymptotically $C_A(t)$ decays as $e^{-t/\Theta}$, where Θ is the autocorrelation time. The short-time behavior of $C_A(t)$ depends on the quantity A, whereas Θ is a characteristic time-scale of the simulation itself. Since measurements separated by a time t do not contain significantly different information if $C_A(t)$ has not decayed appreciably from 1, a simulation with a long Θ is inefficient, even if the acceptance rate may be reasonable. This is the case for algorithms utilizing local updates (involving one or a small number of spins) if the system is close to a second order phase transition, where there are fluctuations on large length scales. Small local updates can only gradually lead to fluctuations of large ordered domains, and a simulation of practical duration may therefore not efficiently sample all types of important configurations. Non-Metropolis methods based on global updates (flips of any-size clusters of spins) have been developed that can overcome this *critical slowing down* problem [10].

Monte Carlo methods such as that outlined above can in principle be used for any classical many-body system, but the details of the updating procedure of course depends on the system under consideration.

III. QUANTUM MONTE CARLO

The purpose of a finite-temperature QMC calculation is to evaluate a quantum mechanical expectation value

$$\langle \hat{A} \rangle = \frac{1}{Z} \text{Tr}\{\hat{A}e^{-\beta \hat{H}}\}, \quad Z = \text{Tr}\{e^{-\beta \hat{H}}\}, \tag{6}$$

using stochastic methods analogous to the classical Monte Carlo algorithm described in the previous section. If the eigenstates and energies of $\hat{H}$ are not known, Monte Carlo cannot be directly applied. One then has to employ some "trick" (transformation) in order to construct a configuration space $\{C_i\}$ with an associated weight function $W(C_i)$, such that the expectation value can be written in the classical form

$$\langle \hat{A} \rangle = \frac{\text{Tr}\{\hat{A}e^{-\beta\hat{H}}\}}{\text{Tr}\{e^{-\beta\hat{H}}\}} \rightarrow (\text{tricks}) \rightarrow \frac{\sum_i A(C_i)W(C_i)}{\sum_i W(C_i)}. \tag{7}$$

The nature of the configuration space depends on the transformation used, and hence there are typically many possible QMC methods for a given Hamiltonian (some more efficient than others).

A fundamental problem in some cases (in fact, in most cases) is that the resulting weight function $W(C_i)$ is not positive definite. Hence, it cannot be used as a probability distribution in a Monte Carlo process. This difficulty is normally referred to as "the sign problem" [1,11,12]. Simulations can in principle still be carried out, using $|W(C_i)|$ for the probability distribution. Denoting the sign associated with the weight $S(C_i) = \pm1$, one can easily verify that Eq. (7) can also be written as

$$\langle A \rangle = \frac{\sum_i A(C_i)S(C_i)|W(C_i)|}{\sum_i |W(C_i)|} \times \frac{1}{\langle S \rangle} = \frac{\langle AS \rangle}{\langle S \rangle}, \tag{8}$$

where the "average sign" $\langle S \rangle$ is given by

$$\langle S \rangle = \frac{\sum_i |W(C_i)|}{\sum_i W(C_i)}. \tag{9}$$

The average sign is measured in the simulation along with the average $\langle AS \rangle$ of $A(C_i)S(C_i)$, and the ratio is calculated after the simulation. This process clearly becomes unstable if $\langle S \rangle \rightarrow 0$ (which also implies $\langle AS \rangle \rightarrow 0$). This is unfortunately the case at low temperatures for many of the most interesting systems, such as frustrated quantum spins and interacting electrons in dimensions higher than one. The sign problem can be avoided (i.e., a transformation is known for which $W(C_i)$ is positive definite), for, e.g., non-frustrated quantum spin systems, 1D fermion systems, electron-phonon systems (also with electron-electron interactions included in $D = 1$), and the half-filled Hubbard model in any number of dimension.

Next, three different QMC algorithms are discussed. The intention here is only to give an impression of the theoretical foundations of the methods. The focus is therefore on the transformations used, and the relationships between the degrees of freedom of the resulting "classical" system and physical observables of the original quantum model. Actual simulations of the corresponding classical systems are only discussed briefly. Details of these important technical points can be found in the references provided.

A. The Worldline Method

The worldline method is the perhaps most straight-forward way of constructing an effective classical system corresponding to a quantum many-body Hamiltonian. It employs a real-space, discrete imaginary-time (Euclidean) path integral representation [13], constructed using the so called *Trotter break-up* of the operator $\exp(-\beta\hat{H})$ [3,14,1].

In general, to construct a path integral corresponding to a Hamiltonian $\hat{H}$, the partition function is first written in the form

$$Z = \mathrm{Tr}\{e^{-\beta\hat{H}}\} = \mathrm{Tr}\{\prod_{l=1}^{L} e^{-\Delta\tau\hat{H}}\}, \quad \Delta\tau = \beta/L, \tag{10}$$

where $\Delta\tau$ is the discretization in imaginary time, or the "time slice width". A basis $\{|\alpha\rangle\}$, is chosen, unit operators $\sum_\alpha |\alpha\rangle\langle\alpha|$ are inserted between the exponentials, and the trace is written as a sum over diagonal matrix elements:

$$Z = \sum_{\{\alpha_i\}} \langle\alpha_0|e^{-\Delta\tau\hat{H}}|\alpha_{L-1}\rangle\langle\alpha_{L-1}|\cdots|e^{-\Delta\tau\hat{H}}|\alpha_1\rangle\langle\alpha_1|e^{-\Delta\tau\hat{H}}|\alpha_0\rangle. \tag{11}$$

The idea is now to calculate the matrix elements approximately, in a way that becomes exact as $\Delta\tau \to 0$. Writing the Hamiltonian as a sum of two terms, $\hat{H}_A$ and $\hat{H}_B$, the Trotter approximation for non-commuting operators,

$$e^{\Delta\tau(\hat{H}_A+\hat{H}_B)} \approx e^{\Delta\tau\hat{H}_A}e^{\Delta\tau\hat{H}_B}, \tag{12}$$

carried out for all exponentials in (11) typically introduces errors in Z and $\langle A\rangle$ of order $(\Delta\tau)^2$ [15]. Each matrix element in Eq. (11) is then written as

$$\langle\alpha_{l+1}|e^{-\Delta\tau\hat{H}}|\alpha_l\rangle \approx \sum_{\beta_l}\langle\alpha_{l+1}|e^{-\Delta\tau\hat{H}_A}|\beta_l\rangle\langle\beta_l|e^{-\Delta\tau\hat{H}_B}|\alpha_l\rangle. \tag{13}$$

The decomposition into $\hat{H}_A$ and $\hat{H}_B$, and the basis states $\{|\alpha\rangle\}$ and $\{|\beta\rangle\}$, must be chosen so that the matrix elements can be evaluated (sometimes a decomposition into more than two operators has to be considered). This way, the problem can be reduced to a multi-dimensional sum over states. The numerator in Eq. (7) is transformed in the same way, and the expectation value has then been cast into the form required for evaluation using the Monte Carlo method. Although the path-integral construction in principle is completely general, its practical use in the context of numerical simulations is limited to cases where the transformation can be done in such a way that all the terms (or at least most of them) are positive. This is the case for some classes of important models, such as one-dimensional interacting fermions, and non-frustrated quantum spin and boson models (in any number of dimensions).

In a numerical simulation the time slice $\Delta\tau$ is kept finite, but small enough for the resulting error to be small. If exact results are needed, a scaling to $\Delta\tau \to 0$ can be performed, based on calculations for different $\Delta\tau$ and the asymptotic $(\Delta\tau)^2$ form for the systematical error.

The term "worldline method" normally refers to the formulation employing a certain real-space decomposition of the Hamiltonian, introduced by Hirsch, Scalapino, Sugar, and Blankenbecler in the context of 1D fermion models [1]. A slightly different construction was previously suggested for quantum spin systems by Suzuki and collaborators [14].

The method will here be discussed within a simple, yet nontrivial example, namely the $S = 1/2$ antiferromagnetic Heisenberg chain. This model is defined by the Hamiltonian

$$\hat{H} = J \sum_{i=1}^{N} \mathbf{S}_i \cdot \mathbf{S}_{i+1} = J \sum_{i=1}^{N} [S_i^z S_{i+1}^z + \tfrac{1}{2}(S_i^+ S_{i+1}^- + S_i^- S_{i+1}^+)], \tag{14}$$

and is, by the Jordan-Wigner transformation, equivalent to a system of spinless fermions. Here the worldline algorithm is discussed using the spin language, but the correspondence with tight-binding fermions is trivial. The spin states $S_i^z = \pm 1/2$ correspond to the presence or absence of a particle at site i ($n_i = 0, 1$), and the spin flipping operators $S_i^+ S_{i+1}^- + S_i^- S_{i+1}^+$ become fermion hopping operators $c_i^+ c_{i+1} + c_{i+1}^+ c_i$. The diagonal $S_i^z S_{i+1}^z$ corresponds to a nearest-neighbor interaction between the fermions. Including other diagonal fermion interactions poses no problems in principle.

For the path-integral construction, the two-body operators are first divided into two terms, acting on the even and odd "bonds", respectively, i.e.,

$$\hat{H} = \sum_{i=1}^{N} \hat{H}_i = \sum_{\text{even } i} \hat{H}_i + \sum_{\text{odd } i} \hat{H}_i = \hat{H}_{\text{even}} + \hat{H}_{\text{odd}}. \tag{15}$$

This is generally referred to as the *checkerboard decomposition*, for reasons that will become evident below. All operators in $\hat{H}_{\text{even}}$ and $\hat{H}_{\text{odd}}$ commute with each other, but $\hat{H}_{\text{even}}$ and $\hat{H}_{\text{odd}}$ do not commute. The Trotter break-up is used to separate the even and odd terms:

$$Z = \text{Tr}\{e^{-\beta(\hat{H}_{\text{even}} + \hat{H}_{\text{odd}})}\} = \text{Tr}\{\prod_{l=1}^{L} e^{-\Delta\tau(\hat{H}_{\text{even}} + \hat{H}_{\text{odd}})}\}$$

$$\approx \text{Tr}\{\prod_{l=1}^{L} e^{-\Delta\tau \hat{H}_{\text{even}}} e^{-\Delta\tau \hat{H}_{\text{odd}}}\}. \tag{16}$$

Choosing the basis $\{|\alpha\rangle\} = \{|S_1^z, S_2^z, \ldots, S_N^z\rangle\}$, this can be written as

$$Z = \sum_{\{\alpha_i\}} \langle\alpha_0|e^{-\Delta\tau\hat{H}_{\text{odd}}}|\alpha_{2L-1}\rangle\langle\alpha_{2L-1}|e^{-\Delta\tau\hat{H}_{\text{even}}}|\alpha_{2L-2}\rangle\langle\alpha_{2L-2}|\cdots$$

$$\cdots|\alpha_2\rangle\langle\alpha_2|e^{-\Delta\tau\hat{H}_{\text{odd}}}|\alpha_1\rangle\langle\alpha_1|e^{-\Delta\tau\hat{H}_{\text{even}}}|\alpha_0\rangle. \tag{17}$$

Since the operators in, e.g., H_{odd} commute,

$$e^{-\Delta\tau\hat{H}_{\text{odd}}} = e^{-\Delta\tau\hat{H}_1}e^{-\Delta\tau\hat{H}_3}e^{-\Delta\tau\hat{H}_5}\ldots e^{-\Delta\tau\hat{H}_{N-1}}, \tag{18}$$

all matrix elements of $e^{-\Delta\tau\hat{H}_{\text{odd}}}$ are products of 2-spin matrix elements (here an even N is assumed),

$$\langle\alpha_{l+1}|e^{-\Delta\tau\hat{H}_{\text{odd}}}|\alpha_l\rangle =$$
$$\langle S_1^z(l+1)S_2^z(l+1)|e^{-\Delta\tau\hat{H}_1}|S_1^z(l)S_2^z(l)\rangle \times$$
$$\langle S_3^z(l+1)S_4^z(l+1)|e^{-\Delta\tau\hat{H}_3}|S_3^z(l)S_4^z(l)\rangle \times \cdots \times$$
$$\langle S_{N-1}^z(l+1)S_N^z(l+1)|e^{-\Delta\tau\hat{H}_{N-1}}|S_{N-1}^z(l)S_N^z(l)\rangle, \tag{19}$$

and can be easily calculated. Since each $\hat{H}_i$ conserves the z-component of the spin on the pair of sites $\langle i, i+1\rangle$, there are only 6 non-zero matrix elements:

$$\langle\uparrow\uparrow|e^{-\Delta\tau H_i}|\uparrow\uparrow\rangle = \langle\downarrow\downarrow|e^{-\Delta\tau H_i}|\downarrow\downarrow\rangle = \exp(-\Delta\tau/4) \tag{20a}$$

$$\langle\uparrow\downarrow|e^{-\Delta\tau H_i}|\uparrow\downarrow\rangle = \langle\downarrow\uparrow|e^{-\Delta\tau H_i}|\downarrow\uparrow\rangle = \exp(\Delta\tau/4)\cosh(\Delta\tau/2) \tag{20b}$$

$$\langle\uparrow\downarrow|e^{-\Delta\tau H_i}|\downarrow\uparrow\rangle = \langle\downarrow\uparrow|e^{-\Delta\tau H_i}|\uparrow\downarrow\rangle = -\exp(\Delta\tau/4)\sinh(\Delta\tau/2). \tag{20c}$$

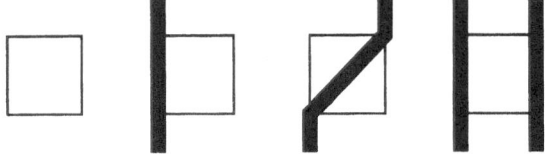

FIG. 1 Graphical representations of the non-zero matrix elements
$\langle S_i^z(l+1)S_{i+1}^z(l+1)|e^{-\Delta\tau\hat{H}_i}|S_i^z(l)S_{i+1}^z(l)\rangle$. These are, from left to right,
$\langle\downarrow\downarrow|e^{-\Delta\tau H_i}|\downarrow\downarrow\rangle$, $\langle\uparrow\downarrow|e^{-\Delta\tau H_i}|\uparrow\downarrow\rangle$, $\langle\uparrow\downarrow|e^{-\Delta\tau H_i}|\downarrow\uparrow\rangle$, $\langle\uparrow\uparrow|e^{-\Delta\tau H_i}|\uparrow\uparrow\rangle$ (two more are obtained by permuting $\uparrow$ and $\downarrow$). The lower and upper corners of the squares represent the spin states l and $l+1$, respectively. The presence of a bold line indicates an $\uparrow$ state, and an empty corner a $\downarrow$ state. A diagonal line indicates that the two spins are flipped, whereas straight lines (or no lines) correspond to diagonal matrix elements.

These can be represented graphically as explained in Fig. III A. A product of matrix elements constituting a non-vanishing term in the partition function (17) can then be graphically represented by a "checkerboard" lattice of 2-spin matrix elements as shown in Fig. III A. In this representation of the configurations, the effective classical system is one of "worldlines" for the up spins (or particles in the fermion language). The weight of a term is given by the product of all the matrix elements, implying that worldlines have to be continuous and periodic in the "time" direction. Even though the matrix elements (20c) corresponding to spin flipping processes are negative, every term in Z is positive, since the total number of spin flips (diagonal worldline segments) must be even (for a periodic system, this is strictly true only if the number of sites is even). There is hence no sign problem, and the system can be studied using standard Monte Carlo sampling techniques. In the fermion case, some configurations may be (depending on N and the number of fermions) associated with negative phase factors due to cyclic permutations of the worldlines between $\tau = 0$ and $\tau = \beta$, if periodic boundary conditions are used. The sign problem resulting from this can be avoided by a modification of the boundary condition [1] (e.g., in some cases antiperiodic boundary conditions are appropriate).

FIG. 2 Worldline representation of a term in the partition function. A pair of rows representing a double matrix element $\langle \alpha_{l+2} | e^{-\Delta\tau \hat{H}_{\text{even}}} | \alpha_{l+1} \rangle$ and $\langle \alpha_{l+1} | e^{-\Delta\tau \hat{H}_{\text{odd}}} | \alpha_l \rangle$ constitutes one time slice. White squares represent 2-spin matrix elements, and must be associated only with worldlines as illustrated in Fig. 1 in order for the configuration to have a non-zero weight. Shaded squares indicate the absence of operators that can flip the corresponding spin pairs, and diagonal worldline segments are therefore not allowed on these. A non-vanishing term must have only un-broken worldlines, and the trace implies that the states at $\tau = 0$ and $\tau = \beta$ must be the same. This configuration has winding number $W = 0$, since there are no cyclic permutations of lines between $\tau = 0$ and $\tau = \beta$.

The configuration updating can be accomplished by small local distortions of the worldlines, as illustrated in Fig. III A. The weight change needed for calculating the acceptance probability associated with such a modification can be easily obtained from Eqs. (20). These updates cannot change the total magnetization, $m = \sum_i S_i^z$, of the system (or the number of particles in the fermion case). Hence, in order to study a grand canonical ensemble, additional updates where whole worldlines are inserted and removed must be considered. The acceptance rate for such updates is typically very low at low temperatures, and the method is therefore better suited for systems in the canonical ensemble. For systems with periodic boundary conditions, there is yet another type of update needed in order for the process to be ergodic. Since the spins (or fermions) are indistinguishable, configurations for which the worldlines are cyclically permuted between $\tau = 0$ and $\tau = \beta$ also contribute. A configuration where the lines are cyclically permuted W steps is said to have a *winding number* W. Local updates conserve the winding number, and one needs to construct a global update for studying a system with fluctuating W. This can be done in principle, but in practice the acceptance rate for changing W becomes very low (for Metropolis-type schemes deviced so far) as the system size grows, and most simulations with the worldline method therefore have to be restricted to the $W = 0$ subspace [1]. This is an approximation for a finite system, but not in the thermodynamic limit since the winding number is associated with the boundary condition.

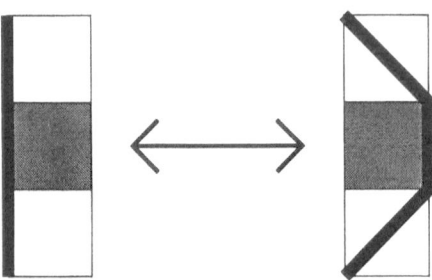

FIG. 3 Example of a local update of a worldline. A series of such updates can lead to any worldline configuration in the canonical ensemble, in a given winding number sector.

Now, consider the numerator $\mathrm{Tr}\{\hat{A}e^{-\beta\hat{H}}\}$ in Eq. (6). For an operator $\hat{A}$ which is diagonal in the real-space basis used, constructing the worldline path-integral for the numerator results in the same set of configurations as that contributing to the denominator Z, with the weight multiplied by the corresponding eigenvalue of $\hat{A}$ in the state $|\alpha_0\rangle$. Hence, the "measuring function" $A(C_i)$ to be averaged during the simulation is simply the eigenvalue. Since the trace is taken, there is nothing special with the state $|\alpha_0\rangle$ (cyclically permuting the states $|\alpha_l\rangle$ an arbitrary number of times leads to an allowed configuration with the same weight), and one can average the eigenvalue over all time slices, which greatly improves the simulation statistics. For example, an equal-time spin correlation function $\langle S_i^z S_j^z \rangle$ (corresponding to the density-density correlation $\langle n_i n_j \rangle$ in the fermion language) is given by

$$\langle S_i^z S_j^z \rangle = \frac{1}{2L} \sum_{l=0}^{2L-1} \langle S_i^z(l) S_j^z(l) \rangle. \tag{21}$$

For periodic systems, correlation functions can of course also be averaged in space. An imaginary-time dependent correlation function is obtained averaging correlations between states with the corresponding time-slice separation:

$$\langle S_i^z(\tau = n\Delta\tau) S_j^z(\tau = 0) \rangle = \frac{1}{2L} \sum_{l=0}^{2L-1} \langle S_i^z(l) S_j^z(l+2n) \rangle. \tag{22}$$

The static susceptibility χ_{ij}, i.e., the linear response dS_i^z/dh_j of S_i^z to a local field h_j coupled to S_j^z (via $h_j S_j^z$), is given by the Kubo formula

$$\chi_{ij} = \int_0^\beta d\tau \langle S_j^z(\tau) S_i^z(0) \rangle, \qquad (23)$$

and can be obtained by numerically integrating Eq. (22) over the discrete set of τ-points available with the discretization used.

Expectation values of off-diagonal operators such as $\langle S_i^-(\tau) S_j^+(0) \rangle$ (corresponding to a single-particle Green's function in the fermion case) cannot be evaluated as easily, since the worldline representation of the numerator in Eq. (6) then involves an additional finite worldline segment between time slices l_1 and l_2, with $\Delta\tau(l_2 - l_1) = \tau$. This segment can follow many different paths between its fixed endpoints, and if the space-time separation is large it is impossible to sum up all contributions exactly. One can in principle obtain an estimate by a stochastic sampling of the paths [1], but in practice this is often too time consuming. An exception is the case $\tau = 0$, $j = i \pm 1$, for which the contributing numerator configurations has only continuous worldlines and hence are the same as those for the partition function. The expectation value can then be evaluated as easily as in the diagonal case. This is true also for off-diagonal expectation values of the form $\langle S_i^-(\tau) S_{i\pm1}^+(\tau) S_j^-(0) S_{j\pm1}^+(0) \rangle$, which appear, e.g., in current-current correlation functions [1].

The checkerboard decomposition can be directly generalized to higher dimensions. For, e.g., a 2D Heisenberg model the Hamiltonian can be decomposed into four terms, each containing only mutually commuting operators. For systems with longer-range interactions the checkerboard decomposition cannot be used, since the operators in $\hat{H}_{\text{even}}$ and $\hat{H}_{\text{odd}}$ (or the corresponding operators in higher dimensions) are then not commuting. Straigh-forward generalizations are possible, however. Unfortunately, for frustrated systems the weight function is not positive definite (frustration, or competing interactions, occurs when closed loops on the lattice can be constructed that contain an odd number of antiferromagnetic bonds, so that all spins on the loop cannot be oriented relative their interacting neighbors as favored by the signs of the interactions). This is due to the negativity of spin-flipping matrix elements, such as (20c), for antiferromagnetic couplings. With interactions extending further than between nearest neighbors, an odd number of such matrix elements can appear in allowed worldline configurations, resulting in a negative weight. This sign problem generally makes simulations of frustrated systems difficult (note, however, that frustration in only the diagonal part of the interaction does not cause a sign problem).

The checkerboard decomposition described above can be used also for spin-1/2 fermions in one dimension, as described in the original paper by Hirsch *et al.* [1]. The effective classical system then has two types of worldlines, representing particles with spin up and down, respectively. In higher dimensions the fermion anticommutation relations always lead to a sign problem, and the worldline method is therefore not useful in this case. This is the case even in 1D if the Hamiltonian includes hopping with range further than between nearest neighbors.

It has recently been noted that the autocorrelation times of worldline simulations can be very long for some models. More efficient sampling techniques utilizing global "loop" updates of the worldlines have been developed that overcomes this problem [16,17] in some important cases. These non-Metropolis *loop algorithms* are generalizations of classical cluster spin algorithms [10], and lead to autocorrelation times up to several orders of magnitude shorter than for the standard local Metropolis method. Additional advantages are that simulations can easily be carried out in the grand canonical ensemble, and with fluctuating winding numbers. Very recently, it was discovered that it is possible to construct an algorithm for the Heisenberg model directly in the $\Delta\tau \to 0$ limit [18]. The Trotter error is then eliminated (this is the case also for the "stochastic series expansion" method discussed below).

B. Fermion Determinant Method

The determinant method [2] is based on the fact that traces over exponentials of bilinear fermion operators can be calculated exactly. After using the Trotter break-up, the partition function can be transformed into a sum (or integral) over such bilinear forms by a so called Hubbard-Stratonovich (HS) transformation. The method is useful for models of the Hubbard type, and will here for simplicity be discussed in the context of the 1D Hubbard model, defined by

$$\hat{H} = -t\sum_{i=1}^{N}\sum_{\sigma=\uparrow,\downarrow}(c_{\sigma,i+1}^{+}c_{\sigma,i} + c_{\sigma,i}^{+}c_{\sigma,i+1}) + U\sum_{i=1}^{N}n_{\uparrow,i}n_{\downarrow,i}. \tag{24}$$

Including a chemical potential μ, the Hamiltonian can be written in the form

$$\hat{H} = -t\sum_{i=1}^{N}\sum_{\sigma=\uparrow,\downarrow}(c_{\sigma,i+1}^{+}c_{\sigma,i} + c_{\sigma,i}^{+}c_{\sigma,i+1}) + \mu\sum_{i=1}^{N}(n_{\uparrow,i} + n_{\downarrow,i}) \tag{25}$$

$$+U\sum_{i=1}^{N}(n_{\uparrow,i} - \tfrac{1}{2})(n_{\downarrow,i} - \tfrac{1}{2}),$$

which is particle-hole symmetric for $\mu = 0$, and hence this corresponds to a half-filled band. Denoting the single-particle terms for up and down spin $K_\uparrow$ and $K_\downarrow$, respectively, and the interaction term V, the Trotter break-up is used to write the partition function as

$$Z = \text{Tr}\{e^{-\beta\hat{H}}\} \approx \text{Tr}\{\prod_{l=1}^{L}e^{-\Delta\tau K_\uparrow}e^{-\Delta\tau K_\downarrow}e^{-\Delta\tau V}\}. \tag{26}$$

There are several types of HS transformations. For the Hubbard model the following discrete variant, deviced by Hirsch [19], has proven useful:

$$e^{-\Delta\tau U(n_\uparrow - 1/2)(n_\downarrow - 1/2)} = B \sum_{\sigma=\pm 1} e^{-\lambda\sigma(n_\uparrow - n_\downarrow)}. \tag{27}$$

One can easily verify that this equality is valid for all combinations of $n_\uparrow \in \{0,1\}$ and $n_\downarrow \in \{0,1\}$ if $B = \frac{1}{2}e^{-\Delta\tau U/4}$ and $\cosh(\lambda) = e^{\Delta\tau U/2}$. Applying the transformation for all sites i and all time slices l leads to a sum over $N \times L$ HS variables $\sigma_i(l) = \pm 1$ (the HS field), with each term containing only exponentials of bilinear operators. Using the fact that the spin up and spin down operators commute gives (a prefactor B^{2NL} is irrelevant, since it appears also in the numerator $\text{Tr}\{\hat{A}e^{-\beta\hat{H}}\}$)

$$Z = \sum_{\{\sigma_i(l)\}} \text{Tr}\Big\{\prod_{l=1}^{L} e^{-\Delta\tau K_\uparrow} e^{-V_\uparrow(l)}\Big\} \text{Tr}\Big\{\prod_{l=1}^{L} e^{-\Delta\tau K_\downarrow} e^{-V_\downarrow(l)}\Big\}, \tag{28}$$

where

$$V_\uparrow(l) = \lambda \sum_{i=1}^{N} \sigma_i(l) n_\uparrow, \quad V_\downarrow(l) = -V_\uparrow(l). \tag{29}$$

The traces can now be calculated for each HS field configuration [2,19]. As a simplified problem illustrating this, consider

$$\text{Tr}\{e^{-\hat{O}}\}, \quad \hat{O} = \sum_{i,j} a_{ij} c_i^+ c_j. \tag{30}$$

Since $\hat{O}$ is bilinear it is diagonal in some basis $\{|k\rangle\}$; $\hat{O} = \sum_k a_k n_k$. One then has

$$\text{Tr}\{e^{-\hat{O}}\} = \text{Tr}\Big\{\prod_k e^{-a_k n_k}\Big\} = \prod_k (1 + e^{-a_k}) = \det[1 + e^{-O}], \tag{31}$$

and since the determinant is independent of the basis it can be evaluated in the original basis where the matrix O is that of the coefficients in Eq. (30):

$$O = \begin{pmatrix} a_{11} & a_{12} & \cdots \\ a_{21} & a_{22} & \cdots \\ \cdots & \cdots & \ddots \end{pmatrix}. \tag{32}$$

It can be shown [2,19] that this result generalizes to the type of exponential product appearing in Eq. (28):

$$\text{Tr}\{e^{-\hat{A}} e^{-\hat{B}} e^{-\hat{C}} \cdots\} = \det[1 + e^{-A} e^{-B} e^{-C} \cdots]. \tag{33}$$

For the 1D Hubbard model, the $N \times N$ matrices corresponding to the operators in the partition function (28) are the "hopping matrices" (including the diagonal chemical potential terms, and setting $t = 1$),

$$K = \Delta\tau K_{\uparrow} = \Delta\tau K_{\uparrow} = -\Delta\tau \begin{pmatrix} -\mu & 1 & 0 & \cdots & 1 \\ 1 & -\mu & 1 & \cdots & 0 \\ 0 & 1 & -\mu & \cdots & 0 \\ \vdots & \vdots & \vdots & \ddots & \vdots \\ 1 & 0 & 0 & \cdots & -\mu \end{pmatrix}, \tag{34}$$

and the HS field matrices

$$V(l) = \lambda \begin{pmatrix} \sigma_1(l) & 0 & \cdots & 0 \\ 0 & \sigma_2(l) & \cdots & 0 \\ \vdots & \vdots & \ddots & \vdots \\ 0 & 0 & \cdots & \sigma_N(l) \end{pmatrix}, \quad \sigma_i(l) = \pm 1. \tag{35}$$

The partition function is hence

$$Z = \sum_{\{\sigma_i(l)\}} \det[M_{\uparrow}]\det[M_{\downarrow}], \tag{36}$$

where

$$M_{\uparrow} = [I + e^{-K}e^{-V(1)}e^{-K}e^{-V(2)}\ldots e^{-K}e^{-V(L)}]$$
$$M_{\downarrow} = [I + e^{-K}e^{V(1)}e^{-K}e^{V(2)}\ldots e^{-K}e^{V(L)}]. \tag{37}$$

Note that the HS matrices are diagonal only because the Hamiltonian used includes just the on-site repulsion. With further-range interactions, off-diagonal matrix elements also appear.

In order to obtain an expression for some expectation value $\langle\hat{A}\rangle$, the above steps have to be repeated for $\text{Tr}\{\hat{A}e^{-\beta\hat{H}}\}$. For $\hat{A} = c_{i,s}^+ c_{j,s}$ (here s denotes the fermion spin; $s = \uparrow, \downarrow$) the resulting expression is

$$\text{Tr}\{c_{i,s}^+ c_{j,s}e^{-\beta\hat{H}}\} = \sum_{\{\sigma_i(l)\}} [M_s]_{ij}^{-1}\det[M_{\uparrow}]\det[M_{\downarrow}], \tag{38}$$

and hence an equal-time single-particle Green's function $\langle c_{i,s}^+ c_{j,s}\rangle$ is obtained by averaging the corresponding matrix elements of the inverses of the fermion matrices M_s. Time dependent Green's functions can also be calculated, and involve slightly more complicated matrices [19]. For any HS field configuration, Wick's theorem for operator products apply, and one can therefore in principle calculate the expectation value of any fermion operator product [2,19].

The sampling of the HS field is normally accomplished by single-spin flips $\sigma_i(l) \rightarrow -\sigma_i(l)$. Calculating the weight ratio associated with such a change requires on the order of N^2 operations, and a sweep through the whole HS lattice therefore requires $\sim \beta N^3$ operations. Nonlocal updates involving whole lines of spins in the imaginary-time direction have also been developed [20], and can sometimes alleviate problems with long autocorrelation times. The matrices M_σ are often ill-conditioned at low temperatures, and stabilization techniques have to be used for the matrix operations. Details of the technical points of the updating of the HS field can be found, e.g., in Ref. [21]. Expressions for time-dependent Green's functions are discussed in detail in Ref. [19].

An important advantage of the determinant over the worldline method is that the weight $W = \det[M_\uparrow]\det[M_\downarrow]$ is positive definite at half filling *in any number of dimensions*. Away from half-filling this is not the case, but the sign problem is considerably less severe than for the worldline method, and many results have been obtained for, e.g., the 2D Hubbard model [21]. Very low temperatures cannot be reached, however.

As noted in the previous section, the single-particle Green's function is not easy to calculate using the worldline method, even in 1D where worldline simulations otherwise are useful. In contrast, it is the perhaps simplest quantity to evaluate within the determinant method. Hence, this is the QMC method of choice for calculating this quantity also in 1D.

C. Stochastic Series Expansion

Already in 1961, well before the first application of the Trotter formula to QMC simulations, Handscomb proposed a Monte Carlo scheme for studies of the $S = 1/2$ ferromagnetic Heisenberg model [4]. This method is based on a power series expansion of $\exp(-\beta\hat{H})$, which for the Heisenberg model can be written as a sum of products of permutation operators. Traces of such products can be calculated analytically (in any number of dimensions), and one can then carry out Monte Carlo simulation in the space of operator products. Handscomb's technique is not associated with any systematical errors, since unlike the Trotter decomposition the series expansion is exact. However, in its original formulation the method can only be applied to the ferromagnetic Heisenberg model. For antiferromagnetic couplings the expansion is not positive definite and for most other models the traces cannot even be evaluated. Due to these limitations, the method was evidently not considered a promising starting point for a more general quantum simulation scheme. A solution to the sign problem for the antiferromagnetic case was presented by Lee et. al. in 1984 [5], but the technique still proved to be inefficient compared to worldline methods. In particular, only systems at relatively high temperatures could be studied [5,22].

The fundamental limitations of Handscomb's method are not present in the more general stochastic series expansion (SSE) techniques first developed in Refs. [6] and [7]. The simple idea behind the generalization is to first write the traces as diagonal matrix elements in a suitably chosen basis. As will be demonstrated below, one can then construct a sampling scheme in a combined space of operator products and basis states. The nature of this SSE configuration space shows a strong resemblance with standard path integrals, but owing to the different treatment of the imaginary-time dimension there are no errors due to discretization. The SSE method is as widely applicable as the worldline method, and is also significantly more efficient for $S = 1/2$ antiferromagnets than the earlier formulations of Handscomb's method. Like the worldline method, the SSE technique is also in practice limited to models for which the sign problem can be avoided, such as non-frustrated spin systems and 1D fermions.

Here the SSE technique is for simplicity only discussed in the context of the $S = 1/2$ antiferromagnetic Heisenberg models in one and two dimensions, but algorithms for, e.g., one-dimensional fermion models are very similar [7]. Since the SSE technique is not yet as widely used as the worldline and determinant methods, and its present formulation not as well described in previous literature, it will be discussed in some more detail here.

For a Heisenberg model on a lattice with N_b interacting nearest-neighbor spin pairs $\langle i(b), j(b) \rangle$ (in any number of dimensions), the Hamiltonian is first written as

$$\hat{H} = -\frac{J}{2} \sum_{b=1}^{N_b} [\hat{H}_{1,b} - \hat{H}_{2,b}] + \frac{N_b J}{4}, \tag{39}$$

where the operators $\hat{H}_{1,b}$ and $\hat{H}_{2,b}$ are defined as

$$\hat{H}_{1,b} = 2[\tfrac{1}{4} - S_{i(b)}^z S_{j(b)}^z], \tag{40a}$$

$$\hat{H}_{2,b} = S_{i(b)}^+ S_{j(b)}^- + S_{i(b)}^- S_{j(b)}^+. \tag{40b}$$

An exact expression for an expectation value is obtained by Taylor expanding $e^{-\beta \hat{H}}$ and writing the traces as sums over diagonal matrix elements in the basis $\{|\alpha\rangle = |S_1^z, \ldots, S_N^z\rangle\}$. The partition function can then be written as

$$Z = \sum_\alpha \sum_n \sum_{S_n} \frac{(-1)^{n_2}}{n!} \left(\frac{\beta}{2}\right)^n \left\langle \alpha \left| \prod_{l=1}^n \hat{H}_{a_l, b_l} \right| \alpha \right\rangle, \tag{41}$$

where S_n denotes a sequence of index pairs defining the operator string $\prod_{l=1}^n \hat{H}_{a_l, b_l}$,

$$S_n = [a_1, b_1][a_2, b_2] \ldots [a_n, b_n], \quad a_i \in \{1, 2\}, \ b_i \in \{1, \ldots, N_b\}, \tag{42}$$

and n_2 denotes the total number of index pairs (operators) $[a_i, b_i]$ with $a_i = 2$. Both $\hat{H}_{1,b}$ and $\hat{H}_{2,b}$ can act only on states where the spins at sites $i(b)$ and $j(b)$ are antiferromagnetically oriented. $\hat{H}_{1,b}$ leaves such a state unchanged, whereas $\hat{H}_{2,b}$ flips the spin pair. A propagated state $|\alpha(p)\rangle = |S_1^z[p], \ldots S_N^z[p]\rangle$ is defined as

$$|\alpha(p)\rangle = \prod_{l=1}^p \hat{H}_{a_l, b_l} |\alpha\rangle, \quad |\alpha(0)\rangle = |\alpha\rangle. \tag{43}$$

A configuration (α, S_n) must clearly satisfy the periodicity condition $|\alpha(L)\rangle = |\alpha(0)\rangle$ in order to contribute to the partition function. For a lattice with L_x^D sites and L_x even, this implies that the total number n_2 of the off-diagonal operators must be even, and hence that all terms in (41) are positive and can be used as relative probabilities in a Monte Carlo simulation (this is true for any non-frustrated system).

For a finite system at finite β, the lengths n of the operator strings (the powers of the expansion) contributing significantly to the partition function are restricted to within a well defined regime, and the sampling space is therefore finite in practice.In order to construct an efficient updating scheme for the index sequence it is useful to explicitly truncate the Taylor expansion at some self-consistently chosen upper bound $n = L$, large enough to cause only an exponentially small, completely negligible error [6]. One can then define a sampling space where the length of the index sequence is *fixed*, by inserting a number $L - n$ of unit operators, denoted $\hat{H}_{0,0}$, in the operator strings. The terms in the partition function (41) must be divided by $\binom{L}{n}$, in order to compensate for the number of equivalent contributions corresponding to the number of different ways of inserting the unit operators. With $[a_i, b_i] = [0, 0]$ as an allowed operator, the summation over n in (41) is then implicitly included in the summation over all sequences S_L of length L. Denoting by $W(\alpha, S_L)$ the weight of (α, S_L), the partition function is then

$$Z = \sum_\alpha \sum_{S_L} W(\alpha, S_L). \tag{44}$$

Since all non-zero matrix elements in (41) are equal to one, the weight of an allowed configuration is

$$W(\alpha, S_L) = \left(\frac{\beta}{2}\right)^n \frac{(L - n)!}{L!}, \tag{45}$$

where now $n = n_1 + n_2$ is the number of non-$[0,0]$ operators in S_L.

Fig. IIIC shows a graphical representation of a typical configuration generated for an 8-site Heisenberg chain at a high temperature, $T = J$. The index sequence S_L is shown, along with all the propagated states $|\alpha(p)\rangle$ and graphic representations of the operators in S_L. Note that all the states have the same magnetization, $m^z = \sum_i S_i^z$, reflecting that this quantity is conserved by the Hamiltonian. This kind of representation suggests that one can think of a configuration somewhat like a term in a path integral formulation, with the spins "moving" on a space-time grid, as in real-space Trotter based methods such as the worldline method. However, it should be kept in mind that a given state $|\alpha(p)\rangle$ is not at a fixed imaginary-time, as will be discussed further below in the context of time-dependent expectation values.

During a simulation, the sequence S_L and one of the states $|\alpha(p)\rangle$ are stored. Other propagated states are generated as needed. The simulation is started with a randomly generated state $|\alpha(0)\rangle$, with an index sequence S_L containing only $[0, 0]$ operators (unit operators), and with some arbitrary (small) L. The length L is adjusted until sufficient to eliminate any detectable truncation errors, as will be discussed further below. The configuration illustrated in Fig. IIIC has a small

$L = 44$, reflecting the small system size and high temperature. Simulations of large systems at low temperatures of course require considerably larger cut-offs. Since only one of the states has to be stored, even cases where $L \sim 10^5$ or larger are practically feasible. The memory requirements are considerably smaller than for corresponding worldline simulations, where the whole space-time spin lattice is stored.

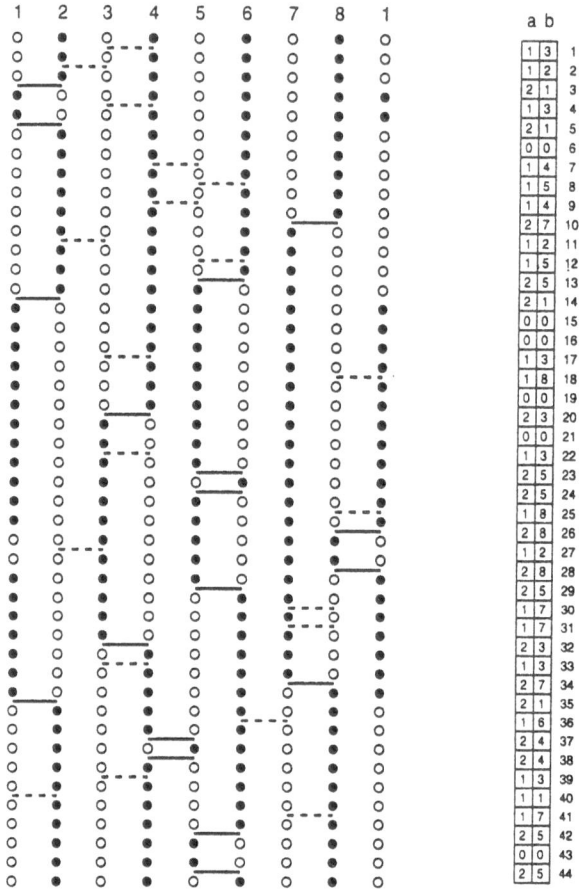

FIG. 4 Graphical representation of a configuration for an 8-site system with periodic boundary conditions at $T = J$. To the right is the operator-index sequence S_L, with $L = 44$ in this case. To the left are the states resulting from propagating the initial state (top row) with the operators in S_L. Up and down spins are represented by solid and open circles, respectively. Dashed bars represent the diagonal operators $[1, b]$ and solid ones the off-diagonal operators $[2, b]$. The unit operators corresponding to the index pairs $[0, 0]$ are not shown; their presence can be inferred from the absence of horizontal bars at the corresponding positions.

The simplest type of update is an exchange of diagonal operators,

$$[0, 0]_p \leftrightarrow [1, b]_p, \tag{46}$$

where b is chosen at random in the $\rightarrow$ direction (the position of the operator in the sequence is here indicates by the subscript p). The update is graphically illustrated in Fig. III C(a). The Metropolis acceptance probabilities required to satisfy detailed balance are easily obtained from Eq. (45), where n in is changed by ± 1. Updates are attempted consecutively at all positions $p = 1, \ldots, L$. The spin states $|\alpha_{p-1}\rangle$, which are needed to determine whether exchanges $[0,0]_p \rightarrow [1,b]_p$ are allowed or not, are generated one by one during the process, starting from the stored $|\alpha_0\rangle$.

Updates involving the off-diagonal operators $[2,b]$ are carried out with n fixed. The simplest involves two identical operators at positions p_1 and p_2 in S_L:

$$[1,b]_{p_1}, [1,b]_{p_2} \leftrightarrow [2,b]_{p_1}, [2,b]_{p_2}. \tag{47}$$

As illustrated in Fig. III C(b), such an update can be carried out provided that there are no other operators between p_1 and p_2 acting on either of the two sites connected by bond b. An efficient way to carry out this type of update is to first partition S_L into a set of subsequences $S_{L_i}^{(i)}$, each containing only the L_i operators acting on a single bond b_i (along with information on the presence of operators at the neighboring bonds, which act as constraints for updating $S_{L_i}^{(i)}$). If b_i and b_j are not connected to each other (i.e., act on different spins only), the subsequences $S_{L_i}^{(i)}$ and $S_{L_j}^{(j)}$ can be updated independently of each other. In 1D two different partitions have to be made, in 2D four, e.t.c.. When all the subsequences of a partition have been updated, they are recombined into a new full S_L, which then can be split into another partition, e.t.c.. Searching for possibilities to carry out substitutions of the type (47) is a fast process within the subsequences, which have a typical length $L_i \sim L/N \sim \beta$. However, care has to be taken that the operators to be substituted are selected in such a way that detailed balance is satisfied [7]. Note that the configuration weight does not change in an allowed pair-substitution.

[a]

[b]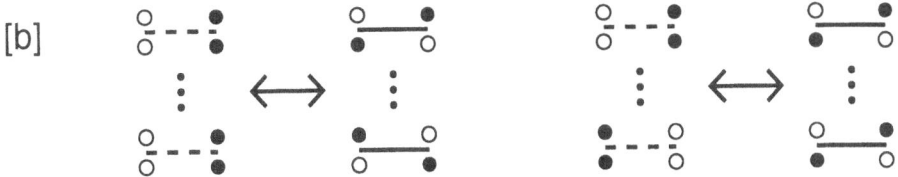

FIG. 5 The two simplest sequence updates shown in the graphical representation used in Fig. III C. (a) represents a single-operator substitution at a bond b, Eq. (46). This type of diagonal update can be carried out at a position p if the spins at sites $b, b+1$ in the state $|\alpha(p-1)\rangle$ are antiferromagnetically oriented. (b) represents a pair substitution, Eq. (47). This kind of substitution can be carried out provided that there are no operators $[a, b-1]$ or $[a, b+1]$ ($a = 1, 2$)

between positions p_1 and p_2. Operators $[a, b]$ are allowed in this interwall, however. The left part of (b) illustrates the case where there is an even number (or zero) of operators $[2, b]$ between p_1 and p_2, and the right graph the case of an odd number of such operators. .

The sequence updates (46) and (47) are the only ones needed for a 1D system with open boundary conditions. For a 2D system, configurations with spin flips around a plaquette (four bonds forming a square) are possible, and require an additional type of update. With $\{b_1, b_2, b_3, b_4\}$ defining a permutation of the four bonds of a plaquette, this can be accomplished by substitutions of the type

$$[2, b_1]_{p_1}, [2, b_2]_{p_2} \leftrightarrow [2, b_3]_{p_1}, [2, b_4]_{p_2}, \tag{48}$$

where not prohibited due to constraints imposed by other operators. For actually carrying out these updates, the sequence is again partitioned, now into subsequences each containing operators acting on a different plaquette. Four different partitions must be considered in 2D, so that bonds of different plaquettes do not share spins.

For systems with periodic boundary conditions, updates involving spin flips on loops wrapping around the whole system are required in order to sample different winding number sectors, which cannot be accomplished by the local sequence alterations discussed above. For a systam of linear dimension L_x, the winding number can be changed by substituting $L_x/2$ operators according to

$$[2, b_1]_{p_1} \ldots [2, b_{L_x/2}]_{p_{L_x/2}} \leftrightarrow [2, b_{L_x/2+1}]_{p_1} \ldots [2, b_{L_x}]_{p_{L_x/2}}, \tag{49}$$

where the set of bonds $\{b_1, \ldots, b_{L_x}\}$ is a permutation of links forming a closed ring around the system (in the $x-$ or y-direction in 2D). In 1D, the whole sequence S_L must be used for carrying out this update, since all bonds of the ring are involved. In 2D, the sequence can be split into subsequences acting on lines in the x- or y-direction (two different partitions). As is also the case for simulations with the worldline method, the acceptance rate for updates changing the winding number decreases rapidly as the system size grows, and in practice calculations for large systems (typically $L_x \gtrsim 16$) have to be restricted to the zero winding number sector.

Updating the index sequence with the operator substitutions described above suffices for generating all possible configurations within a sector of fixed magnetization $m^z = \sum_{i=1}^{N} S_i^z$, without performing explicit spin flips in the states $|\alpha(p)\rangle$. Since the ground state is a singlet, calculations aimed at $T = 0$ properties can be restricted to $m = 0$ (the temperature used then of course has to be much lower than the lowest excitation energy of the finite system considered).

A Monte Carlo step is defined as a sequence of all the updates discussed above. The number of operations needed for one Monte Carlo step (excluding the winding number update) scales as $N\beta$, which is the same as for the worldline method.

However, for the worldline method the prefactor grows with decreasing time slice width as $1/\Delta\tau$, and therefore the SSE technique can be expected to be more efficient if very accurate results are needed.

The truncation L is determined by monitoring the powers n sampled during the equilibration part of the simulation. If the largest n sampled exceeds some threshold $L - \Delta_L/2$, L is increased by Δ_L: $L \to L + \Delta_L$. The increment Δ_L is largely arbitrary. In practice $\Delta_L \approx 0.05 - 0.1 \times L$ works well, and rapidly produces an L which saturates, and is sufficient to cause no detectable truncation errors. The growth of L during equilibration is illustrated by data generated for a 4×4 lattice in Fig. III C. The distribution of n during a subsequent simulation is also shown, and clearly demonstrates that the truncation of the Taylor expansion is no approximation in practice — in a simulation of practical duration n would not exceed L, even if allowed to.

Measurements of observables of interest can be carried out using the index sequences S_n obtained by omitting the $[0,0]$ operators in S_L. One can show that the internal energy per site is simply given by the average of n as (with the constant term in Eq. (39) neglected) [4,6]

$$E = -\frac{\langle n \rangle}{N\beta}. \tag{50}$$

This result shows that the average power $\langle n \rangle$, and hence the cut-off L, scales as $N\beta$ at low temperatures. The spin correlation function is obtained averaging the correlations in the propagated states $|\alpha(p)\rangle$ defined in Eq. (43) [6]:

$$\langle S_i^z S_j^z \rangle = \left\langle \frac{1}{n+1} \sum_{p=0}^{n} S_i^z[p] S_j^z[p] \right\rangle. \tag{51}$$

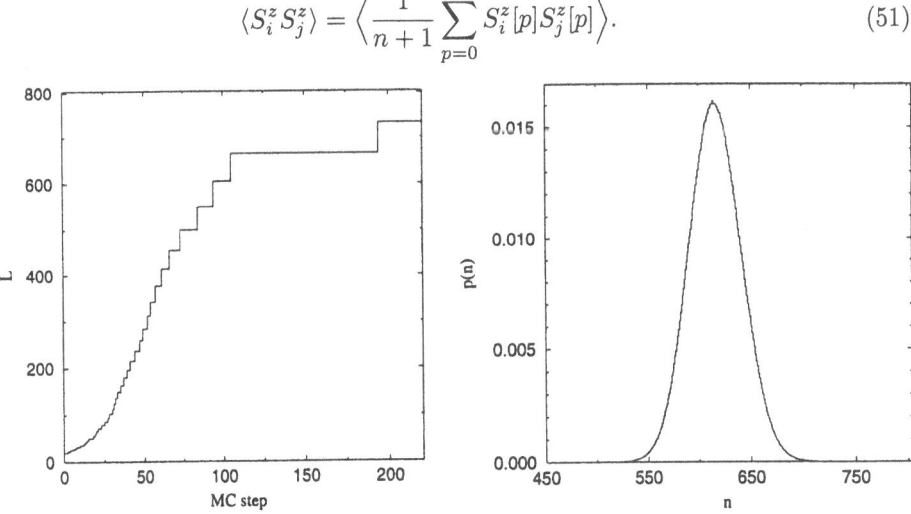

FIG. 6 Left: The truncation L vs. the number of Monte Carlo steps for a 4×4 Heisenberg model at $\beta = 32$. The final L after 10^5 equilibration steps was $L = 804$. Right: The distribution of the power n during a subsequent simulation consisting of 10^6 MC steps.

Imaginary-time dependent correlation functions such as $C_{ij}(\tau) = \langle S_i^z(\tau)S_j^z(0)\rangle$ can also be easily evaluated. Taylor expanding the exponentials, one can show that [7]

$$C_{ij}(\tau) = \left\langle \sum_{m=0}^{n} \frac{\tau^m(\beta-\tau)^{n-m}n!}{\beta^n(n-m)!m!}\bar{C}_{ij}(m)\right\rangle, \tag{52}$$

where $\bar{C}_{ij}(m)$ is a correlator between states separated by m positions:

$$\bar{C}_{ij}(m) = \frac{1}{n+1}\sum_{p=0}^{n} S_i^z[p]S_j^z[p+m]. \tag{53}$$

The periodicity of the propagated states imply that $S_i^z[p+n] = S_i^z[p]$. The corresponding static susceptibility, defined in Eq. (23), can be directly obtained by integrating over τ in (52). The result is [6,7]

$$\chi_{ij} = \left\langle \frac{\beta}{n(n+1)}\left(\sum_{p=0}^{n-1} S_i^z[p]\right)\left(\sum_{p=0}^{n-1} S_j^z[p]\right) + \beta\bar{C}_{ij}(0)\right\rangle. \tag{54}$$

Eq. (52) shows the relationship between the "propagation index" p in $|\alpha(p)\rangle$ and imaginary-time: A time separation τ corresponds to a distribution of separations m between propagated states, peaked around $m = n\tau/\beta$. Hence, the SSE simulation method can be seen as a path-integral formulation where the additional dimension resulting from the mapping corresponds to a fluctuating (non-sharp) imaginary-time.

The absence of detectable systematical errors in SSE calculations have been verified by carrying out long simulations for 4×4 and 6×6 lattices [23], for which exact diagonalization results are available [24]. For example, the results for the energy per site is $E = -0.701778(9)$ and $-0.678871(5)$ for $L = 4$ and $L = 6$, respectively, where the numbers within parenthesis indicate the statistical errors in the last digits of the results. The corresponding exact energies are $E = -0.701780$ and $E = -0.678872$, and hence the SSE results are accurate to withing a relative statistical error as small as $\sim 10^{-5}$. These simulations were carried out at inverse temperatures $\beta = 8L$ in the $m^z = 0$ sector. Results of the same relative accuracy have been obtained for systems with up to 16×16 spins [23].

In terms of the models that can be studied and the types of expectation values that can be evaluated, the SSE technique is comparable to the worldline method. The advantage is the absence of systematical errors. As already noted in Sec. IIIA, a worldline loop algorithm for the Heisenberg model has recently been deviced that directly simulates the $\Delta\tau \to 0$ limit [18], and hence also eliminates the Trotter error. The loop algorithm has the additional advantage that problems with long autocorrelation times are avoided. However, in some cases loop algorithms do not perform well, and the SSE method may then be more efficient. It may also be possible to construct a loop-type updating scheme for the SSE method. It was also recently noted that covariance effects can sometimes be used to significantly enhance the accuracy of results obtained with the SSE method (most likely, with other methods as well) [25]. In particular, very accurate results for spin correlations of Heisenberg spin systems can be obtained using this new way of analyzing the data [23].

IV. DISCUSSION OF APPLICATIONS

Here some applications of the QMC methods described in the previous sections are summarized. The choice of topics is mainly based on the interests of the author, and the list of references is by no means complete. The intention is primarily to provide novices in the field with a representative subset of references where QMC methods are discussed within the context of the physical systems for the study of which they were developed.

The 2D $S = 1/2$ Heisenberg antiferromagnet with nearest-neighbor interactions on a square lattice has been studied extensively using several QMC techniques. These include variants of the worldline method [26,27], as described here as well as in the formulation by Suzuki and co-workers, and also the recently developed variant based on global loop updates [17,18]. The early formulations of Handscomb's method have also been used [22], as well as the more efficient generalization of Handscomb's method (stochastic series expansion) discussed above [23,28]. The efforts focused on the 2D Heisenberg and related models were (and still are) largely motivated by the discovery of the high-T_c cuprate superconductors, the insulating parent compounds of which are good physical realizations of this model. QMC studies have confirmed that the 2D Heisenberg model with only nearest-neighbor interactions has long-range order at $T = 0$ [26], and have also been crucial in establishing the validity of the continuum field theory description of the model in terms of the $O(3)$ nonlinear σ model [29,27]. In addition, studies of the double-layer Heisenberg model [30], which undergoes an order-disorder transition at $T = 0$ as a function of the coupling between the planes, have confirmed the universal finite-temperature behavior predicted for near-critical 2D antiferromagnets [31].

Spin chains with both $S = 1/2$ and higher spins have also been studied extensively [32,6], and have provided ample verifications of the so called Haldane conjecture, which states that integer spin Heisenberg chains have gapped excitation spectra whereas half-integer ones are gapless. Systems of a few coupled $S = 1/2$ chains have attracted much interest lately [33], and have also been studied usig several QMC techniques [34].

Early after its development, the worldline method was applied to the 1D Hubbard model with various interactions included [35], and also to systems with electron-phonon interactions, such as the Su-Schrieffer-Heeger model [40]. These models are of interest, e.g., in the context of conducting polymeres and other quasi-1D organic conductors. They are also of general interest as "toy" models where the physics of strongly interacting systems can be explored more easily than in higher dimensions. More recent worldline studies of 1D fermion systems can be found, e.g., in the papers listed in Ref. [36]. The stochastic series expansion technique has also been applied to several 1D fermion models, including the standard Hubbard model [37], disordered Hubbard models [38], and a two-band model representing a Cu-O linear chain [39].

The worldline method is also directly applicable to boson models, where the sign problem can be avoided in any number of dimensions, provided that the hoppings are not frustrated (no closed loops with odd numbers of positive boson hopping matrix elements) [41]. An interesting application is the study of the insulator-superfluid transition, occurring as a function of the band filling [42], and how it is affected by the presence of disorder [43].

Due to the severity of the sign problem in worldline studies, QMC results for 2D and 3D fermion systems have been obtained primarily using the determinant method. Extensive studies of 2D models such as the standard one-band Hubbard model, and the more complicated 3-band model, have been motivated by their suggested relevance to understanding the mechanism of high-T_c superconductivity [44]. At half filling (corresponding to insulating undoped cuprates), the standard one-band Hubbard model does not have a sign problem in any D, but away from half-filling the sign problem rapidly becomes severe as the temperature is lowered. This has so far prevented accurate studies at temperatures where one would expect superconductivity to take place [45]. However, studies at higher temperatures show that many of the normal state properties of the cuprates are qualitatively captured by this model [45]. Studies of the 3-band model have attempted to quantitatively determine the interaction parameters relevant to the cuprate materials. In this case the sign problem is even more severe than for the one-band model, and no conclusive evidence for superconductivity has been presented [46].

The sign problem is absent in determinant QMC for the Hubbard model with attractive interactions (in any D). The finite-temperature superconducting transition (of the Kosterlitz-Thouless type) occuring in this model in 2D have been studied using this method [47].

The determinant method can also be used for electron-phonon models, such as the Holstein model [48]. In the absence of electron-electron interactions there is no sign problem (in any D). A variant of the determinant method deviced by Hirsch and Fye for impurity problems [49] have found recent use in studies of fermion systems in the $D \to \infty$ limit [50].

QMC cannot directly provide results for dynamic quantities, which are of course very important experimentally. Instead, the corresponding imaginary-time correlation function has to be evaluated and analytically continued to real frequency. The continuation has to be done numerically, which is an inherently unstable procedure when the imaginary-time data is affected by statistical errors unavoidable in QMC calculations. Recently there has been significant progress in the development of methods for performing the analytic continuation. In particular, the so-called maximum-entropy method has proven useful [51], and has been applied to many of the models discussed above.

Thanks to improved algorithms and increasing computer power, the class of problems that can be efficiently attacked using QMC methods is growing quickly. Unfortunately, the sign problem is still an obstacle in studies of some of the most challenging models, among them interacting fermions in $D > 1$ and frustrated quantum spin systems. The sign problem is at the heart of quantum mechanics, since it is due to interference between different processes. This is particularly clear for methods formulated in real-space, such as the worldline method and the stochastic series expansion technique. In a sense, in models for which the sign problem is absent in these path-integral formulations, quantum mechanical effects are not manifested to the fullest extent possible.

It is probably no exaggeration to say that a solution of the sign problem would lead to a veritable revolution in the field of strongly correlated systems. Hence, quoting D. J. Scalapino, even though the probability p of solving the sign problem is very low ($p \ll 1$), the potential gain G derived from its solution makes the product pG large enough that some effort should be devoted to this problem every now and then.

---

[1] J. E. Hirsch, R. L. Sugar, D. J. Scalapino and R. Blankenbecler, Phys. Rev. B **26**, 5033 (1982).

[2] R. Blankenbecler, D. J. Scalapino and R. L. Sugar, Phys. Rev. D **24**, 2278 (1981).

[3] M. Suzuki, Prog. Theor. Phys. **56**, 1454 (1976).

[4] D. C. Handscomb, Proc. Cambridge Philos. Soc. **58**, 594 (1962); **60**, 115 (1964).

[5] D. H. Lee, J. D. Joannopoulos, and J. W. Negele, Phys. Rev. B **30**, 1599 (1984).

[6] A. W. Sandvik and J. Kurkijärvi, Phys. Rev. B **43**, 5950 (1991).

[7] A. W. Sandvik, J. Phys. A **25**, 3667 (1992).

[8] For a more extensive pedagogical introduction to classical Monte Carlo, see, e.g., K. Binder and D. W. Heermann, *Monte Carlo Simulation in Statistical Physics* (Springer Verlag, Berlin, 1992).

[9] N. Metropolis, A. Rosenbluth, M. Rosenbluth, A. H. Teller, and E. Teller, J. Chem. Phys. **21**, 1087 (1953).

[10] R.-H. Swendsen and J.-S. Wang, Phys. Rev. Lett. **58**, 86 (1987).

[11] S. Miyashita, M. Takasu, and M. Suzuki, in *Quantum Monte Carlo methods in Equilibrium and Nonequilibrium Systems*, ed. M. Suzuki (Springer-Verlag Berlin, 1987).

[12] E. Y. Loh Jr., J. E. Gubernatis, R. T. Scalettar, S. R. White, D. J. Scalapino and R. L. Sugar, Phys. Rev. B **41**, 9301 (1990).

[13] For an introduction to path integrals, see, e.g., R. P. Feynman and A. R. Hibbs, *Quantum Mechanics and Path Integrals* (McGraw-Hill, New York, 1965).

[14] M. Suzuki, S. Miyashita, and A. Kuroda, Prog. Theor. Phys. **58**, 1377 (1977).

[15] R. M. Fye, Phys. Rev. B **33**, 6271 (1986).

[16] H. G. Evertz, G. Lana, and M. Marcu, Phys. Rev. Lett. **70**, 875 (1993); N. Kawashima, J. E. Gubernatis, and H. G. Evertz, Phys. Rev. B **50**, 136 (1994).

[17] U.-J. Wiese and H.-P. Ying, Z. Phys. B **93**, 147 (1994).

[18] B. B. Beard and U. J. Wiese, preprint (1996).

[19] J. E. Hirsch, Phys. Rev. B **31**, 4403 (1985).

[20] R. T. Scalettar, R. M. Noack, and R. R. P. Singh, Phys. Rev. B **44**, 10502 (1991).

[21] S. R. White, D. J. Scalapino, R. L. Sugar, E. Y. Loh Jr., J. E. Gubernatis, and R. T. Scalettar, Phys. Rev. B **40**, 506 (1989).

[22] E. Manousakis and R. Salvador, Phys. Rev. B **39**, 575 (1989); G. Gomez-Santos, J. D. Joannopoulos, and J. W. Negele, Phys. Rev. B **39**, 4435 (1989).

[23] A. W. Sandvik, Phys. Rev. B (submitted).

[24] Results for 6×6 sites were obtained by H. J. Schulz, T. A. L. Ziman, and D. Poilblanc, J. Physique I **6**, 675 (1996).

[25] A. W. Sandvik, Phys. Rev. B (in press).

[26] J. D. Reger and A. P. Young, Phys. Rev. B **37**, 5978 (1988).

[27] H.-Q. Ding and M. S. Makivić, Phys. Rev. Lett. **64**, 1449, (1990); M. S. Makivić and H.-Q. Ding, Phys. Rev B **43**, 3562 (1991).

[28] A. W. Sandvik and D. J. Scalapino, Phys. Rev. B **51**, 9403 (1995); A. W. Sandvik and D. J. Scalapino, Phys. Rev. B **53**, R526 (1996).

[29] S. Chakravarty, B. I. Halperin, and D. R. Nelson, Phys. Rev. Lett. **60** 1057 (1988); Phys. Rev. B **39**, 2344 (1989).

[30] A. W. Sandvik and D. J. Scalapino, Phys. Rev. Lett. **72**, 2777 (1994); A. W. Sandvik, A. V. Chubukov, and S. Sachdev, Phys. Rev. B **51**, 16483 (1995).

[31] S. Sachdev and J. Ye, Phys. Rev. Lett. **69**, 2411 (1992); A. V. Chubukov and S. Sachdev, Phys. Rev. Lett. **71**, 169 (1993).

[32] K. Nomura, Phys. Rev. B **40**, 2421 (1989); S. Yamamoto and S. Miyashita, J. Phys. Soc. Jpn. **63**, 2866 (1994).

[33] S. R. White, R. M. Noack, and D. J. Scalapino, Phys. Rev. Lett. **73**, 886 (1994). E. Dagotto and T. M. Rice, Science **271**, 618 (1996).

[34] M. Troyer, H. Tsunetsugu, and D. Würtz, Phys. Rev. B **50**, 13515 (1994); A. W. Sandvik, E. Dagotto, and D. J. Scalapino, Phys. Rev. B **53**, R2934 (1996); M. Greven, M. Birgeneau, and U.-J. Wiese, Phys. Rev. Lett. **77**, 1865 (1996). B. Frischmuth, B. Ammon, and M. Troyer, Phys. Rev. B **54**, R3714 (1996).

[35] J. E. Hirsch and D. J. Scalapino, Phys. Rev. B **27**, 7169 (1983); **29**, 5554 (1984); J. E. Hirsch, Phys. Rev. Lett. **53**, 2327 (1984).

[36] W. R. Somsky et al., Synthetic Metals **43**, 3531 (1991); F. F. Assaad and D. Würtz, Phys. Rev. B **44**, 2681 (1991); Y. Iino and M. Imada, J. Phys. Soc. Jpn **64**, 4392 (1995); J. Schulte and M. C. Bohm, Phys. Rev. B **53**, 15385 (1996).

[37] A. W. Sandvik, D. J. Scalapino, and C. Singh, Phys. Rev. B **48**, 2112 (1993).

[38] A. W. Sandvik and D. J. Scalapino, Phys. Rev. B **47**, 10090 (1993); A. W. Sandvik, D. J. Scalapino, and P. Henelius, Phys. Rev. B **50**, 10474 (1994).

[39] A. W. Sandvik and A. Sudbø, Phys. Rev. B **54**, R3746 (1996).

[40] J. E. Hirsch and E. Fradkin, Phys. Rev. Lett. **49**, 402 (1982); E. Fradkin and J. E. Hirsch, Phys. Rev. B **27**, 1680 (1983).

[41] G. G. Batrouni and R. T. Scalettar, Phys. Rev. B **46**, 9051 (1992).

[42] G. G. Batrouni, R. T. Scalettar, and G. T. Zimanyi, Phys. Rev. Lett. **65**, 1765 (1990). G. T. Zimanyi, P. A. Crowell, R. T. Scalettar, and G. G. Batrouni, Phys. Rev. B **50**, 6515 (1994).

[43] R. T. Scalettar, G. G. Batrouni, and G. T. Zimanyi, Phys. Rev. Lett. **66**, 3144 (1991). G. G. Batrouni, et al., Phys. Rev. B **48**, 9628 (1993).

[44] For recent reviews, see, e.g, E. Dagotto, Rev. Mod. Phys. **66**, 763 (1994); D. J. Scalapino, Phys. Rep. **250**, 330 (1995).

[45] A. Moreo et al., Phys. Rev. B **41**, 2313 (1990); D. J. Scalapino, S. R. White and S. C. Zhang, Phys. Rev. B **47**, 6157 (1993); N. Bulut, D. J. Scalapino, and S. R. White, Phys. Rev. B **47**, 14599 (1993); D. F. Assaad, W. Hanke, and D. J. Scalapino, Phys. Rev. B **50**, 12835 (1994).

[46] R. T. Scalettar, D. J. Scalapino, R. L. Sugar, and S. R. White, Phys. Rev. B **44**, 770 (1991); G. Dopf, A. Muramatsu, and W. Hanke, Phys. Rev. Lett. **17**, 559 (1992).

[47] A. Moreo and D. J. Scalapino, Phys. Rev. Lett. **66**, 946 (1991). M. Randeria, N. Trivedi, A. Moreo, and R. T. Scalettar, Phys. Rev. Lett. **69**, 2001 (1992); F. F. Assaad, W. Hanke, and D. J. Scalapino, Phys. Rev. B **49**, 4327 (1994).

[48] R. M. Noack, D. J. Scalapino, and R. T. Scalettar, Phys. Rev. Lett. **66**, 778 (1991).

[49] J. E. Hirsch and R. M. Fye, Phys. Rev. Lett. **56**, 2521 (1986).

[50] M. Jarrell, Phys. Rev. Lett. **69**, 168 (1992); J. K. Freericks, M. Jarrell, and D. J. Scalapino, Europhys. Lett. **25**, 37 (1994).

[51] M. Jarrell and J. E. Gubernatis, Phys. Rep. **269**, 133 (1996), and references therein.

Coupled Luttinger Liquids

H.J. Schulz

Laboratoire de Physique des Solides, Université Paris–Sud, 91405 Orsay, France

Many one-dimensional quantum systems, in particular interacting electron and spin systems, can be described a Luttinger liquids. Here, some basic ideas of this picture of one-dimensional systems are briefly reviewed. I then discuss the effect of interchain coupling for a finite number of parallel chains. In the case of spin chains coupled by exchange interactions, the low-energy properties are radically different according to whether the number of coupled chains is even or odd: even number of chains have a gap in the spin excitations, whereas odd numbers of chains are gapless. The effect of interchain tunneling is analyzed for two and three coupled chains of itinerant fermions: for repulsive interactions, the two-chain system is "universally" found to be a d-wave superconductor, with a gap in the spin excitation spectrum. On the other hand, for three chains the ground state depends both on the boundary conditions in the transverse direction and on the strength of the interactions. Weak repulsive interactions in all cases lead to dominant superconducting pairing of d-type. An example of a three-leg spin ladder with a spin gap is proposed. A general scheme to keep track of fermion anticommutation in the bosonization technique is developed.

Correlated Fermions and Transport in Mesoscopic Systems, T. Martin, G. Montambaux and J. Tran Thanh Van eds., Editions Frontieres, Gif-sur-Yvette 1996

H.J. Schulz, in "Proceedings of Les Houches Summer School LXI", ed. E. Akkermans, G. Montambaux, J. Pichard, and J. Zinn-Justin (Elsevier, Amsterdam, 1995), p. 533.

On the Application
of the Non-linear Sigma Model
to Spin Chains and Spin Ladders

Germán Sierra*

Instituto de Matemáticas y Física Fundamental
C.S.I.C., Madrid, SPAIN

Abstract

We review the non-linear sigma model approach (NLSM) to spin chains and spin ladders, presenting new results. The generalization of the Haldane's map to ladders in the Hamiltonian approach, give rise to different values of the θ parameter depending on the spin S, the number of legs n_ℓ and the choice of blocks needed to built up the NLSM fields. For rectangular blocks we obtain $\theta = 0$ or $2\pi S$ depending on wether n_ℓ, is even or odd, while for diagonal blocks we obtain $\theta = 2\pi S n_\ell$. Both results agree modulo 2π, and yield the same prediction, namely that even (resp. odd) ladders are gapped (resp. gapless). For even leeged ladders we show that the spin gap collapses exponentially with n_ℓ and we propose a finite size correction to the gap formula recently derived by Chakravarty using the 2+1 NSLM, which gives a good fit of numerical results. We show the existence of a Haldane phase in the two legged ladder using diagonal blocks and finally we consider the phase diagram of dimerized ladders.

*Based on a talk delivered at the Summer School on "Strongly Correlated Magnetic and Superconducting Systems", held in Madrid, Spain, July 1996

Introduction

There have been three major developments in the 80's and 90's in Condensed Matter or more specifically in Strongly Correlated Systems. In historical order they are

- Haldane's conjecture in 1d antiferromagnetic spin chains (1983) [1, 2]

- Discovery of high-T_c superconductivity and antiferromagnetism in doped and undoped cuprate compounds (1986) [3]

- Discovery of Ladder Materials (1987,91) [4, 5]

All these findings have in common various features: low dimensionality, antiferromagnetism, importance of quantum fluctuations and the fact that they constituted theoretical and experimental surprises. Moreover these topics can be studied using sigma model techniques, which establish a methodological link between them. In this talk we shall be mainly concerned with spin chains and ladders.

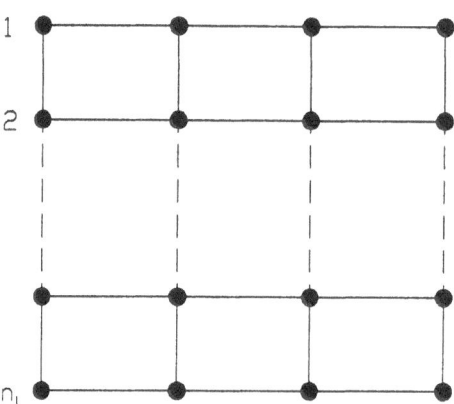

Figure 1 A generic spin ladder with n_ℓ legs.

The first surprise deals with the behaviour of 1d antiferromagnetic Heisenberg spin chains (AFH) as a function of the spin. The Hamiltonian describing the AFH model is given by,

$$H_{\text{chain}} = J \sum_{n=1}^{L} \mathbf{S}(n)\mathbf{S}(n+1) \qquad (1)$$

where $J > 0$ is the exchange coupling constant and $\mathbf{S}(n)$ is a spin-S matrix acting at the nth site of the chain. Haldane's conjecture concerning the spectrum and spin correlations of the Hamiltonian (1) is given in Table 1 ,

$2S$	Spectrum	Correlations
even	gapped	exponential decay
odd	gapless	algebraic decay

Table 1: Behaviour of spin chains

Nowadays there is sufficient theoretical, numerical and experimental evidence to support this conjecture, which should rather be called a "theorem" despite of the lack of a rigorous mathematical proof [6].

The third surprise, in the historical order shown above, deals with the behaviour of spin ladders as function of the number of legs n_ℓ [7] (for a review and references on the subject see [8]). A ladder is an array of n_ℓ spin chains coupled as in Fig. 1. The Hamiltonian of the system is given by,

$$H_{\text{ladder}} = H_{\text{leg}} + H_{\text{rung}}$$
$$H_{\text{leg}} = J \sum_{a=1}^{n_\ell} \sum_{n=1}^{L} \mathbf{S}_a(n)\mathbf{S}_a(n+1) \tag{2}$$
$$H_{\text{rung}} = J' \sum_{a=1}^{n_\ell-1} \mathbf{S}_a(n)\mathbf{S}_{a+1}(n)$$

where $\mathbf{S}_a(n)$ are spin-S matrices located in the ath leg at the position $n = 1, \ldots, L$. We consider periodic boundary conditions alongs the legs, i.e. $\mathbf{S}_a(n) = \mathbf{S}_a(n+L)$, and open BC's along the rungs. The intraleg coupling constant J is positive but the interleg coupling constant J' can be either positive or negative. The qualitative behaviour of spin 1/2 ladders is given in Table 2,

n_ℓ	Spectrum	Correlations
even	gapped	exponential decay
odd	gapless	algebraic decay

Table 2: Behaviour of Spin 1/2 Ladders

This result holds for both signs of J'. The analogy between the integer/half-integer behaviour of spin chains and the even/odd behaviour of spin 1/2 ladders is evident. In the limit where J' goes to minus infinity, the spin ladder Hamiltonian (2) becomes equivalent to the spin chain Hamiltonian (1) for an effective spin equal to $n_\ell/2$, and hence the ladder behaviour given in Table 2 follows from the behaviour of the spin chains. What is not so obvious is that this behaviour holds not just for strong ferromagnetic rung couplings but also for antiferromagnetic ones, regardless their magnitude. As for the Haldane conjecture there is by now sufficient evidence to support the "ladder conjecture" coming from Quantum Monte Carlo, exact diagonalization, mean field theory, experiments on ladder materials like VOPO or cuprates, bosonization, finite size analysis, sigma model, etc [8].

In this talk we shall give a unified description of the behaviour of spin chains and spin ladders utilizing the non-linear sigma model (NLSM) [9, 10, 11]. Let us first recall the logic underlying Haldane conjecture. It is based in the following map[1, 2],

Low Energy Modes of the Spin Chain $\longrightarrow$ Non Linear Sigma Model   (3)

which is obtained in the semiclassical limit where the spin S becomes very large, although (3) can be derived on more general grounds based on symmetry arguments [12]. However these type of arguments miss the theta term in the action, which plays a crucial role in determining the physics of the model.

Hence using the properties of the (NLSM) one derives those of the low energy spectrum of the spin chain.

To arrive at the results of Table 2 for the ladders we shall follow the same logic as for spin chains, that is, we shall construct a map

Low Energy Modes of the Spin Ladder $\longrightarrow$ Non Linear Sigma Model   (4)

and use the properties of the NLSM. In this way the NLSM provides a unified and economic approach to both problems. Actually, the 2d Antiferromagnetic Heisenberg model, can also be mapped into the O(3) NLSM in 2+1 dimensions [13]. The RG analysis at finite temperature of the later model made in [14] has unravelled many of the properties of the undoped cuprate compounds (second surprise). On the other hand the NLSM's that appear in the r.h.s. of the maps (3) and (4) are defined in 1+1 dimensions. This poses the problem of the crossover between 1d and 2d Heisenberg systems through spin ladders [15].

The Non-linear Sigma Model: A Primer

Before we construct the maps (3) and (4), it will be convenient to review the basics of the O(3) NLSM. This model was proposed as a toy version of QCD, as can be seen by enumerating its main features: asymptotic freedom, dynamical mass generation, instantons, skyrmions, integrability , etc (for a review see [16, 17]). For these and other reasons the NLSM still attract the interest of many physicists and mathematicians. In the case of spin chains and ladders the NLSM becomes not just a toy model but a reallistic model describing the low energy physics!.

The O(3) NLSM is a relativistic quantum field theory in 1+1 dimensions whose field Φ is a three component vector living on the 2d-sphere S^2,

$$\Phi^2 = 1, \tag{5}$$

The euclidean action of the model is given by,

$$S = \int d^2x \left[-\frac{1}{2g} \left(\partial_\mu \Phi \right)^2 + i \frac{\theta}{8\pi} \, \epsilon_{\mu\nu} \, \Phi \cdot \left(\partial_\mu \Phi \times \partial_\nu \Phi \right) \right] \tag{6}$$

where $g > 0$ is the sigma model coupling constant, and $\epsilon_{\mu\nu}$ the 2d Levi-Civita symbol. The quantity

$$W = \frac{1}{8\pi} \int d^2x \epsilon_{\mu\nu} \mathbf{\Phi} \cdot (\partial_\mu \mathbf{\Phi} \times \partial_\nu \mathbf{\Phi}) \in \mathcal{Z} \tag{7}$$

takes an integer value for field configurations $\mathbf{\Phi}(\mathbf{x})$ which go to a fixed value, say $\mathbf{\Phi}_0$, when $|\mathbf{x}| \to \infty$ (this condition is required to have a finite action). Compactifying the space-time into the sphere S^2, the integral (7) gives the winding number of the map S^2 (space-time) $\to S^2$ (target space). The parameter θ enters in the partition function as $e^{i\theta W}$, and therefore the integer character of W implies that θ is defined modulo 2π. The theta term in the action is a truly topological term, which leads to dramatic non perturbative effects.

Notice that W changes its sign under a parity (or time reversal) transformations. Hence the topological term of the action breaks explicitly parity (or time reversal) unless $\theta = 0$ or π. Indeed if $\theta = 0$ (mod 2π) the topological term is completely absent while if $\theta = \pi$ the winding number contributes to the action with a sign, i.e. $(-1)^W$. We expect from these properties that the AFH-spin chains and ladders which are parity invariant will be associated with $\theta = 0$ or π.

The basic properties of the NLSM for these values of θ are given in Table 3 [18, 19, 20, 1, 21].

θ (mod 2π)	Spectrum	Correlations
0	gapped	exponential decay
π	gapless	algebraic decay

Table 3: Behaviour of the O(3) NLSM

This behaviour, which is independent of the magnitude of the coupling constant g, will allow us to make a quick derivation of Table 3 in the strong coupling limit. Before we do that, let's write the Hamiltonian that follows from the action (6),

$$H_{\mathrm{NLSM}} = \frac{c}{2} \int dx \left[g \left(1 - \frac{\theta}{4\pi}\mathbf{\Phi}'\right)^2 + \frac{1}{g}\mathbf{\Phi}'^2 \right] \tag{8}$$

where $\mathbf{\Phi}'(x) = \partial_x \mathbf{\Phi}(x)$, c is the "speed of light" and $l(x)$ is the angular momentum density, which is defined as,

$$\mathbf{l} = \mathbf{\Phi} \times \frac{d\mathbf{\Phi}}{dt} \tag{9}$$

Besides the constraint (5) the fields $\mathbf{\Phi}$ and l satisfy,

$$\mathbf{l}(x) \cdot \mathbf{\Phi}(x) = 0 \tag{10}$$

and the cannonical equal-time commutation relations,

$$[l^a(x), l^b(y)] = i\epsilon^{abc}\delta(x-y)l^c(x)$$
$$[l^a(x), \Phi^b(y)] = i\epsilon^{abc}\delta(x-y)\Phi^c(x) \tag{11}$$
$$[\Phi^a(x), \Phi^b(y)] = 0$$

It is very illustrative to prove the statements given in Table 3 in the strong coupling limit ($g >> 1$), for which we shall use a regularized lattice version of the Hamiltonian (8) [20, 21],

$$H_{NLSM}^{lattice} = \frac{c}{2} \sum_n \left(g \, l_n^2 - \frac{2}{g} \, \Phi_n \Phi_{n+1} \right) \tag{12}$$

At every site of the lattice there is a "particle" moving on a sphere ($\Phi_n^2 = 1$), with angular momenta l_n. If $\theta = 0$ the angular momenta takes all possible integer values, $l_n = 0, 1, \ldots \infty$ [20]. However if $\theta = \pi$ there is a monopole of charge 1 at the center of every sphere, which implies that the possible values of the angular momentum are restricted to half-integers values, $l_n = 1/2, 3/2, \ldots \infty$ [21].

In the limit $g >> 1$ the kinetic term $g \sum_n l_n^2$ dominates over the potential term $-\frac{2}{g} \sum \Phi_n \Phi_{n+1}$, and the ground state is obtained choosing the smallest possible value of l_n at every site.

If $\theta = 0$ the optimal choice is given by $l_n = 0$, $\forall n$, which yields a unique ground state. The first excited states are obtained by choosing the irrep $l = 1$ in one site and $l = 0$ in the rest of the chain. Since this can be done at every site, there is a huge degeneracy, which is broken by the potential term, which delocalizes the $l = 1$ excitations. The 3L degenerate first excited states become a band of $l = 1$ magnons, separated from the ground state by a gap, which can be computed in perturbation theory. In [20] this gap was computed up to 6th order in $1/g^2$. The first three terms read,

$$\Delta = cg \left(1 - \frac{2}{3}\frac{1}{g^2} + 0.074\frac{1}{g^4} + O(\frac{1}{g^6}) \right) \tag{13}$$

In the weak coupling limit ($g << 1$) a perturbative RG analysis shows that the gap vanishes exponentially as [18]

$$\Delta \sim \frac{c}{g} e^{-2\pi/g} \tag{14}$$

The proportionality constant depends on the regularization of the model [22] (see also [23]) Combining the strong and weak coupling analysis and using Pade aproximants the authors of [20] concluded that the gap should never vanishes as long as g is non zero.

For $\theta = \pi$ the kinetic energy is minimized by the choice $l_n = 1/2$ $\forall n$, which still leaves a huge degeneracy for the ground state. This degeneracy is lifted by the potential term, which leads to an effective AFH model when restricted to the subspace $l_n = 1/2$ [21]. Since the $S = 1/2$ AFH model is massless one gets that the NLSM at $\theta = \pi$ is also massless. In reference [21] it was argued that this gapless behaviour persists to all values of g.

Haldane Map for Spin Chains

The map from the Heisenberg model into the NLSM can be done in various ways: using coherent states in the path integral formalism, generalizing the Hubbard-Stratonovich formula in the partition function, or applying gradient expansions in the Hamiltonian formalism. We shall mainly use the later one [2], but we shall also briefly explain the path integral approach for ladders.

The starting point of the construction is the spin wave analysis of the AFH model. This consist in the linearization of the evolution equations of the spin operators $\mathbf{S}(n)$, around the classical Neel configuration. The equations of motion of the spin operators read,

$$\frac{d\mathbf{S}(n)}{dt} = i[H_{\text{chain}}, \mathbf{S}(n)] \tag{15}$$
$$= -J\mathbf{S}(n) \times [\mathbf{S}(n+1) + \mathbf{S}(n-1)]$$

The basic asumption is that $\mathbf{S}(n)$ deviates by a small amount from the alternating Neel configuration,

$$\mathbf{S}(n) = (-1)^n S \, \mathbf{z} + \mathbf{s}(n) \tag{16}$$

where S is the magnitude of the spin and $\mathbf{z}$ is the unit vector in the vertical direction. In the linearized approximation eq.(15) becomes,

$$\frac{d\zeta(n)}{dt} = i(-1)^{n+1} JS[\zeta(n+1) + \zeta(n-1) + 2\zeta(n)] \tag{17}$$

where $\zeta(n) = s^x(n) + is^y(n)$. The spin waves are the plane wave solutions of eq.(17),

$$\zeta(n) = e^{i(\omega t + kn)}(\psi(k) + (-1)^{n+1}\phi(n)) \tag{18}$$

We are interested in the low energy and long wavelength solutions which are given by,

$$\omega = vk$$
$$\psi(k) \sim Ak \tag{19}$$
$$\phi(k) \sim B$$

Equation (17) is satisfied by the ansatz (18) if $2A = B$. The spin wave velocity v is given by,

$$v = 2JS \tag{20}$$

The two transverse spin wave solutions (18) can be identified with the massless goldstone bosons associated with the breaking of the rotational symmetry of the AFH Hamiltonian by the classical Neel state. Eq(18) implies that the spin variables have two slowly varying components around the momenta $k = 0$ (field l) and $k = \pi$ (field Φ), which correspond to the local spin density and the local staggered magnetization respectively.

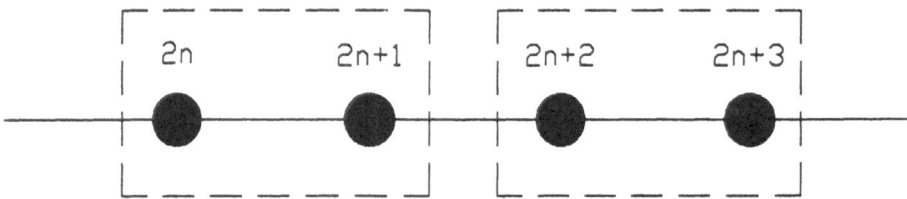

Figure 2 Blocking of the spin chain needed to define the NLSM variables out of the spin ones.

This elementary spin wave analysis is very useful in order to construct the map from the Heisenberg model into the NLSM. This map can be mathematically formulated as a change of variables [2]. First of all, one has to divide the chain into blocks of two sites (see Fig. 2). The block, say (2n, 2n+1), is given the coordinate x= 2n+1/2 and for every block one performs the following change of variables,

$$\mathbf{S}(2n) = \mathbf{l}(x) - S\boldsymbol{\Phi}(x) \tag{21}$$
$$\mathbf{S}(2n+1) = \mathbf{l}(x) + S\boldsymbol{\Phi}(x)$$

The inverse of eqs.(21) are

$$\mathbf{l}(x) = \tfrac{1}{2}[\mathbf{S}(2n+1) + \mathbf{S}(2n)] \tag{22}$$
$$\boldsymbol{\Phi}(x) = \tfrac{1}{2S}[\mathbf{S}(2n+1) - \mathbf{S}(2n)]$$

These relations imply that l is a local spin density and $\boldsymbol{\Phi}$ is a local staggered magnetization. Using (21) and the eqs. $\mathbf{S}^2(n) = S(S+1)$ one gets,

$$\mathbf{l}^2(x) + S^2\boldsymbol{\Phi}^2(x) = S(S+1) \tag{23}$$
$$\mathbf{l} \cdot \boldsymbol{\Phi} = 0$$

In the semiclassical limit $S \gg 1$ eqs.(23) become the sigma model conditions (5) and (10). The commutators (11) can also be obtained in the limit $S \gg 1$ from the commutation relations of the spin matrices, with the following identification of the Dirac's delta function and the Kronecker's delta symbol in the limit where the lattice spacing δ goes to zero,

$$\delta(x - y) = \lim_{\delta \to 0} \frac{1}{2\delta} \, \delta_{x,y} \tag{24}$$

The term 2δ in the denominator of eq.(24) is the lattice spacing in the x-variables (from now on we shall set $\delta = 1$).

The AFH Hamiltonian (1) reads in terms of the new variables l and Φ,

$$H_{\text{chain}} = \sum_x [-S(S + 1) + 2l^2(x) + l(x) \, l(x + 2) \tag{25}$$
$$+S(\Phi(x) \, l(x + 2) - l(x) \, \Phi(x + 2)) - S^2\Phi(x) \, \Phi(x + 2)]$$

Making the asumption that the fields $l(x)$ and $\Phi(x)$ vary slowly in the x coordinate, we can perform a gradient expansion and truncation of the higher derivative terms. Keeping terms with at most two spatial or time derivatives in the field Φ, we obtain the following Hamiltonian,

$$H_{\text{chain}}^{\text{(Continuum)}} = \int dx \, [2l^2 - S(l\Phi' + \Phi'l) + S^2(\Phi')^2] \tag{26}$$

which coincides with the NLSM Hamiltonian given in (8) upon the identifications,

$$\theta = 2\pi S, \quad g = \frac{2}{S}, \quad c = 2JS \tag{27}$$

This is the desired result which permits to derive Table 1 from Table 3. From eqs. (14) and (27) we can estimate the value of the spin gap and the correlation length as functions of the spin,

$$\Delta_S \sim JS^2 e^{-\pi S}, \quad \xi \sim \frac{2}{S} e^{\pi S} \tag{28}$$

In Table 4 we show the numerical values of the gap, correlation length and spin velocity of the $S = 1$ and 2 Heisenberg chains, obtained using Quantum Monte Carlo [24, 25] and DMRG methods [26, 27].

Spin	Δ/J	ξ	c/J
$S = 1$	0.4107	~ 6	2.47
$S = 2$	0.049-0.085	49	4.16

Table 4: Spin gap of the $S = 1$ and 2 chains.

This table shows the semi-quantitative agreement between the numerical (exact) results and the NLSM estimates (28). In the semiclassical limit where $S \to \infty$ the NLSM predicts that the gap should go to zero exponentially fast, recovering in that way the classical gapless behaviour. This shows that the existence of the gap is a truly quantum mechanical effect.

To end up our review of the spin chains, we would like to make some comments on the non uniqueness of the map (3). Indeed eqs. (21) and (22), which give the map between the spin and the sigma model variables, depend on the choice of the 2 site blocks. The other possible choice, given by the blocks (2n+1, 2n+2), leads two another couple of variables $\tilde{l}$ and $\tilde{\Phi}$, which are linearly related to l and Φ, by the eqs,

$$\tilde{l} = l - S \; \Phi' \tag{29}$$
$$\tilde{\Phi} = \Phi$$

These equations leave invariant the constraints (5), (10), and in fact they can be obtained by the following transformation [2],

$$\tilde{l} = e^{-iS \int dx \, \beta' \, \cos\alpha} \; l \; e^{iS \int dx \, \beta' \, \cos\alpha} \tag{30}$$
$$\tilde{\Phi} = e^{-iS \int dx \, \beta' \, \cos\alpha} \; \Phi \; e^{iS \int dx \, \beta' \, \cos\alpha}$$

where α and β are the spherical coordinates that parametrize the staggered field

$$\Phi = (\sin\alpha \, \cos\beta, \sin\alpha \, \sin\beta, \cos\alpha) \tag{31}$$

However the Hamiltonian (26) is not left invariant, but the change affects only the theta parameter which for the new variables becomes,

$$\tilde{\theta} = -2\pi S \tag{32}$$

This change in the value of θ, depending on the blocking, can also be seen as due to a parity transformation. Recall that parity changes the sign of the topological term W. The two blockings (2n, 2n +1) and (2n+1, 2n+2) are indeed related by parity. In the next section we shall consider more general blockings which will lead to changes in θ but that are not related by parity transformations. If $\theta = \pi$ the change (32) leaves invariant the sign factor $(-1)^W$ appearing in the partition function.

Haldane Map for Spin Ladders

There are two ways to study the problem of spin ladders using NLSM methods. The first one consists in the application of the Haldane's map to every chain forming the ladder, obtaining a system of n_ℓ sigma model fields coupled by rung interactions. This approach is similar in spirit to the bosonization studies of spin ladders made in references [28], and its applicability is reasonable in the weak coupling regime $J'/J << 1$. The other possible approach is to built up a unique sigma model field describing the low energy modes of the spin ladder as a whole [9, 10, 11]. This approach should be appropiate to study the intermediate coupling regime ($J'/J \sim O(1)$). The strong coupling regime ($J'/J >> 1$) has been mainly studied using perturbative [29, 30, 31] and mean field methods [32]. We shall see that the later approaches are related to the discrete version of the NLSM based on the Hamiltonian (12).

The map from the ladder into the NLSM follows the same steps as that of the spin chain. We shall review below the approach of reference [10].

Spin Wave Analysis of Ladders

This analysis shows that at the linearized level there are two massless modes corresponding to two Goldstone bosons. These modes describe the gapless deviations from the classical Neel solution, which in the case of AF-coupling along the rungs reads,

$$\mathbf{S}_a^{class}(n) = (-1)^{a+n} S \tag{33}$$

In the rest of the talk we shall confine ourselves to this case. The transverse components of the spin waves, i.e. $\zeta_a(n) = s_a^x(n) + i s_a^y(n)$, are given by

$$\zeta_a(n) = e^{i(\omega t + kn)} \left(A_a k + (-1)^{a+n+1} B \right) \tag{34}$$

where $\omega = vk$. The quantities A_a represent the fraction of total spin carried out by the ath-leg, and we shall normalize them as,

$$\sum_a A_a = 1 \tag{35}$$

The spin wave velocity v and the $A_a's$ are given by,

$$\left(\tfrac{v}{S}\right)^2 = J\, n_\ell / \sum_{u,b} I_{u,b}^{-1} \tag{36}$$
$$A_a = \sum_b L_{a,b}^{-1} / \sum_{c,d} L_{c,d}^{-1}$$

where L^{-1} is the inverse of a matrix L defined as follows,

$$L_{a,b} = \begin{cases} 4J + J' & a = b = 1 \text{ or } n_\ell \\ 4J + 2J' & 1 < a = b < n_\ell \\ J' & |a - b| = 1 \end{cases} \tag{37}$$

Besides the 2 Goldstone bosons there are, at the linearized level, $2(n_\ell - 1)$ massive modes. As an example we give below the values of their masses for the $n_\ell = 2$ and 3 ladders,

$$m_{n_\ell=2} = S\sqrt{8JJ'} \tag{38}$$
$$m_{n_\ell=3} = S\sqrt{J'(J + 4J')} \, , S\sqrt{J'(J + 12J')}$$

Of course in the limit when $J' \to 0$ the masses of the massive modes (38) dissapear and all the modes become massless, as should be the case for a set of n_ℓ uncoupled chains. The basic asumption in the construction of [10] is the truncation of the massive modes, keeping only the gapless ones, which is justified if the energy scales, generated non perturbatively, are lower than the gap of the massive modes. One may expect that this condition is satisfied in the intermediate and strong coupling regimes.

Map: Spin Ladders $\to NLSM$

The spin wave analysis serves as a preparation for the more complicated job of finding the map from the spin ladder into the sigma model. In the Hamiltonian formulation the first step is to split the ladder into blocks of n_B sites, defining for each block an average angular momentum $\mathbf{l}$ and staggered magnetization $\mathbf{\Phi}$ as follows,

$$\mathbf{l}(\mathbf{x}) = \frac{n_\ell}{n_B} \sum_{(a,n)\in B(x)} \mathbf{S}_a(n) \tag{39}$$
$$\mathbf{\Phi}(x) = \frac{1}{Sn_B} \sum_{(a,n)\in B(x)} (-1)^{a+n} \mathbf{S}_a(n)$$

where x denotes the center of mass position of the block $B(x)$ along the leg axis, and the prefactor n_ℓ/n_B is required in order that $\mathbf{l}$ satisfy the cannonical commutation relations (11) in the continuum limit. The relation (24) reads in the more general case,

$$\delta(x-y) = \lim_{\delta \to 0} \frac{n_\ell}{n_B \delta} \delta_{x,y} \tag{40}$$

For single chains there are only two types of blockings related by parity. However for ladders there are many different types of blockings, which may in principle lead to different results. In order to choose those blockings that are physically acceptable we shall impose the following conditions

- The blocks must have an even number of sites.

- Every block must contain a site (or more) of every leg the same number of times, which implies that n_B should be a multiple of n_ℓ.

- Every site in a block must have a nearest neighbour belonging to the same block.

Motivated by the spin wave solution (34), we shall propose the following ansatz for the spin operators in terms of the sigma model fields,

$$\mathbf{S}_a(n) = A_a \, \mathbf{l}(x) + (-1)^{a+n} S \, \mathbf{\Phi}(x) \text{ for } (a,n) \in B(x) \tag{41}$$

The consistency between eqs.(39) and (41) is guaranteed by the following identities,

$$\sum_{(a,n)\in B(x)}(-1)^{a+n} = 0$$
$$\sum_{(a,n)\in B(x)} A_a = \frac{n_B}{n_\ell} \qquad (42)$$
$$\sum_{(a,n)\in B(x)}(-1)^{a+n} A_a = 0$$

which can be proved from the conditions on the allowed blocks given above and the symmetry properties of A_a.

Strictely speaking we should add to the r.h.s. of (41) a set of fields describing the massive modes whose existence we discussed in the spin wave analysis. Their inclusion makes the map (41) more rigorous from a mathematical point of view, allowing a careful derivation of the effective Hamiltonian governing the dynamics of the fields l and Φ [10]. Having found the effective Hamiltonian the massive fields are discarded, so that we shall not consider them from now on.

Let us now consider a few examples ladder's blockings.

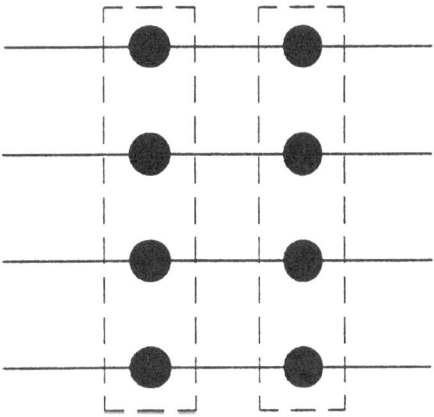

Figure 3 Columnar Blocking $(n_\ell = 4)$

Columnar Blocks $(n_B = n_\ell)$ For ladders with an even numbers of legs the smallest possible blocks, satisfying the conditions given above, coincide with the rungs (see Fig 3), i.e. $n_B = n_\ell$. The partition of the ladders into rungs is what one effectively does in the study of the strong coupling limit of ladders [29, 30, 31]. In that sense one may expect to find relations between the NSLM approach and the strong coupling analysis.

Replacing the ansatz (41) into the ladder Hamiltonian (2), and taking the continuum limit we get the following expression,

$$H = \int dx \left[\frac{1}{2} L_{a,b} A_a A_b \, l^2 + \frac{1}{2} J S^2 \, n_\ell \, \Phi'^2\right] \qquad (43)$$

Comparing (43) with the NLSM Hamiltonian (8) we find,

$$\theta = \quad 0$$
$$\left(\tfrac{c}{S}\right)^2 = \quad Jn_\ell/\sum_{a,b} L_{a,b}^{-1}$$
$$g = \quad 1/\left(S\sqrt{Jn_\ell \sum_{a,b} L_{a,b}^{-1}}\right) \tag{44}$$

Since $\theta = 0$ we see that the NLSM is gapped, which implies that even ladders should always be gapped for any value of the spin. The velocity c coincides with the spin wave velocity v (36). Finally, the coupling constant g has a non trivial dependence of the number of legs and the ratio J'/J. This is interesting because we can derive from the eq.(44) the dependence of the spin gap on the ladder's parameters.

For odd ladders the minimal blocks satisfying the conditions above must have at least $n_B = 2n_\ell$ sites. Unlike the blocks with $n_B = n_\ell$ there are now more geometries. We shall in what follows consider the cases of even and odd values of n_ℓ. In Figs. 4 and 5 we show two of them. The choice of Fig. 4 was the one studied in [10]. We shall give for completeness the results obtained there. Later on we shall study the blocks of Fig. 5.

Rectangular Blocks $(n_B = 2n_\ell)$ The Hamiltonian that one obtains after taking the continuum limit is (fig 4),

$$H = \int \frac{dx}{2}\left[\sum_{a,b} L_{a,b} A_a A_b \, 1^2 + 2JS^2 n_\ell \, \Phi'^2 + 2SJ \sum_{a=1}^{n_\ell}(-1)^a A_a \left(\Phi' \, 1 + 1\Phi'\right)\right] \tag{45}$$

In [10] it was shown that the value of θ corresponding to (45) is given by,

$$\theta = \begin{cases} 0 & n_\ell : \text{even} \\ 2\pi S & n_\ell : \text{odd} \end{cases} \tag{46}$$

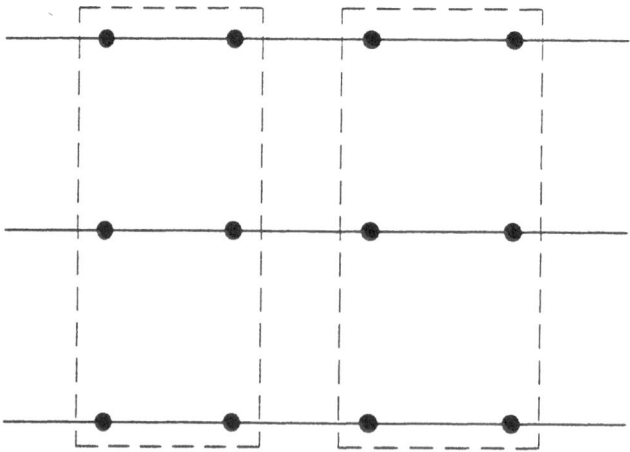

Figure 4 Rectangular Blocking $(n_\ell = 3)$.

The vanishing of θ for even ladders follows simply from the symmetry property $A_a = A_{n_\ell+1-a}$, while the value $\theta = 2\pi S$ for odd ladders requires some more work [10]. The other NLSM parameters of (45) are given by,

$$\left(\frac{c}{S}\right)^2 = 2\frac{Jn_\ell}{\sum_{a,b} L_{a,b}^{-1}} - \delta_{n_\ell,\text{odd}}\frac{1}{\left(2\sum_{a,b} L_{a,b}^{-1}\right)^2}$$

$$g = 1/\left(S\sqrt{2Jn_\ell \sum_{a,b} L_{a,b}^{-1} - \tfrac{1}{4}\delta_{n_\ell,\text{odd}}}\right)$$

(47)

where $\delta_{n_\ell,\text{odd}} = 1$ (resp. 0) if n_ℓ is odd (resp. even). For n_ℓ even the delta terms appearing in these eqs. are absent and we recover (44), except for a renormalization of c and g ($c_{\text{rect}} = \sqrt{2}\, c_{\text{col}}$, $g_{\text{rect}} = g_{\text{col}}/\sqrt{2}$).

For n_ℓ odd, the formula (47) does not coincide with the spin wave velocity (36), except for the case of the spin chains ($n_\ell = 1$)!. For large ladders the velocity approaches the value $4JS$ ($J' = J$) which differs by a factor of $\sqrt{2}$ with respect to the 2d value which is $2\sqrt{2}JS$.

Diagonal Blocks ($n_B = 2n_\ell$)   Within the gradient approximation which we are using, it is consistent to assign to all the points within a diagonal block (fig. 5) the same coordinate x. Under this asumption we get the following Hamiltonian,

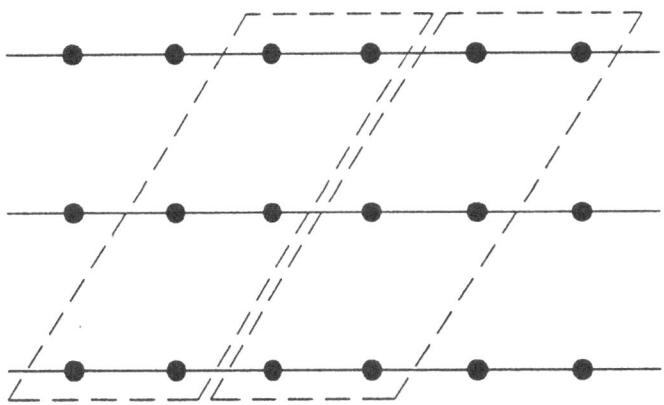

Figure 5 Diagonal Blocking ($n_\ell = 3$).

$$H = \int \frac{dx}{2}\left[\sum_{a,b} L_{a,b} A_a A_b\, \mathbf{l}^2 + 2S^2 \sum_{a=1}^{n_\ell} p_a \mathbf{\Phi}'^2 + 2S \sum_{a=1}^{n_\ell} A_a p_a \left(\mathbf{\Phi}'\,\mathbf{l} + \mathbf{l}\mathbf{\Phi}'\right)\right]$$

(48)

where p_a is defined as

$$p_a = \begin{cases} J + J'/2 & a = 1 \text{ or } n_\ell \\ J + J' & \text{else} \end{cases}$$

(49)

The value of θ that comes out from (48) is given by,

$$\theta = 8\pi S \frac{\sum_a A_a p_a}{\sum_{b,c} L_{b,c} A_b A_c} = 8\pi S \sum_{a,b} p_a L_{a,b}^{-1} \tag{50}$$

To evaluate θ it is convenient to write (49) in matrix form as follows,

$$\theta = 8\pi S n_\ell < F|PL^{-1}|F > \tag{51}$$

where $|F >$ denotes a normalized n_ℓ-component vector with all its entries equal to $1/\sqrt{n_\ell}$ and P is a diagonal matrix with its entries given by p_a. The trick is now to write the matrix L in the form $L = 4P - K^-$, where K^- is defined as,

$$K_{a,b}^- = J' \times \begin{cases} 1 & a = b = 1 \text{ or } n_\ell \\ 2 & a = b \neq 1 \text{ or } n_\ell \\ -1 & |a - b| = 1 \end{cases} \tag{52}$$

Making a "Dyson" decomposition of L^{-1}

$$L = \frac{1}{4P} - \frac{1}{4P} K^- \frac{1}{4P - K^-} \tag{53}$$

and using the fact that K^- annihilates the vector $|F >$ we finally get,

$$\theta = 2\pi S n_\ell \tag{54}$$

It is quite remarkable that all the coupling constants in the expression (50) just combined in order to produce a result which only depends on "global" data S and n_ℓ. Eq.(54) agrees with (46) mod 2π, implying that the choice of block does not affect this parameter. Let us see what are the values of g and c for the Hamiltonian (48),

$$\left(\frac{c}{S}\right)^2 = 2\frac{\sum_a p_a}{\sum_{b,c} L_{b,c}^{-1}} - 4\left(\frac{\sum_{a,b} p_a L_{a,b}^{-1}}{\sum_{c,d} L_{c,d}}\right)^2 \tag{55}$$

$$g = 1/S\sqrt{2\sum_{a,b,c} p_a L_{b,c}^{-1} - \left(2\sum_{a,b} p_a L_{a,b}^{-1}\right)^2}$$

Path Integral Derivation of Haldane Map for Ladders

In this section we shall apply coherent state techniques to derive the map from the spin ladders into the NLSM (for a review see reference [33]). The map for the 2 legged ladder was first worked out in [9]. Our results agree with the result of this reference and extend them to generic values of n_ℓ (see also [11] for an alternative derivation of the results given below).

The action of the spin ladder system in the path integral formulation can be written in the following form,

$$S = \int d\tau \left\{ i \sum_n \sum_{a=1}^{n_\ell} \mathbf{A}(\mathbf{S}_a(n)) \cdot \frac{d\mathbf{S}_a(n)}{d\tau} \right. \tag{56}$$

$$\left. -J \sum_n \sum_{a=1}^{n_\ell} \mathbf{S}_a(n)\mathbf{S}_a(n+1) - J' \sum_n \sum_{a=1}^{n_\ell-1} \mathbf{S}_a(n)\mathbf{S}_{a+1}(n) \right\}$$

where $\mathbf{S}_a(n)$ is a classical spin variable satisfying $\mathbf{S}_a(n)^2 = S^2$ and $\mathbf{A}(\mathbf{S}_a(n))$ is the vector potential that fullfills the constraint $\mathbf{rot}\mathbf{A}(\mathbf{n}) = \mathbf{n}$ $(\mathbf{n} = \mathbf{S}/S)$.

The long wavelength limit of the ladder's spin variables $\mathbf{S}_a(n)$ is given, according to (41), by

$$\mathbf{S}_a(n) = \delta A_a \mathbf{l}(n) + (-1)^{a+n} S \mathbf{\Phi}(n) \left(1 - \frac{\delta^2}{S^2} A_a^2 \mathbf{l}^2 \right)^{1/2} \tag{57}$$

where the term with the square root is needed for the correct normalization of the variable $\mathbf{S}_a(n)$. For spin chains this later term can be replaced by 1, but for ladders it gives a non trivial contribution when considering the coupling between rungs.

Introducing (57) into (56) and performing the standard gradient expansions one finds the following NLSM parameters

$$\begin{aligned} \theta &= 2\pi S \, \delta_{n_\ell,\text{odd}} \\ \left(\tfrac{c}{S} \right)^2 &= Jn_\ell / \sum_{a,b} L_{a,b}^{-1} \\ g &= 1/\left(S\sqrt{Jn_\ell \sum_{a,b} L_{a,b}^{-1}} \right) \end{aligned} \tag{58}$$

which coincide with those obtained using columnar blocks (44) for even legged ladders. For odd ladders the results of the path integral and those obtained with rectangular blocks differ, except in the case of spin chains $(n_\ell = 1)$. This poses the problem between the relation between the Hamiltonian formalism and the path integral formalism for this type of ladders. In any case, notice that the path integral approach gives for even and odd legged ladders a value of c identical to the spin wave velocity (36).

Before we extract more consequences from the mapping of the ladders into the NLSM, it is convenient to express c and g in terms of

The Function f_{n_ℓ}

Let us define f_{n_ℓ} as [10],

$$f_{n_\ell}(J'/J) = \frac{4J}{n_\ell} \sum_{a,b} L_{a,b}^{-1} \tag{59}$$

$$= \frac{1}{n_\ell^2} \left[\delta_{n_\ell,\text{odd}} + 2 \sum_{m=1,3,\dots,n_\ell-1} \sin^{-2}\left(\frac{\pi m}{2n_\ell} \right) \left(1 + \frac{J'}{J} \cos^2 \frac{\pi m}{2n_\ell} \right)^{-1} \right]$$

Its explicit expression and numerical values for $z = 1$ in the cases $n_\ell = 1, 2, 3, 4$ and ∞ is given in Table 5.

$f_{n_\ell}(z)$ is a monotonically decreasing function in z, which varies from 1 to $1/n_\ell^2 \, \delta_{n_\ell,\mathrm{odd}}$ as z varies from 0 to ∞. Interesting enough, the sum appearing in eq.(59) can be performed yielding,

$$f_{n_\ell}(z) = \frac{1}{1+z}\left[1 + \frac{z}{n_\ell\,(1+z)^{1/2}}\frac{\left(1+\frac{2}{z}\left(1+\sqrt{1+z}\right)\right)^{n_\ell} - (-1)^{n_\ell}}{\left(1+\frac{2}{z}\left(1+\sqrt{1+z}\right)\right)^{n_\ell} + (-1)^{n_\ell}}\right] \qquad (60)$$

In the isotropic case $(z = 1)$ we get,

$$f_{n_\ell}(1) = \frac{1}{2}\left(1 + \frac{1}{n_\ell\sqrt{2}}\frac{\left(3+2\sqrt{2}\right)^{n_\ell} - (-1)^{n_\ell}}{\left(3+2\sqrt{2}\right)^{n_\ell} + (-1)^{n_\ell}}\right) \qquad (61)$$

For $n_\ell > 2$, a good approximation of (61) is given by (see Table 5),

$$f_{n_\ell}(1) \sim \frac{1}{2}\left(1 + \frac{1}{n_\ell\sqrt{2}}\right) \qquad (62)$$

In eqs. (60) and (61) the difference between even and odd ladders dissapears exponentially as n_ℓ increases.

This ends the review of the properties of f_{n_ℓ} and we return now to our general discussion.

Summary of Results and Conclusions

In Table 6 we summarize the results obtained so far for the three types of blocks we have considered so far.

From this table we can extract the following conclusions,

- The values of θ, c and g are block dependent. For example, for n_ℓ even we get $g_{\mathrm{rect}} = \frac{1}{\sqrt{2}}\,g_{\mathrm{col}}$. Since the size of both blocks is different we could interpreted the previous relation as the RG relation bewtween $g's$ at two different length scales $(n_B(\mathrm{rect})/n_B(\mathrm{col}) = 2)$. Curiously enough the value of g_{rec} is smaller than the value of g_{col}, suggesting that this "one step RG flow" is similar to the RG flow of the 2+1 NLSM, where the coupling constant g decreases at longer distances in the "renormalized classical region" [14]. After this first RG step, which truncates the ladder's degrees of freedom down to those of the 1+1 NLSM, the value of g will start growing as a consequence of the 1+1 NLSM RG equations, so that the physics of the system will be dominated by the strong coupling regime.

  From Table 6 we find two "RG-invariant" quantities: the θ parameter and the perpendicular spin susceptibility.

- The invariance of the θ parameter relies on its periodicity, which implies that for all purposes we may take $\theta = 2\pi S n_\ell$. This is really a topological result for it only depends on the "global data" S and n_ℓ, and not on the values of the ladder coupling constants J and J'. For odd ladders this result must be a consequence of the generalization, due to Affleck [35], of the well known theorem by Lieb-Schultz-Mattis [36], that asserts that in the infinite length limit the odd ladders must either have a degenerate ground state or else there are gapless excitations.

- The value of the bare perpendicular susceptibility χ^0 is given for all the block choices by,

$$\chi^0 = \frac{1}{c\,g} = \frac{n_\ell f_{n_\ell}}{4J} \tag{63}$$

Notice that for large n_ℓ the susceptibility per site χ^0/n_ℓ goes to a finite value.

It is quite interesting to compare (63) and the spin wave velocity (36),

$$c = \frac{2JS}{\sqrt{f_{n_\ell}}} \tag{64}$$

with the corresponding expressions of the bare perpendicular spin susceptibility and spin wave velocities of a d dimensional sigma model [37],

$$\chi^0 = \frac{1}{4dJa^d} \;,\quad c = 2\sqrt{d}JSa \tag{65}$$

where a denotes the lattice spacing.

The comparison of (63), (64) and (65) suggest somehow that ladders behave as d_{ladder} dimensional spin systems with,

$$d_{\text{ladder}} = \frac{1}{f_{n_\ell}} \tag{66}$$

This naive definition of "fractal" dimension of ladders helps to explain some numerical facts. First of all, if we choose $J' = 0$, then $f_{n_\ell}(0) = 1$ and we get $d_{\text{ladder}} = 1$, which indeed corresponds to 1d chains. For the isotropic models and n_ℓ large we get,

$$d_{\text{ladder}} \sim \frac{2}{1 + \frac{1}{n_\ell\sqrt{2}}} \tag{67}$$

which converges towards $d = 2$ (plane) from below.

A less heuristic proposal is to associate f_{n_ℓ} with a "finite size" effect in the renormalization constants Z_c and Z_χ of the ladder. Combining the recent work of Chakravarty [15], together with the results presented above, we shall propose a finite size correction of the spin susceptibility, spin velocity and spin-stiffness renormalization constants of ladders as follows,

$$\begin{aligned}
Z_\chi(S, n_\ell) &= 2f_{n_\ell}Z_\chi(S) \\
Z_c(S, n_\ell) &= Z_c(S)/\sqrt{2f_{n_\ell}} \\
Z_{\rho_s}(S, n_\ell) &= Z_{\rho_s}(S)
\end{aligned} \tag{68}$$

where $Z_\chi(S), Z_c(S)$ and $Z_{\rho_s}(S)$ are those of the 2d spin system (i.e. $n_\ell = \infty$). The last eq. in (68) follows from the first two thanks to the relation $Z_{\rho_s} = Z_\chi Z_c^2$. We shall give some numerical support to (68) in the next section.

The general formulation we have introduced above, will allow us to discuss certain important issues concerning the ladders.

Spin Gap of Even Ladders

Choosing the columnar-block description of the even ladders we deduce from Table 6 and eq.(14) the following expression for the spin gap,

$$\Delta_{n_\ell}^{(1+1)} \sim JS^2 \, n_\ell \exp\left(-\pi S n_\ell \sqrt{f_{n_\ell}}\right) \tag{69}$$

which predicts a exponential decay of the gap as a function on the number of legs [10]. This implies in particular that the spin gaps of the 2, 4 and 6 legged ladders should be related. Indeed one finds from numerical results of the isotropic ladders

$$\frac{\Delta_2 \Delta_6}{\Delta_4^2} \sim 1 \tag{70}$$

where we have used the data of Table 7 together the value $\Delta_2/J = 0.504$ [29, 38].

In agreement with (69), Chakravarty has recently derived the exponential fall off of the gap with n_ℓ using the 2+1 NLSM [15]. In his approach a spin ladder of width n_ℓ at zero temperature and periodic boundary conditions along the rungs, is equivalent to a Heisenberg plane of infinite extent at a finite temperature inversely proportional to n_ℓ. This allows the use of the 2+1 NLSM results to study ladder systems. In particular the expressions for the spin gap and correlation length are given by (isotropic ladders $J' = J$), [15]

$$\Delta_{n_\ell}^{(2+1)} = \frac{16\pi}{e} JS^2 Z_{\rho S} \exp\left(-\frac{\pi S}{\sqrt{2}} \frac{Z_{\rho S}}{Z_c} \frac{L}{a}\right) \left(1 - \frac{1}{\pi\sqrt{2}S} \frac{Z_c}{Z_{\rho S}} \frac{a}{L}\right)^{-1} \tag{71}$$

$$\xi_{n_\ell}^{(2+1)} = \frac{e}{4\pi\sqrt{2}} \frac{J Z_c}{S Z_{\rho_S}} \exp\left(\frac{\pi S}{\sqrt{2}} \frac{Z_{\rho S}}{Z_c} \frac{L}{a}\right) \left(1 - \frac{1}{\pi\sqrt{2}S} \frac{Z_c}{Z_{\rho S}} \frac{a}{L}\right)$$

where L and a are the width and lattice spacing of the ladder ($L/a = n_\ell$). The exponential behaviour of eqs. (69) and (71) agree in the classical limit $S \to \infty$ provided we choose Z_c and Z_{ρ_s} as the renormalization constants $Z_c(S, n_\ell)$ and $Z_{\rho_s}(S, n_\ell)$ defined in (68). This gives further support to the finite size correction propose in (68).

n_ℓ	1	2	3	4	∞
$f_{n_\ell}(z)$	1	$1/\left(1+\frac{z}{2}\right)$	$\left(1+\frac{z}{12}\right)/\left(1+\frac{3z}{4}\right)$	$\left(1+\frac{z}{4}\right)/\left(1+z+\frac{z^2}{8}\right)$	$\frac{1}{1+z}$
$f_{n_\ell}(1)$	1	0.666	0.6190	0.5882	0.5
$\frac{1}{2}\left(1+\frac{1}{n_\ell\sqrt{2}}\right)$	0.853	0.677	0.6178	0.5884	0.5

Table 5: Expressions for the function $f_{nl}(z)$ and numerical values.

Block	n_ℓ	θ	c	g
Columnar	even	0	$2JS/\sqrt{f_{n_\ell}}$	$2/Sn_\ell\sqrt{f_{n_\ell}}$
Rectangular	even	0	$2\sqrt{2}JS\sqrt{f_{n_\ell}}$	$\sqrt{2}/Sn_\ell\sqrt{f_{n_\ell}}$
Rectangular	odd	$2\pi S$	$\frac{2\sqrt{2}JS}{\sqrt{f_{n_\ell}}}\left(1-\frac{1}{2n_\ell^2 f_{n_\ell}^2}\right)^{1/2}$	$\frac{\sqrt{2}}{Sn_\ell\sqrt{f_{n_\ell}}}\left(1-\frac{1}{2n_\ell^2 f_{n_\ell}}\right)^{-1/2}$
Diagonal	both	$2\pi Sr_\ell$	$\frac{2JS}{f_{n_\ell}}\left[2f_{n_\ell}(1+(1-\frac{1}{n_\ell})\frac{J'}{J})-1\right]^{1/2}$	$\frac{2}{Sn_\ell}\left[2f_{n_\ell}(1+(1-\frac{1}{n_\ell})\frac{J'}{J})-1\right]^{-1/2}$
Path Int.	both	$2\pi S\delta_{n_\ell,\text{ odd}}$	$2JS/\sqrt{f_{n_\ell}}$	$2/Sn_\ell\sqrt{f_{n_\ell}}$

Table 6: Haldane-Map parameters for different choices of blocks.

In Tables 7 and 8 we give the values of the gap and correlation length of the 4 and 6 legged ladders obtained using: i) numerical methods (Quantum Monte Carlo [39, 40] and DMRG [38]), ii) the NLSM in 2+1 ([15]) and iii) the finite size correction of the NLSM results of [15]. For the two later set of data the values of the renormalization constant for S=1/2 are choosen as $Z_c = 1.18, Z_{\rho_s} = 0.724$ [41].

n_ℓ	DMRG	QMC	$NLSM(2+1)$	$NLSM(2+1)$+finite size
4	0.190	0.16 - 0.17	0.268	0.209
6	?	0.055 - 0.05	0.064	0.050

Table 7: Ladder's spin gap.

n_ℓ	DMRG	QMC	$NLSM(2+1)$	$NSLM(2+1)$+finite size
4	5-6	10.3	6.23	7.37
6	?	~ 30	26.2	31.6

Table 8: Ladder's correlation length.

We observe from Tables 7 and 8 that the finite size modification of the Chakravarty formulas (71) seems to give a rather good agreement with the numerical results. Further work needs to be done to settle this matter.

Limits of Applicability of the Ladder's Map

We mentioned at the beginning of the construction of the map from the ladder into a unique NSLM field, that it would be valid for the intermediate coupling region ($J'/J \sim O(1)$). We shall next explain this point in more detail.

First of all let us consider the weak coupling region $J'/J << 1$. As shown in eqs.(38) the masses of the higher modes of the ladder, at the linearized level, are of order $\sqrt{JJ'}$. On the other hand the mass generated non perturbatively is given by (69). A consistent truncation of the massive modes then requires that the mass of these modes should be larger than the mass generated non perturbatively, which leads to,

$$Ae^{-Bn_\ell} << \frac{J'}{J} \tag{72}$$

This equation implies that there is a lower critical value, $(J'/J)_c$ below which the truncation of the high energy modes, at least in the way it is done here, is not valid. In particular, in the weak coupling region the spin gap is approximately proportional to J' (for $n_\ell = 2, \Delta \sim 0.41J'$)[29, 42, 40]. This behaviour is not consistent with (69). On the other hand eq.(72) suggests that the range of applicability of our model is bigger as n_ℓ increases.

Let us now consider the strong coupling regime ($J'/J >> 1$), and choose as an example the two legged ladder. For the columnar block, the map (41) reads,

$$S_1(n) = \tfrac{1}{2}\mathbf{l}(n) + S(-1)^n \Phi(n)$$
$$S_2(n) = \tfrac{1}{2}\mathbf{l}(n) - S(-1)^n \Phi(n) \tag{73}$$

which plugged into the ladder Hamiltonian leads, without making any approximation, to

$$H_{\text{ladder}} = \sum_n \frac{J'}{2} \mathbf{l}(n)^2 + J \left(\frac{1}{2}\mathbf{l}(n)\,\mathbf{l}(n+1) - 2S^2 \Phi(n)\,\Phi(n+1) \right) \tag{74}$$

If we now apply a gradient expansion in the fields $\mathbf{l}$ and Φ, we obtain the results given in (44) (see also Table 9), for the case $n_\ell = 2$. It is interesting to compare (74) with the lattice NLSM Hamiltonian (12). We see that the gradient expansion and truncation of the $\mathbf{l}'s$ field is crucial in order that (74) becomes a NLSM Hamiltonian. In the strong coupling regime, as can be seen from the mean field analysis of [32], what is pertinent is to drop the term proportional to $\mathbf{l}(n)\mathbf{l}(n+1)$ in (74). If we do this approximation then (74) become the discrete NLSM Hamiltonian with the following identification of g and c,

$$c_{\text{latt}} = S\sqrt{2JJ'}$$
$$g_{\text{latt}} = \tfrac{1}{S}\sqrt{\tfrac{J'}{2J}} \tag{75}$$

If we replace these parameters into the gap formula obtained in the strong coupling regime (13) we get

$$\Delta/J' = 1 - \frac{4S^2}{3}\frac{J}{J'} + 0.296S^4 \left(\frac{J}{J'} \right)^2 \tag{76}$$

If we replace in (76) S^2 by $S(S+1)$ and we particularize to the case $S = 1/2$, then we get the correct behaviour of the gap in perturbation theory to order J/J' [30],

$$\Delta/J' = 1 - \frac{J}{J'} + \frac{1}{2}\left(\frac{J}{J'}\right)^2 + \frac{1}{4}\left(\frac{J}{J'}\right)^3 + O(\left(\frac{J}{J'}\right)^4) \tag{77}$$

The previous discussion illustrates that in the strong coupling limit one has to be careful in making gradient expansions of the NLSM fields. A more detailed analysis is needed to clarify this matter.

Haldane Phase in the 2-Legged Ladder

An important question concerning even spin ladders is wether these systems are in a fundamentally new state or they are in a more familiar state, as for example the integer spin chains. This problem has been addressed by various authors arriving at different conclusions [45, 44]. In this section we shall apply the NLSM techniques to clarify this issue, finding support for the view of [44] that the 2 legged ladder is in the same phase than the $S = 1$ chain [43]. A more detailed discussion is delayed to the future.

The $n_\ell = 2$ ladder can be studied using columnar, rectangular and diagonal blocks. The values of the corresponding NLSM parameters are given in Table 9.

Block	θ	c	g
Columnar	0	$2JS\left(1 + \frac{J'}{2J}\right)^{1/2}$	$\frac{1}{S}\left(1 + \frac{J'}{2J}\right)^{1/2}$
Rectangular	0	$2\sqrt{2}JS\left(1 + \frac{J'}{2J}\right)^{1/2}$	$\frac{1}{S\sqrt{2}}\left(1 + \frac{J'}{2J}\right)^{1/2}$
Diagonal	$4\pi S$	$2JS\left(1 + \frac{J'}{2J}\right)$	$\frac{1}{S}$

Table 9: NLSM parameters of the 2 legged ladder

From Table 9 we see that the parameters obtained using the diagonal blocks coincide with those of an effective spin chain with spin and exchange coupling given by,

$$S_{\text{eff}} = 2S, \quad J_{\text{eff}} = \frac{1}{4}(2J + J') \tag{78}$$

This is not an accidental fact. An alternative way to arrive to (78) is to construct an effective model in terms of the spin 2S states formed out by symmetrization of two spin S states located in diagonal positions of the ladder (see fig. 5) [44], i.e.

$$\mathbf{S}_{\text{eff}}(n) = \mathbf{S}_1(n) + \mathbf{S}_2(n-1) \tag{79}$$

The Hamiltonian governing this effective spins is given, to lowest order in perturbation theory by the chain Hamiltonian,

$$H_{\text{eff}} = \frac{1}{4}(2J + J') \sum_n \mathbf{S}_{\text{eff}}(n)\, \mathbf{S}_{\text{eff}}(n+1) \tag{80}$$

The NLSM parameters corresponding to this model (27) are precisely the ones given in Table 9 for the diagonal blocks. Hence the 2 legged ladder is mapped into the same NLSM model as the S=1 chain, which suggests that both systems are in the same phase. The relationship between the "standard" phase of the ladder, given by the RVB picture of [38], and that of the spin 1 chain appears, from the point of view of the NLSM, as the relation between two different types of blockings, namely the rectangular and the diagonal ones. This relation is given by,

$$\tilde{\mathbf{l}} = \mathbf{l} + S\mathbf{\Phi}', \quad \tilde{\mathbf{\Phi}} = \mathbf{\Phi} \tag{81}$$

where $\mathbf{l}, \mathbf{\Phi}$ (resp. $\tilde{\mathbf{l}}, \tilde{\mathbf{\Phi}}$) are the NLSM fields associated to rectangular (resp. diagonal) blocks. Eq.(81) is a cannonical transformation, identical to (29), which changes θ from 0 into $4\pi S$, leaving invariant the values of c and g. This poses a puzzle since the later parameters are different for rectangular and diagonal blocks, except in the case $J' = 2J$. This discrepancy must be understood along the lines of our discussion about the different values of g for columnar and rectangular blocks, and very likely do not affect the conclusion about the equivalence between the 2 legged ladder and the spin 1 chain.

Spin Ladders with Dimerization

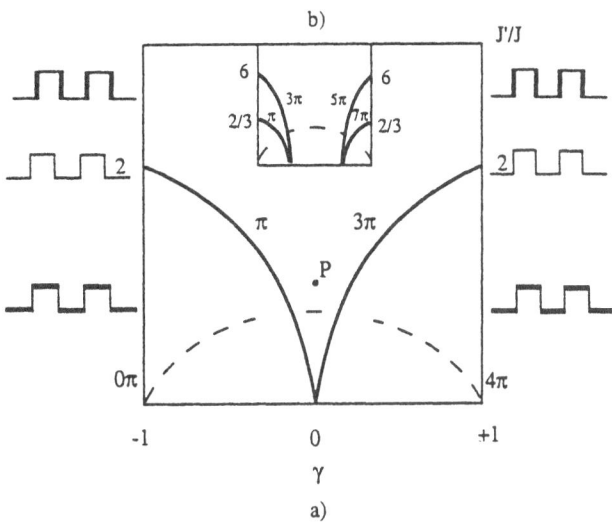

Figure 6 Phase diagrams of the 2 legged ladder with staggered dimerization for S=1/2 and S=1 (inset). If $|\gamma| = 1$ the ladders degenerate into the "snake" chains which are depicted in the margins of the figure.

The main property of uniform spin ladders is that there are no phase transitions as one varies the ratio J'/J. It is thus interesting to investigate the existence of new phases by enlarging the parameter space of the model. There are plenty of possibilites at hand: dimerization, frustration, spin deffects, etc. We shall review here the first one. There are also many possible types of dimerization in a ladder. We shall consider below dimerizations only along the legs. In this case we may distinguish between columnar and staggered dimerizations, for which the intraleg coupling constant is given by,

$$J_a(n) = \begin{cases} J(1 + (-1)^n\gamma) & \text{(columnar)} \\ J(1 + (-1)^{n+a}\gamma) & \text{(staggered)} \end{cases} \tag{82}$$

The dimerization parameter γ will be choosen to vary in the interval $(1, -1)$ in order not to change the AF character of the legs. The rung coupling constant J' may be positive or negative, so all together there are 4 types of models. In references [46] were studied the models with columnar dimerization and ferromagnetic rung coupling. If $J' \to -\infty$ the dimerized ladder becomes effectively a dimerized chain and one can apply the results known for chains.

A chain with spin S and alternation parameter γ can be mapped into a NLSM with θ given by [2, 47],

$$\theta = 2\pi S(1 + \gamma) \tag{83}$$

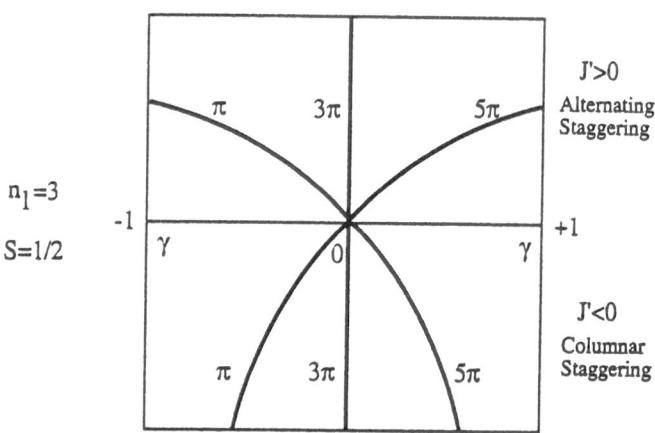

Figure 7 Phase diagram of the 3 legged ladder with columnar (lower half plane) and staggered dimerizations (upper half plane). Observe the existence of 3 critical lines that emerge from the origin.

The criteria that the NLSM models with $\theta = \pi$ (mod 2π) are masless [1, 21] implies the existence of 2S critical points [47]. Indeed, as γ varies from -1 to 1, the parameter θ, given in (83), passes 2S times through π. If S=1/2 there is a single critical point corresponding to the non dimerized chain (i.e. $\gamma = 0$). If S=1 there should exists two critical points for $\gamma_c = \pm 1/2$. Numerical computations show that there are indeed two critical points located at $\gamma_c \sim \pm 0.25$ [48]. Hence the NLSM predicts the existence of these critical points but it is not precise about their localization. The NLSM prediction has also been confirmed for S= 3/2 [49] and S=2 [50].

Returning to ladders with columnar dimerization we expect, from the above discussion, that there should exist two critical lines in the plane $(\gamma, J'/J)$. In the example of the spin 1/2 two legged ladder there should exist two of these lines emanating at $J'/J = -\infty, \gamma_c = \pm 0.25$ and ending at the origin (i.e. $J'/J = 0, \gamma = 0$) [46]. This critical lines separate the Haldane phase associated to the strong ferromagnetic ladder and a dimerized phase of weakly coupled chains.

The staggered dimerization with AF rung couplings has been studied in [51]. The behaviour of these ladders is very interesting and it is based on the map into the NLSM. The value of the θ parameter of a staggered ladder with AF rung coupling is given by [51],

$$\theta = 2\pi S n_\ell (1 + \gamma \, f_{n_\ell}) \tag{84}$$

This formula contains (83) as a particular case (for $n_\ell = 1$). Using the properties of f_{n_ℓ} one can conjecture from (84) the existence of $2S n_\ell$ critical lines in the plane $(\gamma, J'/J)$ [51]. In the particular case of the spin 1/2 two legged ladder there should exist two critical lines separating also the RVB phase of the uniform ladders and a dimerized phase of weakly couple chains (see Fig. 6). The pahse diagram of the 3 legged ladder (S=1/2) contains besides the uniform critical line ($\gamma = 0$) two more critical lines ending at the walls $|\gamma| = 1$ (see Fig. 7).

The two cases analysed above (i.e. columnar/F-rung and staggered/AF-rung) have a very similar phase diagram which suggests some kind of relationship between them [52]. Other types of dimerizations will be considered in [52].

Acknowledgements I would like to thank T.M. Rice for introducing me to the subject of ladders, M.A. Martin-Delgado and R. Shankar for reading this manuscript and E. Dagotto, S. Haas, B. Frischmuth, S. White, H.J. Schulz and S. Sachdev and J. Dukelski for conversations, advises and suggestions.

References

[1] F.D. Haldane, Phys.Rev.Lett. 50, 1153 (1983); Phys.Lett. 93A, 464 (1983).

[2] I.Affleck, Les Houches Lecture Notes, in: Fields, Strings, and Critical Phenomena, ed. E. Brezin and J. Zinn-Justin (North-Holland, Amsterdam, 1988).

[3] J.G. Bednorz and K.A. Muller, Z. Phys. B 64, 188 (1986).

[4] D.C. Johnston, J.W. Johnson, D.P. Goshorn and A.J. Jacobsen, Phys.Rev. B 35, 219 (1987).

[5] Z. Hiroi, M. Azuma, M. Takano and Y. Bando, J. Solid State Chem. 95, 230 (1991).

[6] R. Botet and R. Jullien, Phys. Rev. B27, 613 (1983);
R. Botet, R. Jullien and M. Kolb, Phys. Rev. B28, 3914 (1983);
M. P. Nightingale and H.W.J. Blote, Phys. Rev. B33, 659 (1986);
H.J. Shulz and T.A.L. Ziman, Phys. Rev. B33, 6546 (1986).

[7] E. Dagotto, J. Riera and D. Scalapino, Phys.Rev. B45, 5744 (1992).

[8] E. Dagotto and T.M. Rice, Science 271, 618 (1996).

[9] D.Senechal, Phys. Rev. B52, 15319 (1995).

[10] G. Sierra, J. Math. Phys. A29, 3299 (1996).

[11] S. Dell'Aringa, E. Ercolessi, G. Morandi, P. Pieri and M. Roncaglia, "Effective actions for spin ladders", cond-mat/9610148.

[12] S. Weinberg, Phys. Rev. Lett. 17, 616 (1966);
C. G. Callan, S. Coleman, J. Wess and B. Zumino, Phys. Rev. 177, 2247, (1969).

[13] F.D.M. Haldane, Phys. Rev. Lett. 61, 1029 (1988).

[14] S. Chakravarty, B.I. Halperin and D.R. Nelson, Phys. Rev. Lett. 60, 1057 (1988); Phys. Rev. B39, 2344 (1989).

[15] S. Chakravarty, "Dimensional Crossover in Quantum Antiferromagnets", cond-mat 9608124.

[16] E. Fradkin, "Field Theories of Condensed Matter Systems", Frontiers in Physics, Addison-Wesley (1991).

[17] A. Auerbach, "Interacting Electrons and Quantum Magnetism", Springer-Verlag (1994).

[18] A.M. Polyakov, Phys.Lett. B59,79 (1975).

[19] A. Polyakov and P.B. Weigman, Phys.Lett. B131, 121 (1983).
P.B. Weigman, Phys.Lett. B152, 209 (1985).

[20] C.J. Hamer, J.B. Kogut and L. Susskind, Phys. Rev.D19, 3091 (1979).

[21] R. Shankar and N. Read, Nucl.Phys. B336, 457 (1990).

[22] S.H. Shenker and J. Tobochnik, Phys. Rev. B 22, 4462 (1980).

[23] P. Hasenfratz, M. Maggiore and F. Niedermayer, Phys. Lett. B245, 522 (1990).

[24] M. Takahashi, Phys. Rev. Lett. 62, 2313 (1989).

[25] S. Yamamoto, Phys. Rev. Lett. 75, 3348 (1995).

[26] S.R. White and D.A. Huse, Phys. Rev. B 48, 3844 (1993);
E.S. Sorensen and I. Affleck, Phys. Rev. Lett. 71, 1633 (1993).

[27] U. Schollwock, O. Golinelli and T. Jolicoeur, Phys. Rev. B 54, 4038 (1996).

[28] H.J. Schulz, Phys.Rev. B34, 6372 (1986); "Coupled Luttinger Liquids", cond-mat/9605075
D.G. Shelton, A.A. Nersesyan and A.M. Tsvelik, Phys. Rev. B 53, 8521, (1996).
S.P. Strong and A.J. Millis, Phys. Rev. Lett. 69, 2419 (1992).

[29] T. Barnes, E. Dagotto, J. Riera and E. Swanson, Phys.Rev. B47, 3196 (1993).

[30] M. Reigrotzki, H. Tsunetsugu and T.M. Rice, J. Phys. C: Cond. Matt. 6, 9325 (1994).

[31] B. Frischmuth, S. Haas, G. Sierra and T.M. Rice," "Low-Energy Properties of Antiferromagnetic Spin-1/2 Heisenberg Ladders with an Odd Number of Legs", cond-mat/9606183

[32] S. Gopalan, T.M. Rice and M. Sigrist, Phys.Rev. B49, 8901 (1994).

[33] R. Shankar, Nucl. Phys. B 330, 433, (1990).

[34] D.V. Khveshchenko, Phys.Rev. B50, 380 (1994).

[35] I. Affleck, Phys. Rev. B37, 5186 (1988).

[36] E. Lieb, T. Schultz and D. Mattis, Ann. Phys. (N.Y.) 16, 407 (1961).

[37] E. Manousakis, Rev.Mod.Phys.63, 1 (1991)

[38] S. White, R. Noack and D. Scalapino, Phys.Rev.Lett. 73, 886 (1994).

[39] B. Frischmuth, B. Ammon and M. Troyer , "Susceptibility and low temperature thermodynamics of spin 1/2 Heisenberg ladders", cond-mat/9601025.

[40] M. Greven, R.J. Birgeneau and U.-J. Wiese, Phys. Rev. Lett.77, 1865 (1996).

[41] R.R.P. Singh, Phys. Rev. B 39, 9760 (1989); R.R.P. Singh and D. Huse, Phys. Rev. B 40, 7247 (1989).

[42] N. Hatano, Y. Nishiyama and M. Suzuki, J. Phys. A: Math. Gen. 27, 6077 (1994).

[43] I thank E. Dagotto, for suggesting the application of the NLSM method to study this problem.

[44] S. White, Phys. Rev. B 53, 52 (1996).

[45] K. Hida, Phys. Rev. B 45, 2207 (1992);
K. Hida and S. Takada, J. Phys. Soc. Jpn. 61, 1879 (1992).

[46] K. Totsuka and M. Suzuki, J.Phys.: Condens. Matter 7, 6079 (1995).

[47] I.A. Affleck and F.D.M. Haldane, Phys.Rev. B36, 5291 (1987).

[48] Y. Kato and A. Tanaka, J. Phys. Soc. Jpn. 63, 1277 (1994);
S. Yamamoto, J. Phys. Soc. Jpn. 63, 4327 (1994).

[49] M. Yajima and M. Takahashi, J. Phys. Soc. Jpn. 65, 39 (1996).

[50] M. Yamanaka, M. Oshikawa and S. Miyashita, "Hidden Order and Dimerization Transition in S=2 Chains", cond-mat/9604107.

[51] M.A. Martin-Delgado, R. Shankar and G. Sierra, "Phase Transitions in Staggered Spin Ladders", cond-mat/9605035, to appear in Phys. Rev. Lett.

[52] M.A. Martin-Delgado, R. Shankar and G. Sierra, in preparation.

.

Density Matrix and Renormalization for Classical Lattice Models

T. Nishino[1] and K. Okunishi[2]

1 Department of Physics, Graduate School of Science,
Kobe University, Rokko-dai, Kobe 657, Japan
e-mail: nishino@phys560.phys.kobe-u.ac.jp
2 Department of Physics, Graduate School of Science,
Osaka University, Toyonaka, Osaka 560, Japan

Abstract

The density matrix renormalization group is a variational approximation method that maximizes the partition function — or minimize the ground state energy — of quantum lattice systems. The variational relation is expressed as $Z = \text{Tr}\,\rho \geq \text{Tr}\,(\tilde{1}\rho)$, where ρ is the density submatrix of the system, and $\tilde{1}$ is a projection operator. In this report we apply the variational relation to two-dimensional (2D) classical lattice models, where the density submatrix ρ is obtained as a product of the corner transfer matrices. The obtained renormalization group method for 2D classical lattice model, the corner transfer matrix renormalization group method, is applied to the $q = 2 \sim 5$ Potts models. With the help of the finite size scaling, critical exponents ($q = 2, 3$) and the latent heat ($q = 5$) are precisely obtained.

1   Introduction

The basic procedure in the renormalization group (RG) is to keep relevant information of a physical system, and neglect (or integrate out) irrelevant one. [1, 2, 3] The density matrix renormalization group (DMRG) introduced by White [4] greatly enhances the applicability of the numerical RG, because the method automatically keeps a fixed numbers ($= m$) of the relevant basis; DMRG present the best approximation within the limited numerical resource that we can use. The DMRG has been applied to a number of one-dimensional (1D) quantum lattice systems, such as the spin chain, [5, 6] ladder, [7, 8] Bethe lattice system, [9] strongly correlated electron systems, [11, 12, 13, 14] models in momentum space. [10] Not only the numerical superiority, but also the formulation of DMRG have attracted theoretical interests. Östlund and Rommer [15] have shown that DMRG is a variational method, where the ground state is expressed as a product of 3-index tensors. [16, 17] Martìn-Delgado and Sierra have investigated the analytic formulation of DMRG, and have formulated the correlated block RG. [18, 19] Recently White have refined the finite system algorithm of DMRG, and extend the applicability of DMRG to 2D quantum systems. [20] Quite recently Xiang have reported DMRG study of 1D quantum system at finite temperature, [21] using the quantum transfer matrix formulation [22] and DMRG applied to the transfer matrix. [23]

Another RG approach has been development for 2D classical lattice models: Baxter's method of the corner transfer matrix. (CTM) [24] The method is a generalization of Kramers-Wannier approximation, [25, 26] and therefore Baxter's method is based on a variational principle for the partition function. It should be noted that Baxter's variational relation is in principle the same as the variational relation in DMRG. [27] The purpose of this report is to explain how the concept of DMRG is applied to 2D classical lattice models. We start from a short review of the variational relation in DMRG in the next section.

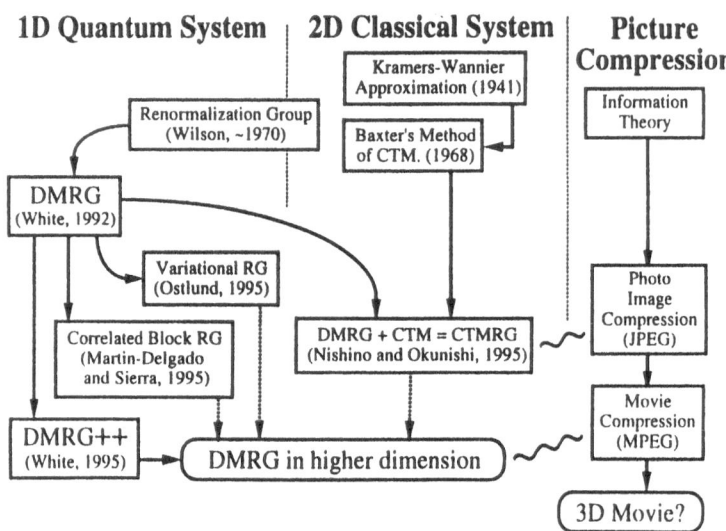

Figure 1 Historical overview of DMRG and related fields.

It is worth looking at a practical use of the RG method as the 2D photo image compression. [28, 29] A photo image in our computer is normally compressed before it is stored, in order to decrease the file size. The compression algorithm is related to the block RG method, [1, 3] since a small region — a pixel — in a 2D picture has strong correlation with its environment. (Computer scientists may insist that their findings about the photo image compression are efficient for RG formulation in physics.) At present, compression of movies — TV pictures — are in progress in the world of computation; [30, 31] the algorithm may be a good reference for the RG study of 3D classical systems and 2D quantum systems.

The development of RG, DMRG, Baxter's CTM method, and the photo image compression is summarized in Fig.1. Originally these methods are proposed independently, however, now it is apparent that their background is in common. It is, to approximate a system (or the *objects*) within a limited number of freedom.

2 Variational Principle in DMRG

We start from a short review of the variational principle in DMRG. We consider the antiferromagnetic $S = 1/2$ Heisenberg spin chain as an example of 1D quantum systems. The spin Hamiltonian is

$$H = J \sum \mathbf{S}_i \cdot \mathbf{S}_{i+1}, \tag{1}$$

where $\mathbf{S}_i$ represents the quantum spin at i-site, and the parameter J is positive. The Hamiltonian H is real-symmetric, and so is the density matrix $\rho = e^{-\beta H}$. (In the following discussion, ρ does not always have to be real-symmetric, but should be positive definite.)

L   **R**

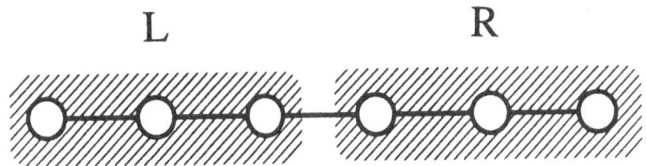

Figure 2 The $S = 1/2$ Heisenberg spin chain with open boundary conditions. We divide it into the local system $[L]$ and the reserver $[R]$.

We consider an open spin chain, (Fig.2) which consists of the left half $[L]$ (= the local system) and the right half $[R]$ (= the reserver). The terms 'local system' and 'reserver' are rather formal, since $[R]$ is not always longer than $[L]$. (In Fig.2 both $[L]$ and $[R]$ has the same size.) The Hilbert space of the whole system is spanned by the real-space basis $|l\rangle|r\rangle$, where $|l\rangle$ and $|r\rangle$ corresponds to the spin configuration for $[L]$ and $[R]$, respectively. The matrix element of ρ is given by

$$\rho_{lr,l'r'} = \langle l\,|\langle r|e^{-\beta H}|r'\rangle|l'\,\rangle. \tag{2}$$

In the context of DMRG, what is called the *density matrix* is actually the density submatrix (DSM)

$$\rho^L = \sum_{ll'} |l\rangle\, \rho^L_{ll'}\, \langle l'| \equiv \sum_{ll'} |l\rangle \left(\sum_r \rho_{lr,l'r} \right) \langle l'| \tag{3}$$

that contains the information only about the local system $[L]$. The trace of ρ^L is equal to the partition function.

The relevant state selection — renormalization — in DMRG is performed through the diagonalization of the DSM

$$O^T \rho^L Q = \mathrm{diag}\{\lambda_1, \lambda_2, \ldots\}, \tag{4}$$

where $\lambda_1 \geq \lambda_2 \geq \ldots \geq 0$ are eigenvalues in decreasing order, $Q = (\mathbf{q_1}, \mathbf{q_2}, \ldots)$ are the set of the corresponding right eigenvectors, $O = (\mathbf{o_1}, \mathbf{o_2}, \ldots)$ are that of the left eigenvectors. The matrices O and Q satisfies the dual orthogonal relation

$QO^T = 1$. When ρ^L is real-symmetric, Q is equal to O, and both of them are orthogonal matrices. The first m column vectors $\mathbf{q}_\alpha$ $(1 \leq \alpha \leq m)$ in Q represent the RG transformation from the original basis $\{|l\rangle|r\rangle\}$ to the renormalized basis $\{|\alpha\rangle\}$. The irrelevant states are thrown away. [4]

We can check the validity of the above basis state selection by observing the inequality

$$Z = \sum_l \lambda_l \geq \sum_{l=1}^m \lambda_l. \tag{5}$$

The quantity $\tilde{Z} = \sum_{l=1}^m \lambda_l$ is the approximate partition function, which is smaller than Z by $\sum_{l>m} \lambda_l$. It has been known that if the system has a finite excitation gap, the eigenvalue λ_i decays exponentially with respect to i, and therefore $\tilde{Z}$ is a good approximation for Z when m is sufficiently large. The approximate partition function is equal to the trace of the m-dimensional diagonal matrix

$$\tilde{\rho}^L \equiv \mathrm{diag}\{\lambda_1, \lambda_2, \ldots, \lambda_m\} = \tilde{O}^T \rho^L \tilde{Q}, \tag{6}$$

where $\tilde{Q}$ is the rectangular matrix $(\mathbf{q}_1, \mathbf{q}_2, \ldots, \mathbf{q_m})$, and $\tilde{O}$ is $(\mathbf{o}_1, \mathbf{o}_2, \ldots, \mathbf{o_m})$. We can regard the matrix operation of $\tilde{O}^T$ and $\tilde{Q}$ to ρ^L as the RG transformation (or the block spin transformation), and $\tilde{\rho}^L$ as the renormalized DSM.

Substituting Eq.(6) into Eq.(5), we obtain a variational relation in the matrix form

$$Z = \mathrm{Tr}\, \rho^L \geq \mathrm{Tr}\, \tilde{\rho}^L = \mathrm{Tr}\left(\tilde{Q}\tilde{O}^T \rho^L\right), \tag{7}$$

where the matrix product $\tilde{1} \equiv \tilde{Q}\tilde{O}^T$ has the property of the projection operator $(\tilde{Q}\tilde{O}^T)^2 = \tilde{Q}\tilde{O}^T$. The projection operator $\tilde{Q}\tilde{O}^T$ is *optimal* in the sense that it gives maximum of $\tilde{Z} = \mathrm{Tr}\,(\tilde{1}\rho)$ under the constraint $\tilde{1}^2 = \tilde{1}$ and $\mathrm{Tr}\,\tilde{1} = m$. In other word, the DMRG minimize the free energy of the system within the restricted degree of freedom. At the zero temperature $(\beta \to \infty)$, DMRG minimize the total energy.

The optimal projection operator $\tilde{1}$ for the local system $[L]$ is dependent on the size of the reserver $[R]$. However, the dependence is not conspicuous when $[R]$ is sufficiently large. The infinite system DMRG algorithm uses the insensitivity of $\tilde{1}$ against the reserver size. The finite system DMRG algorithm is more accurate than the infinite algorithm, because the former correctly takes into account of the reserver-size dependence.

3 From Quantum System to Classical System

In order to apply the variational relation in DMRG to 2D classical lattice system, we define the DSM for 2D systems. We use the fact that the density matrix $e^{-\beta H}$ of a d-dimensional quantum system can be expressed as a partition function of a $d+1$-dimensional classical system with special boundary conditions; the relation is known as the Trotter-Suzuki formula. [32, 33]

The density matrix of the Heisenberg chain in Fig.2 is approximated as

$$e^{-\beta H} = \left(e^{-\frac{\beta}{N}H} \right)^M \sim \left(e^{-\frac{\beta}{N}H_A} e^{-\frac{\beta}{N}H_B} \right)^M \tag{8}$$

where $H_A \equiv \sum_i \mathbf{S}_{2i} \cdot \mathbf{S}_{2i+1}$ and $H_B \equiv \sum_i \mathbf{S}_{2i+1} \cdot \mathbf{S}_{2i+2}$ are the partition of H. The matrix element $\rho_{lr,l'r'}$ in Eq.(2) is approximated by the Boltzmann weight of the *chessboard model*, whose boundary spin configurations are fixed to l, r, l' and r'. Figure 3 shows an example when $M = 4$. When we consider the partition function Z, the boundaries l, r, l' and r' plays the role of the tabs for sticking; $Z = \text{Tr}\, e^{-\beta H}$ is approximately equal to the partition function of the cylindrical system shown in Fig.4(a), which is constructed by attaching l and r in Fig.3 to l' and r', respectively. In the same way, the approximate DSM $\rho_{ll'}^L = \sum_r \rho_{lr,l'r}$ is obtained by attaching r to r' as shown in Fig.4(b).

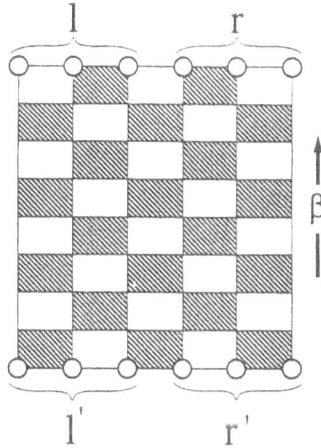

Figure 3 Trotter-Suzuki decomposition of the density matrix in Eq.(8). The label l, r, l' and r' denote the boundary spin configurations.

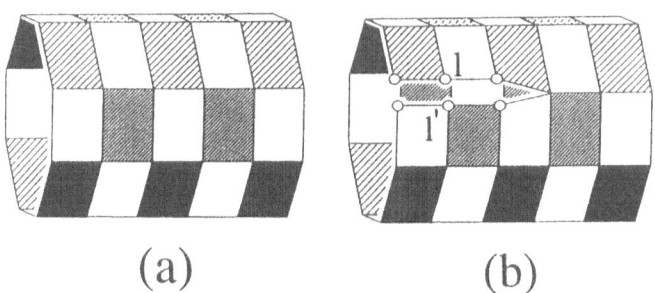

(a)   (b)

Figure 4 The density matrix in the Trotter-Suzuki formula Eq.(8): (a) partition function $\text{Tr}\, \rho$. (b) density sub matrix $\rho_{ll'}^L$.

The expression of ρ^L in Fig.4(b) is a typical example of the DSM for 2D classical lattice models. The ρ^L corresponds to the cylindrical system with a cut L, where the spin configurations around the cut are fixed to l and l'. We generalize this example. Suppose we have a finite size lattice system $[A]$ shown in Fig.5. We then consider an arbitrary line or curve L on $[A]$, and cut $[A]$ along L to derive a new system $[A']$. (The curve L is a kind of string on the 1+1 space time.) The derived system $[A']$ has new boundaries around the cut L, where the boundary spin configurations are represented by the labels l and l'. *The Boltzmann weight of $[A']$, which we write $\rho^L_{ll'}$, is the DSM of $[A]$.*

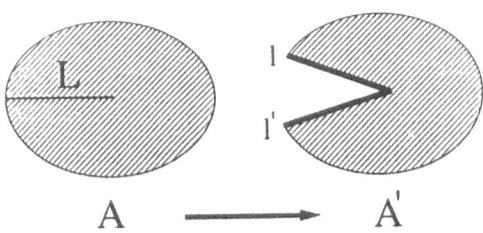

Figure 5 We derive the DSM of the finite size system $[A]$ by cutting it along the curve L. Compared to $[A]$, the system $[A']$ has additional boundaries l and l'.

We have defined ρ^L for 2D classical lattice models. Once we obtain ρ^L for a given system $[A]$, we can perform DMRG along the variational treatment discussed in the previous section, where the group of spins l and l' on $[A]$ are transformed into m-state effective spins. In the next section we present an appropriate choice of $[A]$ and L for a typical 2D classical model.

4 Construction of the Density Matrix via CTM

The Trotter-Suzuki decomposed Heisenberg spin chain in Eq.(8) is an anisotropic 2D lattice model, and the application of DMRG on this system is rather complicated[21]. For the tutorial purpose we consider a simpler 2D system, the symmetric 16-vertex model. [24, 34] The model includes the Ising model [35] as its special case. For simplicity, we assume that the model is ferromagnetic.

The 16-vertex model is defined by the Boltzmann weight

$$W_{abcd} = W_{bcda} = W_{cdab} = W_{dabc} \tag{9}$$

on each vertex (= lattice point) of the simple square lattice, where the spin variables a, b, c, and d take either $+$ (up) or $-$ (down). In the following we consider a square system $[A]$, whose linear dimension is $2N$ or $2N + 1$. We impose fixed boundary condition on $[A]$; we introduce boundary weights

$$P_{abc} = W_{abc+} \tag{10}$$

and

$$C_{ab} = W_{ab++} \tag{11}$$

in order to fix the boundary spins to $+$. (Fig.6) The partition function of $[A]$ is expressed by these weights. For example, the partition function of the system with $2N + 1 = 3$ is expressed as

$$Z = \sum_{ab\ldots l} W_{kheb} P_{abc} C_{cd} P_{def} C_{fg} P_{ghi} C_{ij} P_{jkl} C_{la}, \tag{12}$$

where the position of spin indices a-d is shown in Fig.7.

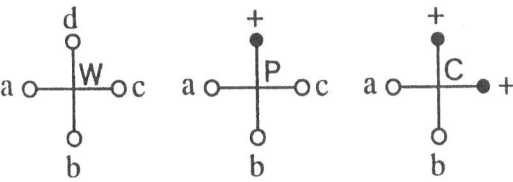

Figure 6 Boltzmann weights of the symmetric 16-vertex model; P and C are the boundary weights in Eq.(10) and Eq.(11), respectively.

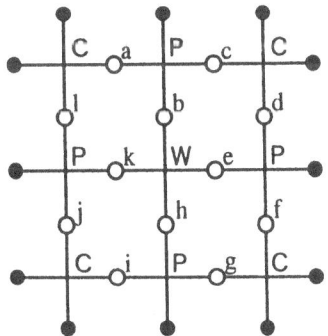

Figure 7 A square system of the linear dimension $2N + 1 = 3$, whose partition function is given by Eq.(12).

In order to generalize Eq.(12) to arbitrary system size, we introduce a half-row transfer matrix, (HRTM) which is a generalization of the boundary weight P in

Eq.(10). The HRTM of length N is defined by the recursion relation

$$P^N_{\mathbf{abc}} = \sum_d W_{a_N\, d\, c_N\, b}\, P^{N-1}_{\mathbf{a'd}\,\mathbf{c'}}, \tag{13}$$

where the label N is the number of vertices in HRTM, $P^1_{\mathbf{abc}}$ is equal to P_{abc} in Eq.(10), and $\mathbf{a}$ represents a group of spins on a row

$$\mathbf{a} = (a_1, a_2, \ldots, a_{N-1}, a_N), \tag{14}$$

which is related to $\mathbf{a'}$ as $\mathbf{a} = (\mathbf{a'}, a_N)$; the same for $\mathbf{c} = (\mathbf{c'}, c_N)$. Figure 8(a) shows an example when $N = 3$. We occasionally drop the vector indices of $P^N_{\mathbf{abc}}$ and write it simply as P^N_b; in that case we think of P^N_b as a 2^N-dimensional matrix $(P^N_b)_{\mathbf{ac}}$. The HRTM is also conventionally called *vertex operator*. [36]

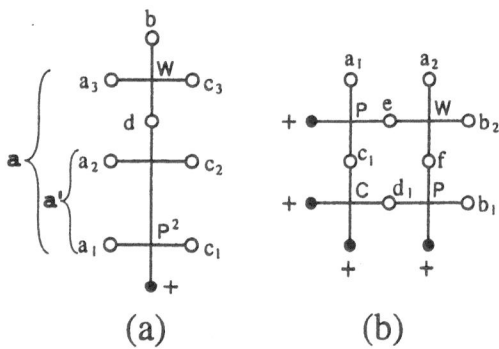

(a) (b)

Figure 8 Recursive definition of (a) P^N in Eq.(13) and (b) C^N in Eq.(15). The shown examples are P^3 and C^2.

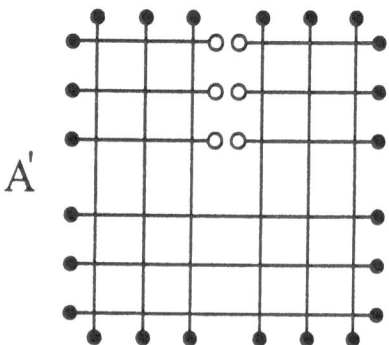

A'

Figure 9 The density submatrix ρ^L of the square system $[A]$ is expressed as a product of four CTMs. (Eq.(16).).

The density submatrix ρ^L of the square system $[A]$ is expressed as a product of corner transfer matrices. (CTMs) [24] The CTM of size N is defined as

$$C_{\mathbf{ab}}^N = \sum_{\mathbf{c'd'}} \left(\sum_{ef} W_{efb_Na_N} P_{\mathbf{a'ec'}}^{N-1} P_{\mathbf{b'fd'}}^{N-1} \right) C_{\mathbf{c'd'}}^{N-1} \tag{15}$$

where we have used the index rule $\mathbf{a} = (\mathbf{a'}, a_N)$, $\mathbf{b} = (\mathbf{b'}, b_N)$, etc. Figure 8(b) shows the example when $N = 2$. The smallest CTM $C_{\mathbf{ab}}^1$ is equal to the boundary weight C_{ab} in Eq.(11). The square system $[A]$ is then constructed by attaching four CTMs. Figure 9 shows the system $[A']$ of the size $2N = 6$, whose Boltzmann weight

$$\rho^L = \left(C^N \right)^4 \tag{16}$$

corresponds to the DSM of the square system $[A]$ of the size $2N = 6$. Such a construction of the DSM was first introduced by Baxter more than 30 years ago in his variational method. [24]

Now we can apply the RG procedure to 2D classical lattice models using ρ^L defined in Eq.(16); following the RG procedure in Sec.2, we combine Baxter's method and DMRG. For the brevity, we call our new RG method *corner transfer matrix renormalization group* (CTMRG) in the following. [27]

5 CTM Renormalization Group

As was discussed in Sec.2, the heart of DMRG is the diagonalization of ρ^L. For the symmetric 16-vertex model, where ρ^L is expressed as $\left(C^N \right)^4$, both ρ^L and C^N have the common eigenvectors. We therefore diagonalize CTM

$$O^T C^N Q = \text{diag}\{\omega_1, \omega_2, \ldots\} \tag{17}$$

instead of ρ^L, where we assume the decreasing order $|\omega_1| \geq |\omega_2|, \ldots \geq 0$. The block-spin transformation is performed by the rectangular matrices $\tilde{O} = (\mathbf{o}_1, \mathbf{o}_2, \ldots, \mathbf{o}_m)$ and $\tilde{Q} = (\mathbf{q}_1, \mathbf{q}_2, \ldots, \mathbf{q}_m)$, where $\mathbf{o}_\alpha$ and $\mathbf{q}_\alpha$ are the left and the right eigenvectors of C^N, respectively. Here after we use greek letters for indices that runs from 1 to m. (Actually, $\tilde{Q}$ is equal to $\tilde{O}$ since C^N is symmetric.) The renormalized CTM

$$\tilde{C}^N \equiv \tilde{O}^T C^N \tilde{Q} = \text{diag}\{\omega_1, \omega_2, \ldots, \omega_m\} \tag{18}$$

is related to $\tilde{\rho}^L \equiv \tilde{O}^T \rho^L \tilde{Q}$ via $\tilde{\rho}^L = (\tilde{C}^N)^4$. The renormalized HRTM is obtained in the same way

$$\tilde{P}_a^N = \tilde{O}^T P_a^N \tilde{Q}, \tag{19}$$

where the tensor elements of $\tilde{P}_a^N$ are $\tilde{P}_{\xi a\eta}^N$.

At this point we remember that P^N and C^N are defined through the recursion relations Eq.(13) and Eq.(15), respectively. The relations are also valid for $\tilde{P}^N$

and $\tilde{C}^N$. For $\tilde{C}^N$, its area is extended by attaching two HRTMs (Fig.10)

$$\bar{C}^{N+1}_{(\alpha,a)(\beta,b)} = \sum_{ef\delta} W_{efba}\, \tilde{P}^N_{\alpha e\delta}\tilde{P}^N_{\beta f\delta}\, \omega_\delta, \tag{20}$$

where $\bar{C}^{N+1}_{(\alpha,a)(\beta,b)}$ is a *partially renormalized* CTM of linear size $N+1$, and the pairs of indices (α,a) and (β,b) represent the row and the column matrix indices for $\bar{C}^{N+1}$. The length of HRTM is simultaneously increased by putting a new vertex at the end point

$$\bar{P}^{N+1}_{(\alpha,a)b(\gamma,c)} = \sum_{d} W_{adcb}\tilde{P}^N_{\alpha d\gamma}. \tag{21}$$

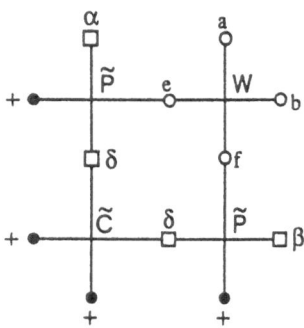

Figure 10 Area extension of the renormalized CTM in Eq.(20).

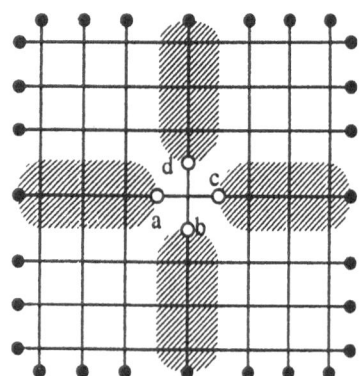

Figure 11 A square system of size $2N+1$ is expressed as a product among $\tilde{C}^N$, $\tilde{P}^N_{a\,(bcd)}$, and W.

The extended CTM in Eq.(20) is not diagonal. As was done in Eqs.(17-18), we diagonalize $\bar{C}^{N+1}$ to obtain the new renormalized CTM $\tilde{C}^{N+1}$ with the matrix dimension m. It is now obvious that we can indefinitely repeat the renormalization process in Eqs.(18-19) and the system size extension in Eqs.(20-21). In this way, we obtain $\tilde{P}^N$ and $\tilde{C}^N$ for arbitrary large N starting from P^1 and C^1. [27]

Approximate thermodynamic functions of $[A]$ can be obtained using $\tilde{C}^N$. The free energy $-k_BT \ln Z$ is estimated from the approximate partition function $\tilde{Z} = \text{Tr}\left(\tilde{C}^N\right)^4 = \sum_i \omega_i^4$. It is also possible to obtain local quantities, such as spin polarization and multi-spin correlation functions, using a combination of P^N and C^N. For example, the local energy is estimated as

$$E(2N+1) = \frac{\sum_{abcd} X_{abcd} \text{Tr}\left(\tilde{P}_a^N \tilde{C}^N \tilde{P}_b^N \tilde{C}^N \tilde{P}_c^N \tilde{C}^N \tilde{P}_d^N \tilde{C}^N\right)}{\sum_{abcd} W_{abcd} \text{Tr}\left(\tilde{P}_a^N \tilde{C}^N \tilde{P}_b^N \tilde{C}^N \tilde{P}_c^N \tilde{C}^N \tilde{P}_d^N \tilde{C}^N\right)}, \tag{22}$$

where the position of the indices a-d are shown in Fig.11, and X_{abcd} is the local energy operator $\ln\left(W_{abcd}\right)$. Since the CTMRG extends the area of the system $[A]$ from the center, local quantities at the center can be calculated most precisely.

The largest matrix element of $\tilde{C}^N$ rapidly grows with respect to N. We should therefore normalize CTM

$$\frac{\tilde{C}^N}{\omega_1} = \text{diag}\left\{1, \frac{\omega_2}{\omega_1}, \ldots, \frac{\omega_m}{\omega_1}\right\} \to \tilde{C}^N \tag{23}$$

in realistic numerical calculations. We should also normalize $\tilde{P}_a^N$ in the same manner. Apart from the critical point, the normalized $\tilde{C}^N$ converges to its thermodynamic limit $\tilde{C}^\infty$ exponentially with respect to N. At criticality the convergence is relatively slow; it is observed that the decay rate at criticality is controlled by the critical exponents. [37]

We have imposed fixed boundary conditions on $[A]$. Since the boundary condition is totally determined by the boundary weight P_a^1 in Eq.(10) and C^1 in Eq.(11), we can choose other boundary conditions by modifying P_a^1 and C^1. For example, the free boundary condition is imposed by the boundary weights

$$P_{abc} - W_{abc+} + W_{abc-} \tag{24}$$

and

$$C_{ab} = W_{ab++} + W_{ab+-} + W_{ab-+} + W_{ab--}. \tag{25}$$

The CTMRG method presented above can be applied to a wide class of 2D classical lattice models, such as the q-state Potts model, IRF model, etc. We have to be careful to anisotropic lattice models whose ρ^L consist of four *different* CTM's. (In Baxter's textbook the DSM are expressed as $\rho = ABCD$. [24]) In such a case, we should not diagonalize each CTM independently, but we should diagonalize ρ^L to perform RG transformation. We should also be careful to antiferromagnetic models, because we have to prepare several sets of $\tilde{P}^N$ and $\tilde{C}^N$ according to the alternating spin order.

6   Numerical Result

The symmetric 16-vertex model includes the square lattice Ising model as its parameter limit. We first look at the critical phenomena of the Ising model using CTMRG. Figure 12 shows the calculated local energy

$$E(2N) = \mathrm{Tr}\left(\sigma\sigma'\tilde{\rho}^L\right)/\tilde{Z} \tag{26}$$

where $\sigma\sigma'$ is a pair of neighboring spin at the center of the square cluster $[A]$ of size $2N$. The data shown by the black dots are $E(\infty)$, that are obtained when $m = 98$. The data deviate from the exact solution at most 10^{-7}. At the critical temperature T_c, we estimate $E_c(\infty)$ by observing its convergence with respect to N. The inset of Fig.11 shows the $1/N$ dependence of $E_c(N)$. A simple $1/N$ fitting gives $E_c(\infty) = 0.707148$, which is close to the exact one $1/\sqrt{2} = 0.707107\ldots$.

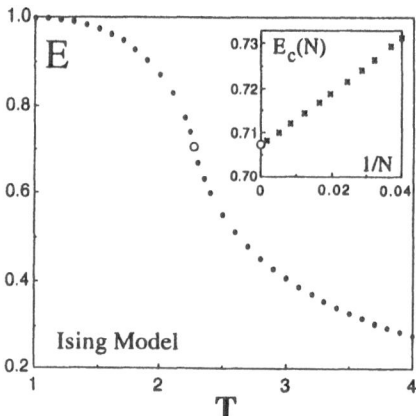

Figure 12 Local Energy $E(\infty)$ of the Ising Model at the center of the square system. The inset shows the size dependence of $E(N)$ at the critical temperature $T_c.$.

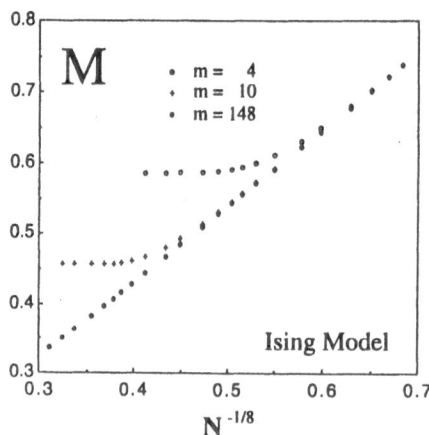

Figure 13 Local Magnetization of the Ising Model at T_c.

Table 1: Critical exponents ν and η of the $q=2,3$ Potts model estimated from the numerical data when $m = 200$. Theoretical values are shown inside the parenthesis.

q	ν (Exact)	η (Exact)
2	1.0006 (1.0000)	0.2501 (0.2500)
3	0.8321 (0.8333)	0.2654 (0.2667)

The $1/N$ dependence of $E_c(N)$ is actually related to the scale invariance at criticality. The finite size scaling (FSS) theory [38, 39] predicts that $E_c(N)$ obeys the scaling form

$$E_c(N) - E_c(\infty) \sim N^{1/\nu - d}, \tag{27}$$

where ν is the correlation-length exponent and $d = 2$ is the spatial dimensionality; in our case $E_c(N)$ is proportional to N^{-1} because ν of the Ising model is equal to unity. Similarly, the local order parameter

$$M(N) = \mathrm{Tr}\left(\sigma \tilde{\rho}^L\right)/\tilde{Z} \tag{28}$$

at the center of $[A]$ obeys

$$M_c(N) \sim N^{-(d-2+\eta)/2} \tag{29}$$

with the anomalous dimension of the spin η. Figure 13 shows the N dependence of calculated magnetization $M_c(N)$ at T_c when $m = 148$. As it is expected from $\eta = 1/4$, the calculated $M_c(N)$ is proportional to $N^{-1/8}$.

The estimation of the critical exponents via FSS is valid for a wide class of 2D classical model. As examples, we apply CTMRG and FSS to the $q=2,3$ Potts models, [40, 41] which can be treated as a symmetric q^4-vertex model. [42, 41] Table I shows the estimated exponents ν and η from the calculated data when $m = 200$ and $100 \leq N \leq 1000$. For comparison, the theoretically determined exponents [41] are shown in the parenthesis. The error in the calculated exponents are less than 0.2%.

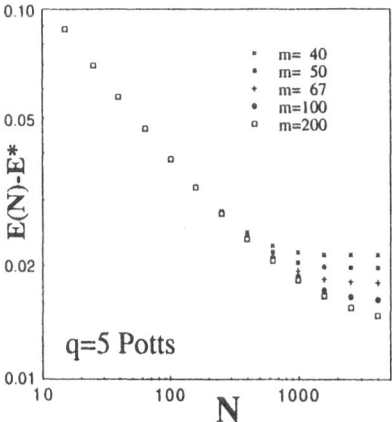

Figure 14 Local Energy of the $q=5$ Potts mode.

Numerical analysis of the case $q = 4$ is in progress. In this case, one has to take care of the logarithmic corrections in the FSS analysis. [43]

We finally show the N dependence of the local energy $E(N)$ of the $q=5$ Potts model at the transition temperature; we measure $E(N)$ from the average $E^* = (E_+ + E_-)/2$, where E_+ and E_- are the energy of the disordered and the ordered phase, respectively, at the transition temperature. (Fig.14) Since the transition is first order, $E(N) - E^*$ does not obey the scaling form in Eq.(27). In this case, it is possible to estimate the latent heat through the scaling analyses with respect to both N and m. The calculated latent heat is 0.0256, where the exact value is 0.0265. [41] The CTMRG is expected to be efficient for the analysis of the weak first order transition. [44]

7 Conclusion and Discussion

We have reviewed the variational principle in DMRG, and explained its application to 2D classical lattice systems. We create the DSM according to Baxter's construction, and perform the RG transformation using DMRG algorithm. As trial calculations, we perform FSS analyses on the q-state Potts models at the transition temperature. It is concluded that the CTMRG is efficient for the determination of the critical exponents or the latent heat.

It is possible to feed-back the CTMRG algorithm to the zero-temperature 1D quantum system, and to obtain a rapid infinite system DMRG algorithm; the product wave function (PWF) RG. [46] The PWFRG is closely related to the improved finite system DMRG algorithm (DMRG++ in Fig.1) recently proposed by White. [20]

Finally, we discuss the applicability of DMRG or CTMRG to 3D classical lattice models. As we have employed a square system for 2D models, we should consider a cubic system for 3D models. We divide it into 8 subcubics, say corner transfer tensor, (CTT) and construct the DSM as their tensor products. We then diagonalize the DSM, and renormalize CTT. Apart from the renormalized CTM in two dimension, renormalized CTT is not diagonal. [45] There is, however, a computational problem that prevents us from practical use of the RG method for 3D system; at present the numerical calculation is too heavy even for a small m.

There are plenty of subject that we can clarify in the field of the numerical RG.

Acknowledgments

The authors would like to express their sincere thanks to Y. Akutsu and M. Kikuchi for valuable discussions. T. N. thank to G. Sierra, M. A. Martìn-Delgado and S. R. White for helpful discussions about the RG method. The numerical result on $q = 5$ Potts model is obtained through the collaboration with A. Yamagata, H. Otsuka and Y. Kato. [44] The present work is partially supported by Kasuya foundation. Most of the numerical calculations were done by NEC SX-3/14R in computer center of Osaka University.

References

[1] L. P. Kadanoff: Physics 2 (1965) 263.

[2] K. G. Wilson and J. Kogut: Phys. Rep. **12 C** (1974) 75.

[3] T. W. Burkhardt and J. M. J. van Leeuwen: *Real-Space Renormalization,* Topics in Current Physics vol.**30**, (Springer, Berlin, 1982), and references therein.

[4] S. R. White: Phys. Rev. Lett. **69** (1992) 2863; Phys. Rev. **B 48** (1993) 10345; see also his article in this volume.

[5] S. R. White and D. A. Huse: Phys. Rev. **B48** 3844.

[6] E. S. Sorensen and I. Affleck: Phys. Rev. Lett **71** (1993) 1633; Phys. Rev. **B49** (1994) 15771.

[7] K. Hida: J. Phys. Soc. Jpn. **64** (1995) 4896.

[8] S. R. White and I. Affleck: to appear Phys. Rev. **B54** (1996).

[9] H. Otsuka: Phys. Rev. **B53** (1996) 14004.

[10] T. Xhiang: Phys. Rev. **53** (1996) R10445.

[11] C. C. Yu and S. R. White: Phys. Rev. Lett. **71** (1993) 3866.

[12] R. M. Noack, S. R. White and D. J. Scalapino: Phys. Rev. Lett **73** (1994) 882.

[13] N. Shibata, T. Nishino, K. Ueda and C. Ishii: Phys. Rev. **B53** (1996) R8828; preprint, cond-mat/9608118.

[14] S. R. White: preprint, cond-mat/9605143.

[15] S. Östlund and S. Rommer: Phys. Rev. Lett **75** (1995) 3537; preprint, cond-mat/9606213.

[16] A. Klümper, A. Schadschneider and J. Zittartz: Z. Phys. **B87** (1992) 281; Europhys. Lett. **24** (1993) 293.

[17] A. Schadschneider and J. Zittartz: Ann. Physik **4** (1995) 157.

[18] M. A. Martín-Delgado and G. Sierra: Int. J. Mod. Phys **A11** (1996) 3145; preprint, UCM/CSIC-96-01 July 1996; See also their article in this volume.

[19] M. A. Martín-Delgado, J. Rodriguez-Laguna and G. Sierra: Nuc. Phys. B473 (1996) 685.

[20] S. R. White: Preprint, cond-mat/9604129; see also his article in this volume.

[21] R. J. Bursill, T. Xiang, G. A. Gehring: preprint, cond-mat/9609001.

[22] M. Suzuki and M. Inoue: Prog. Theor. Phys. **78** (1987) 787.

[23] T. Nishino: J. Phys. Soc. Jpn. **64** (1995) 3598.

[24] R. J. Baxter: J. Math. Phys. **9** (1968) 650; J. Stat. Phys. **19** (1978) 461; *Exactly Solved Models in Statistical Mechanics* (Academic Press, London, 1982) p.363.

[25] H. A. Kramers and G. H. Wannier: Phys. Rev. **60** (1941) 263.

[26] R. Kikuchi: Phys. Rev. **81** (1951) 988.

[27] T. Nishino and K. Okunishi: J. Phys. Soc. Jpn. **65** (1996) 891-894.

[28] K. R. Rao and P. Yip: *Discrete Cosine Transform,* (Academic Press, INC, 1990.)

[29] M. Kaburagi, private communication.

[30] MPEG, CD 11172: *Coding of MovingPictures and AssociatedAudio for Digital Storage Media at Up To About 1.5 Mbps,* (1991).

[31] CCITT Rec. H.261, *Video Codec for Audiovisual Services at p*64 kbit/s,* CCITT COM XV-R 37-E (1990).

[32] H. F. Trotter: Proc. Am. Math. Soc. **10** (1959) 545.

[33] M. Suzuki: Prog. Theor. Phys. **56** (1976) 1454.

[34] M. Jimbo, T. Miwa and A. Nakayashiki: J. Phys. **26A** (1993) 2199.

[35] The origin of the Ising model is reported in S. Kobe, preprint cond-mat/9605174.

[36] T. Miwa, in *Correlation Effects in Low-dimensional Electron Systems*, Solid-State Science **118** ed. A. Okiji and N. Kawakami (Springer, Berlin, 1994) p.68.

[37] T. Nishino, K. Okunishi and M. Kikuchi: Phys. Lett. **A213** (1996) 69-72.

[38] M. E. Fisher, in *Proc. Int. School of Physics 'Enrico Fermi'*, edited by M.S. Green, (Academic Press, New York, 1971), Vol. 51, p. 1.

[39] M. N. Barber, in *Phase Transitions and Critical Phenomena*, edited by C. Domb and J.L. Lebowitz, (Academic Press, New York, 1983), Vol. 8, p. 146. and references therein.

[40] R. B. Potts: Proc. Camb. Phil. Soc. **48** 106.

[41] F. Y. Wu: Rev. Mod. Phys. **54** (1982) 235, and the references there in.

[42] T. Nishino and K. Okunishi: Preprint.

[43] J. Salas, A. D. Sokal: Preprint, lanhep-lat/9607030.

[44] T. Nishino, A Yamagata, H. Otsuka, Y. Kato and K. Okunishi: Preprint.

[45] Baxter: *Exactly Solved Models in Statistical Mechanics* (Academic Press, London, 1982) p.401.

[46] T. Nishino and K. Okunishi: J. Phys. Soc. Jpn. **64** (1995) 4084.

Real-Space Renormalization Group Methods Applied to Quantum Lattice Hamiltonians

Miguel A. Martín-Delgado[1]

[1]*Departamento de Física Teórica I, Universidad Complutense. 28040-Madrid, Spain*
(October 96)

I review recent work and some new results, performed in collaboration with G. Sierra, on the Real-Space Renormalization group method applied to quantum spin lattice systems mainly in spatial dimensions one and two, and to spin ladders which are somehow in between. The first part of these notes is devoted to non-interacting systems in 1D and 2D and the role played by the correlations between blocks. The second part comprises interacting systems in 1D, spin ladders and 2D using the standard BRG method.

Proceedings of the El Escorial Summer School 1996 on STRONGLY CORRE-LATED MAGNETIC AND SUPERCONDUCTING SYSTEMS

I. INTRODUCTION: BRIEF HISTORY OF REAL-SPACE RG METHODS

The Real-Space Renormalization Group Method has undergone a great revival since the Density Matrix RG was introduce by White in 1992 [1], [2]. Nowdays, this method is considered a powerful numerical tool to get non-perturbative results, specially for interacting systems in 1D although some recent advances for 2D systems have been obtained [3], [2]. Despite being numerical, the DMRG has also been a source of inspiration for analytical studies and this is the framework of the present notes. Moreover, much of the El Escorial Summer School has been devoted to real-space RG methods [2], [4], [5].

The Renormalization Group method has become one of the basic concepts in Physics, ranging from areas such as Quantum Field Theory and Statistical Mechanics to Condensed Matter Physics. The many interesting and relevant models encountered in these fields are usually not exactly solvable except for some privileged cases in one dimension. It is then when we resort to the RG method to retrieve the essential features of those systems in order to have a qualitative understanding of what the physics of the model is all about. This understanding is usually recasted in the form of a RG-flow diagram were the different possible behaviours of the model leap to the eyes.

Many authors in the past have contributed significantly to the idea of renormalization and it is out of the scope of these notes to give a detailed account on this issue.

We shall be dealing with the the version of the RG as introduced by Wilson [6] and Anderson [7] in their treatment of the Kondo problem, and subsequent developments of these ideas carried out by Drell et al. at the SLAC group [9] and Pfeuty et al. [10].

Real space Renormalization Group (RG) methods originated from the study of the Kondo problem by Wilson [6]. It was clear from the beginning that one could not hope to achieve the accuracy Wilson obtained for the Kondo problem when dealing with more complicated many-body quantum Hamiltonians. The key difference is that in the Kondo model there exists a *recursion relation* for Hamiltonians at each step of the RG-elimination of degrees of freedom. The existence of such recursion relation facilitates enormously the work, but as it happens it is specific of *impurity problems*.

From the numerical point of view, the Block Renormalization Group (BRG) procedure proved to be not fully reliable in the past particularly in comparison with other numerical approaches, such as the Quantum MonteCarlo method which were being developed at the same time. This was one of the reasons why the BRG methods remained undeveloped during the '80's until the begining of the '90's when they are making a comeback as one of the most powerful numerical tools when dealing with zero temperature properties of many-body systems, a situation where the Quantum MonteCarlo methods happen to be particularly badly behaved as far as fermionic systems is concerned [8].

As it happens, the BRG gives a good qualitative picture of many properties exhibited by quantum lattice Hamiltonians: Fixed points, RG-flow, phases of the system etc. as well as good quantitative results for some properties such as ground state energy and others [9], [10], [11], [20]. However in some important instances the BRG method is off the correct values of critical exponents by a sensible amount.

The origin of the density matrix RG method relies on the special treatment carried out by White and Noack [17] on the 1D tight-binding model, the lattice version of a single particle in a box. It was Wilson [18] the first to point out the relevance of this simple model in understanding the sometimes bad numerical performance of the standard Block Renormalization Group (BRG) method. In reference [17] the authors proposed a method called Combination of Boundary Conditions (CBC) which performs extremely well as compared to the exact known solution of the model. Recently, we have studied the role played by the boundary conditions in the real-space renormalization group method [21] by constructing a new analytical BRG-method which is able to give the exact ground state of the model and the correct $1/N^2$-law for the energy of the first excited state in the large N(size)-limit.

Yet another branch of applications of DMRG inspired ideas is to use the Superblock formalism [17] without resorting to a Density Matrix. For instance, in [19] it has been found that the application of this formalism to the anisotropic Heisenberg model in 1D successfully improves the standard BRG results of Rabin [20].

The Density Matrix RG has been originally devised to deal with quantum lattice Hamiltonians. However, recent new applications have been developed by Nishino and coworkers [4], [24], [27] to address the renormalization of classical lattice models. One of the outcomes of these studies has been to state the relationship between Baxter's corner transfer matrix formalism and the Density Matrix RG of White (in 1D).

The number of new developments on this subject is constantly growing and it is not possible to give a full account of all of them here. More applications can be found in the rest of contributions to these proceedings devoted to real-space RG methods.

II. REVIEW OF STANDARD BLOCK RENORMALIZATION GROUP METHODS (BRG)

For the sake of completeness and to set up the notation used throughout these notes, the block renormalization group method is revisited in this section along the lines of a new and unified reformulation of it based on the idea of the *intertwiner operator T* to be discussed below. This treatment is by all means equivalent to the standard approach presented by S.R. White in his contribution to this volume. This formulation has recently allowed us to introduce the new Variational and Fokker-Planck DMRG methods [12] on equal footing as the standard BRG method. For a more extensive account on this method we refer to [13] and chapter 11 of reference [14] and references therein.

Let us first summarize the main features of the real-space RG. The problem that one faces generically is that of diagonalizing a quantum lattice Hamiltonian H, i.e.,

$$H|\psi >= E|\psi > \tag{1}$$

where $|\psi >$ is a state in the Hilbert space $\mathcal{H}$. If the lattice has N sites and there are k possible states per site then the dimension of $\mathcal{H}$ is simply

$$dim\mathcal{H} = k^N \tag{2}$$

As a matter of illustration we cite the following examples: $k = 4$ (Hubbard model), $k = 3$ (t-J model), $k = 2$ (Heisenberg model) etc.

When N is large enough the eigenvalue problem (1) is out of the capability of any human or computer means unless the model turns out to be integrable which only happens in some instances in $d = 1$.

These facts open the door to a variety of approximate methods among which the RG-approach, specially when combined with other techniques (e.g. numerical, variational etc.), is one of the most relevant. The main idea of the RG-method is the mode elimination or thinning of the degrees of freedom followed by an iteration which reduces the number of variables step by step until a more manageable situation is reached. These intuitive ideas give rise to a well defined mathematical description of the RG-approach to the low lying spectrum of quantum lattice hamiltonians.

To carry out the RG-program it will be useful to introduce the following objects:

- $\mathcal{H}$: Hilbert space of the original problem.

- $\mathcal{H}'$: Hilbert space of the effective degrees of freedom.

- H: Hamiltonian acting in $\mathcal{H}$.

- H': Hamiltonian acting in $\mathcal{H}'$ (effective Hamiltonian).

- T : embedding operator : $\mathcal{H}' \longrightarrow \mathcal{H}$

- $T^\dagger$:truncation operator : $\mathcal{H} \longrightarrow \mathcal{H}'$

The problem now is to relate H, H' and T. The criterium to accomplish this task is that H and H' have in common their low lying spectrum. An exact implementation of this is given by the following equation:

$$HT = TH' \tag{3}$$

which imply that if $\Psi'_{E'}$ is an eigenstate of H' then $T\Psi'_{E'}$ is an eigenstate of H with the same eigenvalue (unless it belongs to the kernel of T: $T\Psi'_{E'} = 0$), indeed,

$$HT\Psi'_{E'} = TH'\Psi'_{E'} = E'T\Psi'_{E'} \tag{4}$$

To avoid the possibility that $T\Psi' = 0$ with $\Psi' \neq 0$, we shall impose on T the condition,

$$T^\dagger T = 1_{\mathcal{H}'} \tag{5}$$

such that

$$\Psi = T\Psi' \Rightarrow \Psi' = T^\dagger \Psi \tag{6}$$

Condition (5) thus stablishes a one to one relation between $\mathcal{H}'$ and $\text{Im}(T)$ in $\mathcal{H}$. Observe that Eq. (3) is nothing but the commutativity of the following diagram:

$$
\begin{array}{ccc}
\mathcal{H}' & \xrightarrow{\ T\ } & \mathcal{H} \\
H' \downarrow & & \downarrow H \\
\mathcal{H}' & \xrightarrow{\ T\ } & \mathcal{H}
\end{array}
$$

Eqs. (3) and (5) characterize what may be called exact renormalization group method (ERG) in the sense that the whole spectrum of H' is mapped onto a part (usually the bottom part) of the spectrum of H. In practical cases though the exact solution of Eqs. (3) and (5) is not possible so that one has to resort to approximations (see later on). Considering Eqs. (3) and (5) we can set up the effective Hamiltonian H' as:

$$H' = T^\dagger HT \tag{7}$$

This equation does not imply that the eigenvectors of H' are mapped onto eigenvectors of H. Notice that Eq.(7) together with (5) does not imply Eq. (3). This happens because the converse of Eq.(5), namely $TT^\dagger \neq 1_{\mathcal{H}}$ is not true, since otherwise this equation together with (5) would imply that the Hilbert spaces $\mathcal{H}$ and $\mathcal{H}'$

are isomorphic while on the other hand the truncation inherent to the RG method assumes that $dim\mathcal{H}' < dim\mathcal{H}$.

What Eq.(7) really implies is that the mean energy of H' for the states Ψ' of $\mathcal{H}'$ coincides with the mean energy of H for those states of $\mathcal{H}$ obtained through the embedding T, namely,

$$< \Psi'|H'|\Psi' >=< T\Psi'|H|T\Psi' > \tag{8}$$

In other words $T\Psi'$ is used as a variational state for the eigenstates of the Hamiltonian H. In particular T should be chosen in such a way that the states truncated in $\mathcal{H}$, which go down to $\mathcal{H}'$, are the ones expected to contribute the most to the ground state of H. Thus Eq. (7) is the basis of the so called variational renormalization group method (VRG) As a matter of fact, the VRG method was the first one to be proposed. The ERG came afterwards as a perturbative extension of the former (see later on).

More generally, any operator $\mathcal{O}$ acting in $\mathcal{H}$ can be "pushed down" or renormalized to a new operator $\mathcal{O}'$ which acts in $\mathcal{H}'$ defined by the formula,

$$\mathcal{O}' = T^{\dagger}\mathcal{O}T \tag{9}$$

Notice that Eq.(7) is a particular case of this equation if choose $\mathcal{O}$ to be the Hamiltonian H.

In so far we have not made use of the all important concept of the block, but a practical implementation of the VRG or ERG methods does require it. The central role played by this concept makes all the real-space RG-methods to be block methods.

Once we have established the main features of the RG-program, there is quite a freedom to implement specifically these fundamentals. We may classify this freedom in two aspects:

- The choice to how to reduce the size of the lattice.

- The choice of how many states to be retained in the truncation procedure.

We shall address the first aspect now. There are mainly two procedures to reduce the size of the lattice:

- by dividing the lattice into blocks with n_B sites each. This is the blocking method introduced by Kadanoff to treat spin lattice systems. See Fig. 1.

- by retrieving site by site of the lattice at each step of the RG-program. This is the procedure used by Wilson in his RG-treatment of the Kondo problem. This method is clearly more suitable when the lattice is one-dimensional.

We shall be dealing with the Kadanoff block methods mainly because they are well suited to perform analytical computations and because they are conceptually easy to be extended to higher dimensions. On the contrary, the DMRG method introduced by White [1] works with the Wilsonian numerical RG-procedure what makes it intrinsically one-dimensional and difficult to be generalized to more dimensions. This situation has changed recently in part as S.R. White has devised a numerical improvement of the DMRG which is applicable to a 1/5-depleted 2D lattice [3]. We have formulated our Variational and Fokker-Planck DMRG procedures as block renormalization methods [12].

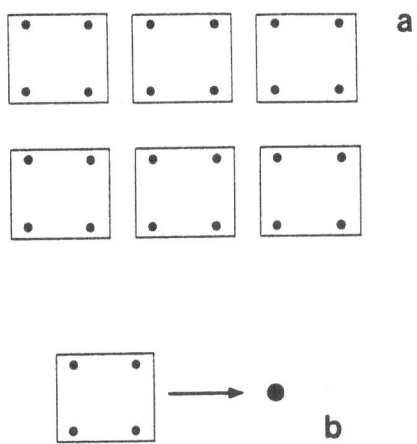

FIG. 1 a) Block decomposition of a square lattice into 4-site blocks. b)Squematic truncation of states in the BRG method associated to the previous lattice decomposition.

Block RG-methods have recently received also renewed attention in one-dimensional problems in connection to what is called a *quantum group* symmetry [15], [16]. Based upon this symmetry we have constructed a new BRG-method that we call q-RG which among other features it is able to predict the exact line of critical XXZ models in the Anisotropic Heisenberg model, unlike the standard BRG-method.

III. THE ROLE OF BOUNDARY CONDITIONS AND REAL-SPACE RG

The first advance in trying to understand the sometimes bad numerical performance of the BRG methods came in the understanding of the effect of *boundary conditions* (BC) on the standard RG procedure [17].

White and Noack [17] pointed out that the standard BRG approach of neglecting all connections to the neighbouring blocks during the diagonalization of the block Hamiltonian H_B introduces large errors which cannot be corrected by any reasonable increase in the number of states kept. Moreover, in order to isolate the origin of this problem they study an extremely simple model: a free particle in a 1D lattice. As a matter of fact, it was Wilson [18] who pointed out the importance of understanding real-space RG in the context of this simple tight-binding model where the standard BRG clearly fails as we are going to show.

The reason for this failure can be traced back to the importance of the boundary conditions in diagonalizing the states of a given block Hamiltonian H_B in which the lattice is decomposed into. Notice that in this fashion we are isolating a given block from the rest of the lattice and this applies a *particular BC* to the block. However, the block is not truly isolated! A statement which is the more relevant the more strongly correlated is the system under consideration. Thus, if the rest of the lattice were there it would apply different BC's to the boundaries of the block. This in turn makes the standard block-diagonalization conceptually not faithfully suited to account for the interaction with the rest of the lattice.

Once the origin of the problem is brought about the solution is also apparent: devise a method to change the boundary conditions in the block in order to mimick the interaction with the rest of the lattice. This is called the Combination of Boundary Conditions (CBC) method which yields very good numerical results. This method has not yet been generalized to interacting systems. However in reference [1] an alternative approach is proposed under the name of Density Matrix Renormalization Group (DMRG) which applies to more general situations and also produces quite accurate results.

In a recent paper [21] we have reconsidered again the role of BC's in the real space RG method for the case of a single-particle problem in a box. The continuum version of this Hamiltonian is simply $H = -\frac{\partial^2}{\partial x^2}$. We shall consider open chains with two types of BC's at the ends:

$$\text{Fixed BC's:} \quad \psi(0) = \psi(L) = 0 \tag{10}$$

$$\text{Free BC's:} \quad \frac{\partial \psi}{\partial x}(0) = \frac{\partial \psi}{\partial x}(L) = 0 \tag{11}$$

The lattice version of H for each type of BC's is given as follows:

$$H_{Fixed} = \begin{pmatrix} 2 & -1 & & & & \\ -1 & 2 & -1 & & & \\ & -1 & 2 & & & \\ & & & \ddots & & \\ & & & & 2 & -1 \\ & & & & -1 & 2 \end{pmatrix},$$

$$H_{Free} = \begin{pmatrix} 1 & -1 & & & & \\ -1 & 2 & -1 & & & \\ & -1 & 2 & & & \\ & & & \ddots & & \\ & & & & 2 & -1 \\ & & & & -1 & 1 \end{pmatrix} \tag{12}$$

The only difference between H_{Fixed} and H_{Free} appear at the first and last diagonal entry $(2 \leftrightarrow 1)$. The exact solution of (12) is very well-known and we give it for completeness:

$$\text{Fixed BC's:} \quad \psi_n(j) = N_n^{Fx} \sin \frac{\pi(n+1)}{N+1} j, \quad E_n = 4\sin^2\left(\frac{\pi(n+1)}{2(N+1)}\right) \tag{13}$$

Free BC's: $\quad \psi_n(j) = N_n^{Fr} \cos \frac{\pi n}{N}(j - \frac{1}{2}), \quad E_n = 4 \sin^2(\frac{\pi n}{2N})$ (14)

$$j = 1, 2, \ldots, N; \quad n = 0, 1, \ldots, N - 1.$$

where the $N_n's$ are normalization constants and N is the number of sites of the chain.

Before getting into the problem of the renormalization of these Hamiltonians, it is worth to pointing out another physical realization of H_{Free}: *A simple magnon above a ferromagnetic background satisfies Free BC's.* See [21] for more details.

Now let us get to the problem of renormalizing the tight-binding Hamiltonians (12).

In Fig. 2 we show the ground state and first excited states of the chain with fixed and free BC's.

It is clear from Fig.2 that *a standard Block RG method is not appropiate to study the ground state of fixed BC's since this state is non-homogeneous while the block truncation does not take into account this fact.*

Each piece of the ground state within each block satisfies BC's which vary from block to block. This is the motivation of reference [17] to consider different BC's in the block method, yielding quite accurate results.

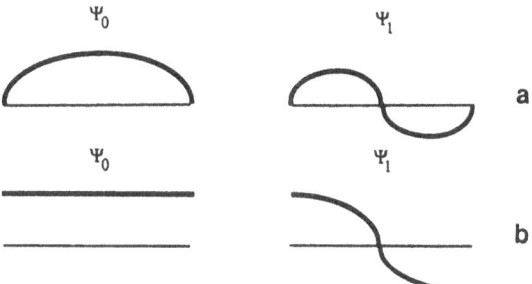

FIG. 2 a) Ground state ψ_0 and first excited state ψ_1 for the Hamiltonian H_{Fixed} with fixed BC's. b) Ground state ψ_0 and first excited state ψ_1 for the Hamiltonian H_{Free} with free BC's.

We observe that the standard RG method performs rather poorly as compared to the CBC method which yields quite the exact results.

The other alternative to the CBC method is the Density Matrix RG method which can be phrased by saying that the rest of the chain produces on every block the appropiate BC's to be applied to its ends, and it has the virtue that can be generalized to other models, something which is not the case as for the CBC method.

On the other hand, the ground state of H_{Free} is an homogeneous state (see Fig.2) which in turn suggests that a standard RG analysis may work for this type of BC's. We shall show that this is indeed the case if the RG procedure is properly defined. The key of our RG-prescription is to notice that H_{Free} has a geometrical meaning: H_{Free} *is the incidence matrix of a graph*, and it is called minus the discrete laplacian $-\Delta$ of that graph. Notice that H_{Fixed} has not such geometrical interpretation, in fact, it concides with the Dynkin diagram of the algebra A_N.

Based on this observation the Kadanoff blocking is nothing but the breaking of the graph into N/n_s disconnected graphs of n_s sites each. We shall choose $n_s = 3$ in our later computation.

The previous geometrical interpretation of H_{Free} suggests that we choose the block Hamiltonian H_B to be the incidence matrix of a disconnected graph, namely,

$$
H_B =
\begin{pmatrix}
1 & -1 & & & & & & \\
-1 & 2 & -1 & & & & & \\
& -1 & 1 & & & & & \\
& & & 1 & -1 & & & \\
& & & -1 & 2 & -1 & & \\
& & & & -1 & 1 & & \\
& & & & & & \ddots &
\end{pmatrix},
$$

$$
H_{BB} =
\begin{pmatrix}
0 & & & & & & & \\
& 0 & & & & & & \\
& & 1 & -1 & & & & \\
& & -1 & 1 & & & & \\
& & & & 0 & & & \\
& & & & & 1 & -1 & \\
& & & & & -1 & 1 & \\
& & & & & & & \ddots
\end{pmatrix}
\tag{15}
$$

and the interblock Hamiltonian H_{BB} above describes the interaction between blocks.

H_{BB} in turn also coincides with the incidence matrix of a graph which contains the missing links which connects consecutive blocks. In a few words: our RG-prescription introduces free BC's at the ends of every block. This condition fixes uniquely the breaking of H_{Free} into the sum $H_B + H_{BB}$. This is the choice we make. It should be emphasized that the splitting of H_{Free} into two parts $H_B + H_{BB}$ is by no means unique, so that different choices may lead to very different results.

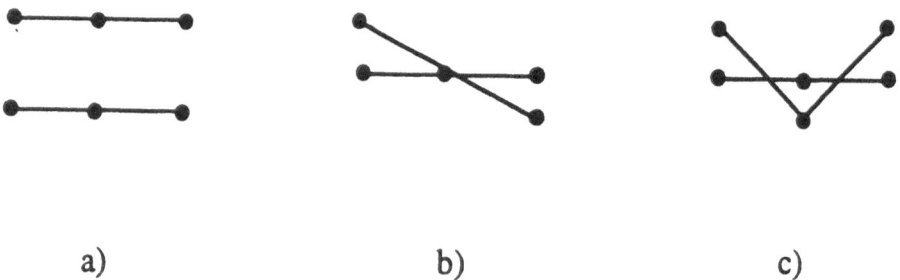

a)   b)   c)

FIG. 3 Building blocks of the 3-site BRG for the tight-binding model in 1D with free BC's. a) Ground state, b) First excited estate, c) Second excited state.

Prior to any computation we notice that the previous RG-prescription should lead to an exact value of the ground state energy, for the ground state of each block is again a constant function. The question is therefore to what extent our method is capable of describing the excited states. We shall concentrate ourselves to the first excited state since computations can be carried out analytically.

First of all we diagonalize H_B within each block of 3 sites, keeping only the ground state $\psi_0^{(0)}$ and the first excited state $\psi_1^{(0)}$ ($3 \rightarrow 2$ truncation). The superscript denotes the initial step in the truncation method. In Fig. 3 we picture the 3 eigenvectors for the 3-site Hamiltonian which will be the building blocks for our BRG, namely the two lowest ones. In the standard RG method we would choose $\psi_0^{(0)}$ and $\psi_1^{(0)}$ as the orthonormal basis for the truncated Hilbert space and obtain the effective Hamiltonian H_B' and H_{BB}'. In our case it is convenient to express these effective Hamiltonians in a basis expanded by the following linear combination:

$$\psi_+^{(0)} = \frac{1}{\sqrt{2}}(\psi_0^{(0)} + \psi_1^{(0)}) \tag{16}$$

$$\psi_-^{(0)} = \frac{1}{\sqrt{2}}(\psi_0^{(0)} - \psi_1^{(0)}) \tag{17}$$

which are also an orthonormal basis of the truncated Hilbert space. In this basis the truncation of H_B reads as follows,

$$H_B \longrightarrow H_B' = \begin{pmatrix} A & 0 & 0 & \\ 0 & A & 0 & \\ 0 & 0 & A & \\ & & & \ddots \\ & & & & A \end{pmatrix}, \quad A = \frac{\epsilon}{2}\begin{pmatrix} 1 & -1 \\ -1 & 1 \end{pmatrix} \tag{18}$$

with ϵ taking on the value $\epsilon^{(0)} = 1$ in the initial step of the RG-method, which is the energy of the state $\psi_1^{(0)}$.

The truncation of H_{BB} is more complicated, the result being:

$$H_{BB} \longrightarrow H_{BB}' = \begin{pmatrix} B & C & 0 & \\ C^t & D+B & C & \\ 0 & C^t & D+B & \\ & & & \ddots \\ & & & & D+B & C \\ & & & & C^t & D \end{pmatrix} \tag{19}$$

$$B = \begin{pmatrix} a^2 & ab \\ ab & b^2 \end{pmatrix} \quad C = \begin{pmatrix} -ab & -a^2 \\ -b^2 & -ab \end{pmatrix} \quad D = \begin{pmatrix} b^2 & ab \\ ab & a^2 \end{pmatrix}$$

with a b taking on the values $a^{(0)} = \frac{1}{\sqrt{6}} - \frac{1}{2}$ and $b^{(0)} = \frac{1}{\sqrt{6}} + \frac{1}{2}$ in the initial step of the RG-method.

The nice feature about the basis (16)-(17) is that all rows and columns of (18) and (19) add up to zero, just like the original Hamiltonians (15), implying that the constant vector is an eigenvector with zero eigenvalue of the renormalized Hamiltonian!

We shall call $H_{N/3}(\epsilon, a, b)$ the sum of the Hamiltonians (18) and (19) for generic values of ϵ, a and b. Next step in our RG-procedure is to form blocks of 4 states of the new Hamiltonian $H_{N/3}(\epsilon, a, b)$ and truncating to the two lowest $\psi_0^{(1)}$ and $\psi_1^{(1)}$ energy states within each 4-block ($4 \to 2$ truncation). The reason for this change in the number of sites per block (from 3 to 4) is motivated by the form of H'_{BB} in (19) and the fact that if we try to make a second step in the RG-method with 3-blocks the method is doomed to failure because the constant state of Fig.2 would no longer be the ground state.

Fortunately enough, with 4-blocks if we define new states $\psi_+^{(1)}$ and $\psi_-^{(1)}$ in the same form as we did in Eq.(12), we obtain that the new effective Hamiltonian is obtained by a redefinition of the parameters, namely,

$$H_{N/3}(\epsilon, a, b) \longrightarrow H_{N/6}(\epsilon', a', b') \tag{20}$$

$$\epsilon' = \frac{\epsilon}{2} + a^2 + b^2 - \Delta \tag{21}$$

$$a' = \frac{1}{2\sqrt{2}} \left[a + b - \frac{a(a^2 - 3b^2 + \Delta) + \frac{b\epsilon}{2}}{\sqrt{\Delta(\Delta + a^2 - b^2)}} \right] \tag{22}$$

$$b' = \frac{1}{2\sqrt{2}} \left[a + b + \frac{a(a^2 - 3b^2 + \Delta) + \frac{b\epsilon}{2}}{\sqrt{\Delta(\Delta + a^2 - b^2)}} \right] \tag{23}$$

$$\Delta \equiv \sqrt{(a^2 - b^2)^2 + (\frac{\epsilon}{2} - 2ab)^2} \tag{24}$$

In this fashion, the constant state of Fig.2 is again the ground state of the model and moreover, upon iteration of Eqs.(20)-(24) there are no level crossing among the excited states. Otherwise stated this means that the level structure of the block Hamiltonian H_B is preserved under the action of our BRG-method based upon the reduction from 4 to 2 states.

The energy $E_1(N)$ of the first excited state of a chain with $N = 3 \times 2^m$ sites can be obtained iterating m times Eqs.(20)-(24):

$$E_1(N = 3 \times 2^m) \equiv \epsilon^{(m)} \tag{25}$$

The initial data are given by:

$$\epsilon^{(0)} = 1, \quad a^{(0)} = \frac{1}{\sqrt{6}} - \frac{1}{2}, \quad b^{(0)} = \frac{1}{\sqrt{6}} + \frac{1}{2} \tag{26}$$

For low values of N the deviation of $\epsilon^{(m)}$ with respect to the exact result is small (see [21]). Recall that we are only keeping two states in our RG-procedure, and that the ground state energy is exactly zero by construction!. But what is more interesting about these (see [21]) is that we are able to obtain the correct size dependence, i.e., $1/N^2$ of $\epsilon^{(m)}$. As a matter of fact, the energy of the first excited state behaves for large N as (14):

$$E_1^{(exact)}(N) \sim c_{exact}/N^2, \quad \text{with } c_{exact} = \pi^2 \tag{27}$$

while our BRG-method gives,

$$E_1^{(BRG)}(N) \sim c_{BRG}/N^2, \quad \text{with } c_{BRG} = 12.6751 \tag{28}$$

The achievement of the $1/N^2$-law is a remarkable result which in turn allows us to match the correct order of magnitude of the energy. For instance, for 10 iterations our RG-method with 2 states kept gives the energy of the order of 10^{-6}, which is precisely the same order of magnitude as for the CBC method but with 8 states kept in the case of Fixed BC's. Recall that the standard BRG performs as bad as a 10^{-2} order of magnitude.

IV. WAVE-FUNCTION RECONSTRUCTION

Insofar we have only used our RG method to "reconstruct" the energies of the lowest states step by step, but we can also use this method to reconstruct the shape of the associate wave function in the real space. This simple fact leads to the consideration of RG applications beyond the original scope for which it was devised. As in this fashion we are making a picture of the wave function, it is natural to use the RG as a image compression method for coding images in order to facilitate its transport through the networks. For more details see the notes by Nishino in these proceedings [4].

We are able to make a reasonable picture of the first excited state wave-function based upon our BRG-procedure when compared with the exact form depicted in Fig.2. As we are working with a real-space realization of the renormalization group method, this is something we have at hand. To do this we need to perform a "reconstruction" of the wave-function. This reconstruction amounts to plot the form of our aproximate wave-function in each and every of the 3-blocks out of the 2^{m+1} in which the original chain is decomposed into under the BRG-procedure. Recall that in the initial step we started out with blocks of 3 states keeping the two lowest states $\psi_0^{(0)}$ and $\psi_1^{(0)}$ ($3 \to 2$ truncation). In the next step we make blocks of 4 states keeping the two lowest states $\psi_0^{(1)}$ and $\psi_1^{(1)}$ ($4 \to 2$ truncation) and then we perfom the iteration procedure over and over. As a result of this procedure we may express the two lowest wave functions of the $m + 1$-th step in terms of those of the previous m-th step by means of the following matricial form:

$$\begin{pmatrix} \psi_0^{(m+1)} \\ \psi_1^{(m+1)} \end{pmatrix} = \frac{1}{\sqrt{2}} \begin{pmatrix} 1 & 0 \\ \alpha_m & \beta_m \end{pmatrix} \begin{pmatrix} \psi_0^{(m)} \\ \psi_1^{(m)} \end{pmatrix}_L + \frac{1}{\sqrt{2}} \begin{pmatrix} 1 & 0 \\ -\alpha_m & \beta_m \end{pmatrix} \begin{pmatrix} \psi_0^{(m)} \\ \psi_1^{(m)} \end{pmatrix}_R \qquad (29)$$

where the LHS of Eq. (29) represents the wave function of $3 \times 2^{m+1}$ sites while in the RHS we have a left-wave-function of 3×2^m sites and another right-wave-function of 3×2^m sites, so that everything squares. The parameters appearing in Eq. (29) turn out to be given by:

$$\alpha_m = \frac{(\frac{\epsilon_m}{2} - 2a_m b_m) + (a_m^2 - b_m^2 + \Delta_m)}{2\sqrt{\Delta_m(\Delta_m + a_m^2 - b_m^2)}} \qquad (30)$$

$$\beta_m = \frac{(\frac{\epsilon_m}{2} - 2a_m b_m) - (a_m^2 - b_m^2 + \Delta_m)}{2\sqrt{\Delta_m(\Delta_m + a_m^2 - b_m^2)}} \qquad (31)$$

with Δ_m as in Eq.(24). Their initial values are $\alpha_0 = 1/\sqrt{10}$ and $\beta_0 = 3/\sqrt{10}$. We may recast Eq. (29) in more compact form by writing:

$$\Psi^{(m+1)} = L_m \Psi_L^{(m)} + R_m \Psi_R^{(m)} \qquad (32)$$

where

$$L_m = \frac{1}{\sqrt{2}} \begin{pmatrix} 1 & 0 \\ \alpha_m & \beta_m \end{pmatrix}, \quad R_m = \frac{1}{\sqrt{2}} \begin{pmatrix} 1 & 0 \\ -\alpha_m & \beta_m \end{pmatrix} \qquad (33)$$

We may call Eq.(32) the *reconstruction equation*. This is the master equation that when iterated "downwards" (reconstruction) allows us to obtain the picture of our approximate BRG-wave-function corresponding to every and each block of 3 sites of the 2^{m+1} blocks in which the chain is decomposed into. At the end of the iteration procedure we end up with expressions for the values of the 3-sites wave-functions in terms of the initial two lowest states $\psi_0^{(0)}$ and $\psi_1^{(0)}$. The first one is a constant function while the second is a straight line of negative slope. Thus, these two states turn out to be the building blocks of our BRG-procedure.

When using the reconstruction equation to obtain the wave function we may use a binary code based upon the labels L (left) and R (right) to keep track of the different 3-sites blocks which make up the chain. Thus, in one dimension the RG-blocks are in a one-to-one correspondence with a binary numerical system. In general, for other dimensions we may state squematically the following correspondence:

$$\text{BRG-prescription} \longleftrightarrow \text{"Number System"}$$

This simple observation is the basis of a coding system for compressing pictures whatever may be its origin. In two dimensions we need more digits to make the coding, but it works likewise and serves for image compression [4].

V. THE CORRELATED BLOCK RENORMALIZATION GROUP (CBRG)

We have already mentioned in the Introduction that the DMRG is not only a powerful computational method but also a source of inspiration for further works concerning the RG. For these reasons, it is worthwhile to explore different options or alternatives to the DMRG which may be useful in situations where the DMRG encounters difficulties, as in the case of 2D quantum systems. The main message of the DMRG is that blocks are correlated. The implementation of this idea by means of the density matrix formalism may be not the unique way to proceed. On the other hand, the "onion-scheme" a la Wilson adopted by the DMRG, while being one of the reasons of its spectacular accuracy, imposes certain limitations. At this stage it is not clear how fundamental are the density-matrix formalism or the onion-scheme for a RG method which takes into account the correlation between blocks. One can indeed combine the Kadanoff block method with the use of a density matrix in the process of truncation, as in reference [12]. More work remains to be done to see wheather there is a real improvement of the standard BRG method by combining it with the DMRG as in [12]. In this section we want to explore another possibility which is to give up both the density matrix and the onion-scheme (see [22]). With this point of view in mind, it would seem that we should come pretty close to the standard BRG method, were it not for the enormous freedom hidden in a Real-Space RG method. This freedom comes from the separation of the Hamiltonian into an intrablock H_B and an interblock H_{BB} Hamiltonian. This is a source of ambiguities which can be sometimes mitigated with the aide of symmetry arguments, but not fully eliminated though. This ambiguity shows up specially for terms in the Hamiltonian acting at the boundaries of the block. There are no general criteria as to how to include this type of terms either into the intrablock or into the interblock Hamiltonians, or into both! For example, in the 1D Ising model in a transverse field (ITF model), a choice which preserves the selfduality of the model attributes some self-couplings to the H_B and others to the interblock H_{BB}, and it yields to an exact value of the critical point and the critical exponent ν [23], [15]. The ambiguity in the splitting of H into the sum $H_B + H_{BB}$ thus affect deeply the truncation procedure itself, which is based on the diagonalization of H_B. Rather than blaming the BRG for its lack of uniqueness, we should use its freedom to allow the blocks to become correlated in the RG procedure. In our present approach this correlation will be taken into account in a "dynamical" way rather than in a "statistical" way as in the DMRG. This will be achieved by the introduction of interblock operators which reflect the "influence" between neighbour blocks and which are defined at the boundary of the block in the first step of our CBRG method.

We have chosen to illustrate our approach the 1D and 2D tight-binding models mainly for simplicity reasons, but we believe that our method could be applied to more complicated problems. In fact, the first step in this direction was already undertaken in reference [21], where only 2 states at each stage of the RG-blocking were retained. This in turn allowed us to obtain the $1/N^2$ scaling law for the size dependence of the first-excited-state energy.

We shall give the general mathematical structure underlying the results of reference [21]. This will allow us to retain more than two states in the RG-truncation and also to consider the two-dimensional tight-binding model. In this fashion, we shall recover the n^2/N^2 scaling law for the n-th excited state of the 1D model and the scaling law $\frac{n_1^2+n_2^2}{N^2}$ in the 2D case. These results will then show that the CBRG method describes correctly the low energy behaviour of the 1D and 2D Laplacian.

VI. THE CBRG METHOD: ONE DIMENSION

The problem we want to study is the one-dimensional Tight-Binding model in an open chain with different boundary conditions at its ends. The Hamiltonian for this system takes the following matricial form,

$$
H_{b,b'} = \begin{pmatrix}
b & -1 & & & & \\
-1 & 2 & -1 & & & \\
& -1 & 2 & & & \\
& & & \ddots & & \\
& & & & 2 & -1 \\
& & & & -1 & b'
\end{pmatrix}
\tag{34}
$$

where b and b' take on the values 1 (or 2) corresponding to Free (or Fixed) BC's respectively. This Hamiltonian is the discrete version of the Laplacian $H = -\partial_x^2$, while the Free or Fixed BC's correspond in the continuum to the vanishing of the wave function (Fixed BC's) or its spatial derivative (Free BC's) at the ends of the chain, i.e.,

$$
\begin{aligned}
b = 2 &\Rightarrow \Psi(0) = 0 \quad \text{Fixed BC} \\
b = 1 &\Rightarrow \frac{\partial \Psi}{\partial x}(0) = 0 \quad \text{Free BC}
\end{aligned}
\tag{35}
$$

and similarly for b' which contains the BC at the other end of the chain.

Hence, altogether there are 4 Hamiltonians of the type in (34), whose eigenstates and eigenvalues are the subject of our RG-techniques.

The first step in the RG method is to divide the lattice into blocks containing n_s sites each and labeled with and index p ($= 1, \ldots, N/n_s$). Let us suppose for a moment that we isolate the pth-block from the rest of the lattice so that its dynamics, as an independent entity, is governed by a Hamiltonian denoted by A_p, which we may call *uncorrelated block Hamiltonian*. The restoration of the block back into the lattice involves two effects. The first one is that the BC's of the p-th block may change under the influence of the $p+1$ and $p-1$ blocks. We describe this change of BC's by the action of *Boundary Operators* denoted by $B_{p,p\pm1}$ on the pth-block. The second effect is the interaction between the pth-block and its neighbours $p+1$ and $p-1$, given by interaction Hamiltonians $C_{p,p\pm1}$ which act on both p and $p+1$ blocks simultaneously. If the problem under consideration is translationally invariant, all the Hamiltonians defined above are independent of the block label p, in which case we denote them by,

$$A_p = A$$
$$B_{p,p+1} = B_R \quad B_{p,p-1} = B_L \tag{36}$$
$$C_{p,p+1} = C \quad C_{p,p-1} = C^\dagger$$

The $H_{Free,Free}$ Hamiltonian (34) gives an example of this as we shall show below. Hence, for the time being, we shall consider the situation described by (36) and leave the more general case after explaining the general ideas.

In the standard BRG method the block Hamiltonian H_B and the interblock Hamiltonian H_{BB} are given, according to our previous definitions, by the following formulas

$$H_B = A + B_L + B_R \tag{37}$$

$$H_{BB} = \begin{pmatrix} 0 & C \\ C^\dagger & 0 \end{pmatrix} \tag{38}$$

The whole Hamiltonian is by all means the sum of H_B and H_{BB} for all the blocks of the chain.

Next step in the RG method is to diagonalize H_B and keep its, say m ($m < n_s$), lowest eigenstates. The truncation is given by a $n_s \times m$ matrix T whose columns are precisely the components of the m lowest eigenstates of H_B. The renormalized Hamiltonian in the new basis is given by,

$$H' = T^\dagger (H_B + H_{BB}) T \tag{39}$$

At first sight from Eq. (37) it would seem that we have taken into account the effect of the BC's on a given block. However, as the examples show, this is quite a bit illusory. On the other hand, the distinction among A, B_L and B_R is rather inmaterial as far as H_B is concerned, and in fact no distinction of this sort is made in the standard BRG formalism. Finally, let us observe that H_B and H_{BB} play rather different roles in the truncation procedure. This asymmetry has been observed as a source of problems by several authors in the past [23], [20].

We shall mention that this asymmetry has recently been related to quantum groups in a fashion which has led to a new RG method called the Renormalization Quantum Group method [15], [16].

Therefore, from various points of view, one is urged to make more explicit the role played by the BC-operators B_L and B_R in our CBRG procedure. For this purpose, we have found convenient to use the concept of superblock already introduced in reference [17]. We shall define a superblock as the set of two consecutive blocks, p and $p+1$ and denoted by $(p, p+1)$. The great advantage of the superblock is that it allows us to materialize the distinction among A, B_L and B_R. In fact, just as the isolation of a single block leads us to the definition of the Hamiltonian A, the isolation of two blocks contained in a superblock allows us to define B_L, B_R and also C through the superblock Hamiltonian H_{sB} as follows,

$$H_{sB} = \begin{pmatrix} A + B_R & C \\ C^\dagger & A + B_L \end{pmatrix} \tag{40}$$

Similarly, the Hamiltonian describing the interaction between superblocks is given by (see Fig.4)

$$H_{sB,sB} = \begin{pmatrix} 0 & & \\ & B_R & C \\ & C^\dagger & B_L \\ & & & 0 \end{pmatrix} \tag{41}$$

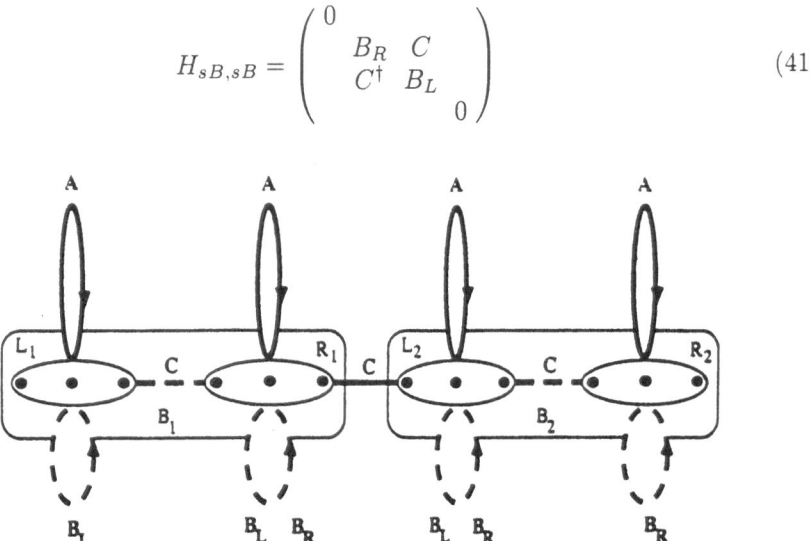

FIG. 4 Pictorical decomposition of a given Hamiltonian H into uncorrelated A-operators, correlation B_L- B_R-operators and interaction C-operators according to the CBRG method. B_1 is a superblock made up of two L_1 and R_1 blocks.

Now instead of diagonalizing H_B in Eq. (37), in the CBRG method we shall diagonalize H_{sB} in Eq. (40), and afterwards keep the $m = n_s$ lowest eigenstates in the tight-binding model. As in the standard BRG method, the change to the truncated basis defines the renormalized operators as follows:

$$H_{sB} \longrightarrow T^\dagger H_{sB} T = A' \tag{42}$$

$$H_{sB,sB} \longrightarrow T^\dagger H_{sB,sB} T = \begin{pmatrix} B'_R & C' \\ C'^\dagger & B'_L \end{pmatrix} \tag{43}$$

where the matrices A', B'_R, B'_L and C' are the renormalized version of the operators A, B_R, B_L and C, and they exhibit the same geometrical interpretation for the renormalized block as their unprimed partners for the original blocks.

If we set $B_R = B_L = 0$ in Eqs. (40) and (41), then after the first RG-step we get $B'_R = B'_L = 0$ and thus the previous RG-scheme coincides with the standard BRG. We may say that uncorrelated blocks are in a sense a fixed point of our method. However, this fixed point may be unstable, and to explore this possibility one has to look for non-vanishing B-operators and their RG-evolution.

Let us address now some examples. We shall first study the Hamiltonian (34) with Free BC's at the ends ($b = b' = 1$). Choosing $n_s = 3$ for example, we see that the choice for the operators A, B_R, B_L and C in the first step of the CBRG procedure is given by,

$$A = \begin{pmatrix} 1 & -1 & 0 \\ -1 & 2 & -1 \\ 0 & -1 & 1 \end{pmatrix}, \quad B_R = \begin{pmatrix} 0 \\ 0 \\ 1 \end{pmatrix},$$

$$B_L = \begin{pmatrix} 1 \\ 0 \\ 0 \end{pmatrix}, \quad C = \begin{pmatrix} 0 & 0 & 0 \\ 0 & 0 & 0 \\ -1 & 0 & 0 \end{pmatrix} \tag{44}$$

This choice is equivalent to the assumption that an isolated block satisfies Free BC's at its ends. The role of B_R and B_L is to *join* these blocks into a single chain. This is the *geometrical* explanation of Eqs. (44). In more general cases one must have to explore which is the best choice. The generalization of Eqs.(44) to blocks with more than 3 sites is obvious. In Table 1 we collect our CBRG-results for the first 5 excited states for a chain of $N = 12 \times 2^6 = 768$ sites. Comparison with exact results gives a good agreement.

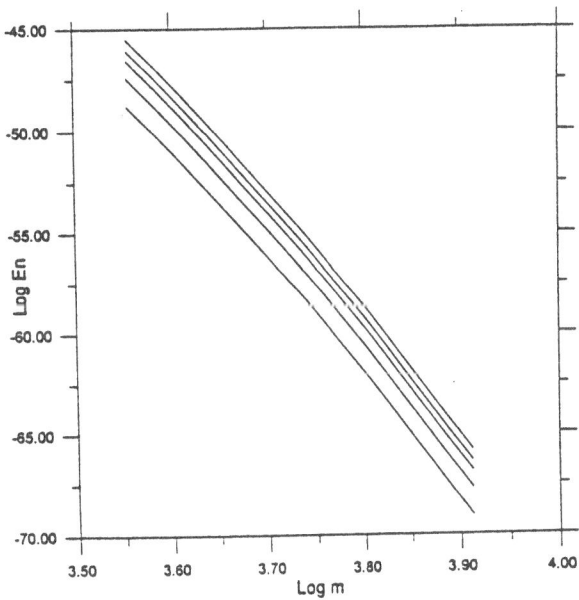

FIG. 5 The n^2/N^2-law for the first 5 excited states of the 1D Tight-Binding Model for a chain of $N = 12 \times 2^m$ sites with Free-Free BC's. This is a $\ln E_n$-$\ln m$ plot. .

An important feature of our CBRG method is that the n^2/N^2- scaling law $(N \longrightarrow \infty)$ for the energy of the n-excited states of a chain made up of N sites, is reproduced correctly (see Fig.5). In Table 2 we show the variation of the first-excited-state energy with the size N of the chain. From those values we can extract the corresponding $1/N^2$-law which turns out to be,

$$E_1^{(CBRG)}(N) = c_{CBRG}^{(1)} \frac{1}{N^2}, \quad c_{CBRG}^{(1)} = 9.8080, \quad (N \longrightarrow \infty) \text{ Free-Free BC's} \quad (45)$$

while the exact value for the proportionality constant c is $c_{\text{exact}} = \pi^2 = 9.86$. This amounts to a 0.6 % error.

Likewise, we have enough data so as to obtain the corresponding n^2/N^2-law for the whole set of 5 excited states. Thus, the scaling law we obtain is,

$$E_n^{(CBRG)}(N) = c_{CBRG} \frac{n^2}{N^2}, \quad c_{CBRG} = 8.4733, \quad (N \longrightarrow \infty) \text{ Free-Free BC's} \quad (46)$$

which now amounts to a 7.34 % error. This is a natural fact from the worse knowledge of the highest excited states of the spectrum in a RG-scheme.

We can make even more explicit the successful achievement of the $1/N^2$-scaling law by leaving as a free adjustable parameter the exponent of $1/N$ in addition to the proportionality constant. Let us denote by θ this critical exponent. Using data from 20 to 50 steps of our CBRG-method for several truncation of states according to our scheme $2n_s \rightarrow n_s$ (namely, $n_s = 10, 13, 20$) we arrive at the following results,

$$E_1^{(CBRG)}(N) = c_{CBRG} \frac{1}{N^\theta}, \quad (N \longrightarrow \infty) \text{ Free-Free BC's} \quad (47)$$

$$\begin{array}{lll} \text{For} & 20 \longrightarrow 10 & \theta = 1.9708 \\ \text{For} & 26 \longrightarrow 13 & \theta = 1.9734 \\ \text{For} & 40 \longrightarrow 20 & \theta = 1.9854 \end{array} \quad (48)$$

These results clearly support the fact that we have correctly reproduced the exact value of $\theta = 2$ for the finite-size critical exponent.

Last, but not least, as was proved in [21] our CBRG method gives the *exact* energy of the ground state for every step of the RG-procedure for Free-Free BC's.

In tables 1 and 2 we also show the results we have obtained with a DMRG analysis following White's method [1]. This analysis is based on the onion scheme of enlarging the lattice site by site á la Wilson. The results coincide with the exact values within the 4 digits precision used here, but they start differing when keeping more digits. Nevertheless, the DMRG is much more time consuming than our CBRG method for it has to build the lattice site by site, while the CBRG reproduces the lattice by blocking which is much more efficient as far as CPU time is concern, and moreover, it applies to two-dimensional situations where the onion scheme fails to reproduce the lattice. We have also performed the DMRG analysis in 1D for Fixed-Fixed BC's in table 4 where the same considerations apply.The CBRG method is also more suitable for analytic formulations [21].

In reference [21] it was shown that one can reproduce easily the wave function of the excited states. This procedure was called *reconstruction* since it works "downwards" in the CBRG method. The basic equation to be used is the *reconstruction equation* [21],

$$\Psi^{(r+1)} = L_r \Psi_L^{(r)} + R_r \Psi_R^{(r)} \quad (49)$$

where $\Psi^{(r)}$ denotes the collection of m lowest eigenstates in the r-step of the CBRG-procedure, and L_r, R_r are the block matrices in terms of which the truncation matrix $T^\dagger$ can be written as $T^\dagger = (L_r, R_r)$.

Our results for a chain of $N = 12 \times 2^6 = 768$ sites and $n_s = 6$ states kept are given in Fig.6 where we have plotted the first 5 excited states and compare them with the exact wave functions. There are some remarkable facts regarding these figures. Firstly, the number of nodes is correctly preserved by our CBRG wave functions. Secondly, the Free-Free type of boundary conditions are also correctly reproduced at the ends of the chain. And lastly, it is worthwhile to point out that the CBRG wave functions "degrade gracefully" as the energy of the excited state raises in accordance with the fact that the lower the energy is, the more reliable are the results.

This ends the results for the Free-Free BC's. In order to address other types of BC's we must come back to the case where the matrices A, B_R, B_L and C depend on each particular block.

Thus, for example, for the Fixed-Free BC's we shall choose as the uncorrelated A-matrix for the block located to the left end of the chain the following form ($n_s = 3$),

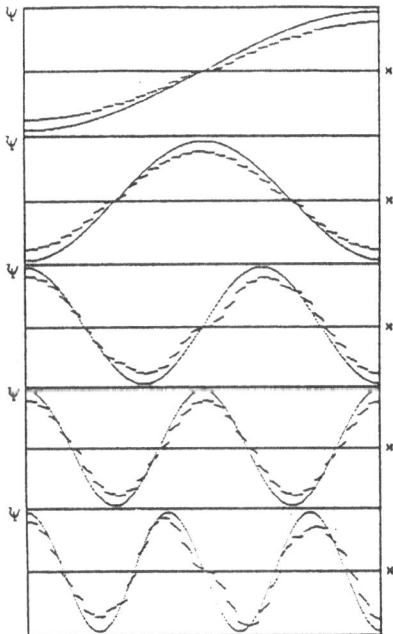

FIG. 6 The wave function reconstruction for the first 5 excited states of the 1D Tight-Binding Model for a chain of $N = 12 \times 2^6 = 768$ sites with Free-Free BC's. We have scaled up the exact results by a factor of 1.23 for clarity. .

$$A_1 = \begin{pmatrix} 2 & -1 & 0 \\ -1 & 2 & -1 \\ 0 & -1 & 1 \end{pmatrix} \quad \text{Fixed-Free BC's} \qquad (50)$$

while the remaining matrices A_n ($n = 2, \ldots, N/3$), will be given by Eqs.(36), (44),

For Free-Fixed BC's, it is the last A-matrix which we have to take different from the others, namely,

$$A_{N/3} = \begin{pmatrix} 1 & -1 & 0 \\ -1 & 2 & -1 \\ 0 & -1 & 2 \end{pmatrix} \quad \text{Free-Fixed BC's} \tag{51}$$

As for the Fixed-Fixed BC's case, we must change the A-matrix at both ends of the chain according to the following prescription,

$$A_1 = \begin{pmatrix} 2 & -1 & 0 \\ -1 & 2 & -1 \\ 0 & -1 & 1 \end{pmatrix}, \quad A_{N/3} = \begin{pmatrix} 1 & -1 & 0 \\ -1 & 2 & -1 \\ 0 & -1 & 2 \end{pmatrix} \quad \text{Fixed-Fixed BC's} \tag{52}$$

Then we follow the same steps as for the Free-Free BC's, taking care that the A, B_R, B_L and C matrices in each CBRG-step may depend on the position of the blocks. This implies in particular that the embedding T-matrices may also vary from block to block.

In Tables 3 and 4 we summarize our results for the Free-Fixed and Fixed-Fixed BC's (Fixed-Free BC's are equivalent to Free-Fixed BC's by parity transformation). In these tables we present our CBRG results for the first 6 lowest lying states for the 1D tight-binding model in a chain of $N = 12 \times 2^5 = 384$ sites with mixed boundary conditions, and they are compared against the exact and standard BRG values. Several remarks are in order. First, we observe that the CBRG method produces a good agreement with the exact results and certainly much more accurate by several orders of magnitude than the old BRG method. Second, the CBRG method is able to reproduce the corresponding n^2/N^2-scaling laws for the spectrum of excited states in each case of mixed BC's. Namely,

- For Free-Fixed BC's and considering just the ground state, we have

$$E_0^{(CBRG)}(N) = c_{CBRG}^{(0)} \frac{1}{4N^2}, \quad c_{CBRG}^{(0)} = 9.072,$$

$$(N \longrightarrow \infty) \quad \text{Free-Fixed BC's} \tag{53}$$

which amounts to a 8 % error with respect to the exact value of $c_{exact} = \pi^2$.

As for the corresponding law for the whole spectrum, we find

$$E_n^{(CBRG)}(N) = c_{CBRG} \frac{(n+1)^2}{4N^2}, \quad c_{CBRG} = 7.6729,$$

$$(N \longrightarrow \infty) \quad \text{Free-Fixed BC's} \tag{54}$$

which represents a 11.5 % error with respect to the exact value of π^2.

- For Free-Fixed BC's and considering just the ground state, we have

$$E_0^{(CBRG)}(N) = c_{CBRG}^{(0)} \frac{1}{N^2}, \quad c_{CBRG}^{(0)} = 8.35,$$

$$(N \longrightarrow \infty) \text{ Fixed-Fixed BC's} \tag{55}$$

which amounts to a 8 % error with respect to the exact value of $c_{exact} = \pi^2$. As for the corresponding law for the whole spectrum, we find

$$E_n^{(CBRG)}(N) = c_{CBRG} \frac{(n+1)^2}{N^2}, \quad c_{CBRG} = 6.9696,$$

$$(N \longrightarrow \infty) \text{ Fixed-Fixed BC's} \tag{56}$$

which represents a 16 % error with respect to the exact value of π^2.

We obtain bigger errors in the determination of these scaling laws as compared with the Free-Free case mainly because we have used less data in our fitting. Nevertherless, we find a good agreement with the exact results. Yet, there is another reason as to why the accuracy in the case of mixed BC's is worse, namely, the ground state wave function Ψ_0 is not homogeneous in space as it is in the Free-Free case [21]. This makes the RG-procedure more involved and a source of extra uncertainties.

Let us mention in passing that we are also able to make a wave function reconstruction in the mixed BC's cases as has been done for the Free-Free BC case.

The outcome of all the results presented so far is that we have succeded in devising a Real-Space RG method capable of reproducing the correct eigenvalues and eigenstates for the tight-binding model as originally envisaged by Wilson, within a certain accuracy which can in principle be improved.

Althoug the model we have employed to test our CBRG-method is a tight-binding model, there are some remarkable facts regarding the fixed-point structure of our CBRG-solution that we would like to stress. Namely, we have found that after enough number of CBRG-iterations, the matrices A, B_L, B_R and C in the Free-Free case scale nicely with the size N of the chain according to the dynamical critical exponent z. To be more precise, let us introduce the fixed point values of those matrices denoted by A^*, B_L^*, B_R^* and C^* which we define as,

$$A^* = N^{-z} a^*, \quad B_L^* = N^{-z} b_L^*,$$

$$B_R^* = N^{-z} b_R^*, \quad C^* = N^{-z} c^*, \quad \text{Fixed-Point values} \tag{57}$$

in terms of the scaled matrices a^*, b_L^*, b_R^* and c^*. For a block of 3 sites ($n_s = 3$) we find the following Fixed-Point structure parametrized by two constants s and t (for bigger n_s we need extra parameters),

$$a^* = 0, \ b_R^* = \begin{pmatrix} 1 & s & s \\ s & t & t \\ s & t & t \end{pmatrix}, \ b_L^* = \begin{pmatrix} 1 & -s & s \\ -s & t & -t \\ s & -t & t \end{pmatrix}, \ c^* = \begin{pmatrix} -1 & s & -s \\ -s & t & -t \\ -s & t & -t \end{pmatrix} \quad (58)$$

with $s = 1.3993$ and $t = 1.9581$. The critical exponent z we obtain is,

$$z = 0.9999 \quad (59)$$

which is indeed very close to the exact value $z = 1$ (actually, it differs in the ninth decimal digit).

The interpretation of this Fixed-Point in the context of the CBRG method is as follows. We pointed out before that when the boundary operators $b_{L,R}$ vanish we recover the standard BRG method in which the blocks are not correlated. Here we find that it is the uncorrelated Hamiltonian which vanish, while the boundary $b_{L,R}$ and interaction c operators do not vanish within the scaling law. This fact may perhaps be interpreted by saying that in the example under study the correlation between blocks is more important than their selfenergy. In references [25], [26], [27] it was shown that the DMRG method leads, in the thermodynamic limit, to a "product form" ansatz for the ground state wave function. In our case we see from Eqs.(57), (58) that we also reach thermodynamical limit, which leads us to ask about the nature of the ansatz for the ground state and excited states implied by the CBRG method. The answer to this question will be addressed in a future publication but it suffices to say that both the DMRG and the CBRG methods seem to yield different ansatzs of the ground state wave function. In a few words, the DMRG is associated with a "vertex picture" while the CBRG is associated with a "string picture".

VII. THE TWO-DIMENSIONAL CBRG-ALGORITHM

The RG-method that we have devised in the one-dimensional problem can be generalized in a natural way to higher dimensions. We shall consider for simplicity the 2D case. First of all, we divide the square lattice into blocks of n_s sites each. Each block will in turn be a square lattice with a minimum of 4 sites ($= 2 \times 2$ block). As in 1D, we shall define the following Hamiltonians to carry out the CBRG-program,

- A_p = self-energy of the p-th block isolated from the lattice.

- $B_{p,q}$ = self-energy of the p-th block induced by the presence of the q-th block.

- $C_{p,q}$ = interaction between the p-th block and the q-th block.

The difference with respect to the 1D case is that each block has now 4 neighbours and therefore there are four different B and C matrices.

Let us consider again the Hamiltonian of a free particle moving in a 2D-box with Free BC's at the boundaries of the box. The 2D Hamiltonian is given again by the incidence matrix of the lattice. As in 1D we shall choose the matrix A as the incidence matrix of the block. Thus, for example, for a 2×2 block we have,

$$A = \begin{pmatrix} 2 & -1 & 0 & -1 \\ -1 & 2 & -1 & 0 \\ 0 & -1 & 2 & -1 \\ -1 & 0 & -1 & 2 \end{pmatrix} \tag{60}$$

The 4 Boundary Operators B have a diagrammatic representation [22] which helps us to keep track of their location in the block H_B and interblock H_{BB} Hamiltonians. Their explicit matricial form is as follows,

$$B_{12} = B_{43} = B_L = \begin{pmatrix} 0 & & & \\ & 1 & & \\ & & 1 & \\ & & & 0 \end{pmatrix}, \quad B_{21} = B_{34} = B_R = \begin{pmatrix} 1 & & & \\ & 0 & & \\ & & 0 & \\ & & & 1 \end{pmatrix} \tag{61}$$

$$B_{14} = B_{23} = B_D = \begin{pmatrix} 0 & & & \\ & 0 & & \\ & & 1 & \\ & & & 1 \end{pmatrix}, \quad B_{41} = B_{32} = B_U = \begin{pmatrix} 1 & & & \\ & 1 & & \\ & & 0 & \\ & & & 0 \end{pmatrix} \tag{62}$$

where the labels denote the position of the neighbouring blocks and we have used the translation invariance of the 2D tight-binding model so that we need only to distinguish between Right and Left, and Up and Down.

As for the Interaction C-Operators [22] we have the following matricial representation, with the same considerations as for the B-operators,

$$C_{12} = C_{43} = C_{LR} = \begin{pmatrix} 0 & 0 & 0 & 0 \\ -1 & 0 & 0 & 0 \\ 0 & 0 & 0 & -1 \\ 0 & 0 & 0 & 0 \end{pmatrix},$$

$$C_{21} = C_{34} = C_{RL} = \begin{pmatrix} 0 & -1 & 0 & 0 \\ 0 & 0 & 0 & 0 \\ 0 & 0 & 0 & 0 \\ 0 & 0 & -1 & 0 \end{pmatrix} \tag{63}$$

$$C_{14} = C_{23} = C_{DU} = \begin{pmatrix} 0 & 0 & 0 & 0 \\ 0 & 0 & 0 & 0 \\ 0 & -1 & 0 & 0 \\ -1 & 0 & 0 & 0 \end{pmatrix},$$

$$C_{41} = C_{32} = C_{UD} = \begin{pmatrix} 0 & 0 & 0 & -1 \\ 0 & 0 & -1 & 0 \\ 0 & 0 & 0 & 0 \\ 0 & 0 & 0 & 0 \end{pmatrix} \tag{64}$$

Thus translation invariance reduces the number of independent CBRG-matrices by a half. These relations are particular of the problem at hand but we must left open the posibility of having all those matrices different from each other in order to handle more complicated problems.

Now that we have all the elements entering in our CBRG-method we proceed to construct the block H_{sB} and interblock $H_{sB,sB}$ Hamiltonians out of them. To this end we have to consider a superblock made up of 4 blocks [22]. Thus, for H_{sB} we have,

$$
H_{sB} = \begin{pmatrix}
A + B_L + B_D & C_{LR} & 0 & C_{DU} \\
C_{RL} & A + B_R + B_D & C_{DU} & 0 \\
0 & C_{UD} & A + B_R + B_U & C_{RL} \\
C_{UD} & 0 & C_{LR} & A + B_L + B_U
\end{pmatrix} \tag{65}
$$

This is a $4n_s \times 4n_s$ matrix made up of $n_s \times n_s$ matrices.

As for the interblock Hamiltonian $H_{sB,sB}$ we have to distinguish between (sB, sB)-couplings of horizontal type denoted by $H_{sB,sB}^{(hor)}$ and vertical type denoted by $H_{sB,sB}^{(ver)}$, which read explicitly as,

$$
H_{sB,sB}^{(hor)} = \begin{pmatrix}
B_R & & & 0 & C_{RL} & 0 & 0 \\
& 0 & & 0 & 0 & 0 & 0 \\
& & 0 & 0 & 0 & 0 & 0 \\
& & & B_R & 0 & 0 & C_{RL} & 0 \\
0 & 0 & 0 & 0 & 0 \\
C_{LR} & 0 & 0 & 0 & & B_L \\
0 & 0 & 0 & C_{LR} & & & B_L \\
0 & 0 & 0 & 0 & & & & 0
\end{pmatrix},
$$

$$
H_{sB,sB}^{(ver)} = \begin{pmatrix}
B_D & & & 0 & 0 & 0 & C_{UD} \\
& B_D & & 0 & 0 & C_{UD} & 0 \\
& & 0 & 0 & 0 & 0 & 0 \\
& & & 0 & 0 & 0 & 0 & 0 \\
0 & 0 & 0 & 0 & 0 \\
0 & 0 & 0 & 0 & 0 \\
0 & C_{DU} & 0 & 0 & & B_U \\
C_{DU} & 0 & 0 & 0 & & & B_U
\end{pmatrix} \tag{66}
$$

where we have made use again of translational invariance.

Once that we have made our choice for the decomposition of the total Hamiltonian of the 2D-tight-binding model into block and interblock Hamiltonians according to our CBRG-prescription, we can carry on with the truncation part of the RG-method. We shall keep n_s states out of $4n_s$ states per superblock so that our truncation scheme may be summarized as,

$$4n_s \text{ (superblock)} \longrightarrow n_s \text{ (new block)}$$

Recall that at each step of the CBRG-method we need to identify the A, B_L, B_R and C operators which define the truncation procedure for the next step of the method. For this purpose, firstly the truncation of the superblock H_{sB} gives rise to the A' uncorrelated self-energy operator for the next RG-step, namely,

$$H_{sB} \;(\; 4n_s \times 4n_s \text{ matrix}) \longrightarrow A' \;(n_s \times n_s \text{ matrix}) \tag{67}$$

To identify the rest of the operators we have to renormalize the interblock Hamiltonian which comes in two types, horizontal and vertical. The renormalization of the $H_{sB,sB}^{(hor)}$ Hamiltonian is given by [22],

$$H_{sB,sB}^{(hor)} \longrightarrow \begin{pmatrix} B'_R & C'_{RL} \\ C'_{LR} & B'_L \end{pmatrix} \tag{68}$$

Likewise, for the $H_{sB,sB}^{(ver)}$ Hamiltonian we have,

$$H_{sB,sB}^{(ver)} \longrightarrow \begin{pmatrix} B'_D & C'_{UD} \\ C'_{DU} & B'_U \end{pmatrix} \tag{69}$$

Now that we have identified all the operators defining the CBRG method at the new stage of the renormalization, we may reconstruct the new superblock Hamiltonian H'_{sB}, which in turn has the same form as the original H_{sB} in Eq.(65) substituting all the operators by their *primed versions*. This statement can be explicitly checked by considering the set of 4 superblocks [22]. Firstly, the new H'_{sB} has a contribution coming from the truncation of each of the 4 superblocks, each of them contributing with an A'-operator as in Eq.(67). Secondly, H'_{sB} picks up two more contributions coming from the horizontal and vertical interaction between superblocks, which we denote by $H_{\leftrightarrow}$ and $H_{\updownarrow}$. Thus, in the CBRG-method H'_{sB} is renormalized as,

$$H'_{sB} = \begin{pmatrix} A' & & & \\ & A' & & \\ & & A' & \\ & & & A' \end{pmatrix} \longleftarrow \text{(single superblock contribution)}$$

$$(H_{\leftrightarrow}) \;\rightarrow\; + \begin{pmatrix} B'_L & C'_{LR} & & \\ C'_{RL} & B'_R & & \\ & & 0 & \\ & & & 0 \end{pmatrix} + \begin{pmatrix} 0 & & & \\ & 0 & & \\ & & B'_R & C'_{RL} \\ & & C'_{LR} & B'_L \end{pmatrix}$$

$$(H_{\updownarrow}) \;\rightarrow\; + \begin{pmatrix} B'_U & & C'_{DU} & \\ & 0 & & \\ & & 0 & \\ C'_{UD} & & & B'_D \end{pmatrix} + \begin{pmatrix} 0 & & & \\ & B'_U & C'_{DU} & \\ & C'_{UD} & B'_D & \\ & & & 0 \end{pmatrix} \tag{70}$$

and altogether we arrive at the previously stated result of Eq.(65). Similarly we may proceed with the renormalized interblock Hamiltonians $H'^{(hor)}_{sB,sB}$ (68) and $H'^{(ver)}_{sB,sB}$ (69) and we end up with the same form for them as the original ones.

This ends the implementation of the CBRG-method for the 2D-tight-binding model.

In Table 5 we collect our CBRG results for the first 4 lowest lying states for a chain of $N = 4 \times 4 \times 4^6 = 65536$ sites. Comparison with the exact results gives a good agreement. We have also data from truncations with blocks of $n_s = 9$ and $n_s = 16$ sites which enforce this statement. Moreover, notice that the first excited state is a doublet as in the exact solution.

Another important result of our CBRG-method is that the $(n_1^2 + n_2^2)/N^2$ scaling law for the energy of the (n_1, n_2)-excited states of a square lattice of length N is reproduced correctly. In fact, from data of the $n_s = 16$ sites truncation for the first-excited-state energy we can extract the corresponding $1/N^2$-scaling law which turns out to be,

$$E_1^{(CBRG)}(N) = c_{CBRG}^{(1)} \frac{1}{N^2}, \quad c_{CBRG}^{(1)} = 9.7365,$$

$$(N \longrightarrow \infty) \text{ D=2 Free BC's} \tag{71}$$

while the exact value of the proportionality constant c is $c_{exact} = \pi^2 = 9.86$. This amounts to a 1.3 % error.

Likewise, we may obtain the full $(n_1^2 + n_2^2)/N^2$ scaling law for the whole set of 15 excited states and we find,

$$E_{(n_1,n_2)}^{(CBRG)}(N) = c_{CBRG}^{(1)} \frac{(n_1^2 + n_2^2)}{N^2}, \quad c_{CBRG} = 7.9074,$$

$$(N \longrightarrow \infty) \text{ D=2 Free BC's} \tag{72}$$

which now amounts to a 10.5 % error.

As in the 1D Free-Free case, we can determine critical scaling exponent θ (47). For a truncation scheme $16 \rightarrow 4$ we find,

$$\theta = 1.99999981 \quad \text{D=2} \tag{73}$$

which clearly supports the scaling laws introduced above. Notice again (see Table 5) that our CBRG method gives the exact (within machine precision) energy of the ground state. This is true for every step of the RG, as was proved in [21] for 1D.

We can also perform the wave function reconstruction of the excited states in the two-dimensional real space. This is achieved by a two-dimensional extension of the reconstruction equation (49). As an illustration of how the CBRG method performs with this matter, see [22]. The qualitative real-space form of the excited-state wave functions are captured by the CBRG procedure.

With this discussion we close the first part of these notes which have been devoted to new develements of the Real-Space RG method revolving around the new ideas brought about by the Density Matrix RG method.

VIII. STANDARD BRG FOR THE 1D AF HEISENBERG MODEL

In the remaining sections we shall return to the standard BRG methods to deal in an analytical controlled fashion with models which include many-body interactions unlike the free models considered in the first part of these notes.

The arquetypical model we shall be dealing with is the Heisenberg model which will be studied in ladder systems and 2D lattices (square, honeycomb). Our point will be that even with the old-fashion BRG can be useful to retrieve the correct qualitative phyisics when properly implemented. To this end we shall be needing some results concerning the Heisenberg model in one dimension which will be basic.

Let us recall the standard BRG method for the AF Heisenberg-Ising model whose Hamiltonian is given by:

$$H_N = J \sum_{j=1}^{N-1} (S_j^x S_{j+1}^x + S_j^y S_{j+1}^y + \Delta S_j^z S_{j+1}^z) \tag{74}$$

where $\Delta \geq 0$ is the anisotropic parameter and $J > 0$ for the antiferromagnetic case. If $\Delta = 1$ one has the AF-Heisenberg model which was solved by Bethe in 1931. If $\Delta = 0$ one has the XX-model which can be trivially solved using a Jordan-Wigner transformation which maps it onto a free fermion model. For the remaing values of Δ the model is also solvable by Bethe ansatz and it is the 1D relative of the 2D statistical mechanical model known as the 6-vertex or XXZ-model.

The region $\Delta > 1$ is massive with a doubly degenerate ground state in the thermodynamic limit $N \to \infty$ characterized by the non-zero value of the staggered magnetization,

$$m_{\text{st}} = \langle \frac{1}{N} \sum_j S_j^z (-1)^j \rangle \tag{75}$$

The region $0 \leq \Delta \leq 1$ is massless and the ground state is non-degenerate with a zero staggered magnetization. The phase transition between the two phases has an essential singularity.

We would like next to show which of these features are captured by a real-space RG-analysis. The rule of thumb for the RG-approach to half-integer spin model or fermion model is to consider blocks with an *odd number of sites*. This allows in principle, although not necessarilly, to obtain effective Hamiltonians with the same form as the original ones. Choosing for (74) blocks of 3 sites we obtain the block Hamiltonian:

$$\frac{1}{J} H = \vec{S}_1 \cdot \vec{S}_2 + \vec{S}_2 \cdot \vec{S}_3 + \epsilon (S_1^z S_2^z + S_2^z S_3^z)$$

$$= \frac{1}{2} \left\{ [\vec{S}_1 + \vec{S}_2 + \vec{S}_3]^2 - (\vec{S}_1 + \vec{S}_3)^2 - 3/4 \right\} + \epsilon (S_1^z S_2^z + S_2^z S_3^z) \tag{76}$$

$\epsilon := \Delta - 1.$

If $\epsilon = 0$ the block Hamiltonian H_B is invariant under the $SU(2)$ group and according to the introduction to this section, we should consider the tensor product decomposition:

$$\frac{1}{2} \otimes \frac{1}{2} \otimes \frac{1}{2} = \frac{1}{2} \oplus \frac{1}{2} \oplus \frac{3}{2} \tag{77}$$

The particular way of writing H_B given in Eq. (76) suggests to compose first $\vec{S}_1$ and $\vec{S}_3$ and then, the resulting spin with $\vec{S}_2$. The result of this of this compositions is given as follows:

$$\left|\frac{3}{2}, \frac{3}{2}\right\rangle = |\uparrow\uparrow\uparrow\rangle \quad E_B = J/2 \tag{78}$$

$$\left|\frac{3}{2}, \frac{1}{2}\right\rangle = \frac{1}{\sqrt{3}}(|\uparrow\downarrow\uparrow\rangle + |\downarrow\uparrow\uparrow\rangle + |\uparrow\uparrow\downarrow\rangle) \quad E_B = J/2 \tag{79}$$

$$\left|\frac{1}{2}, \frac{1}{2}\right\rangle_1 = \frac{1}{\sqrt{2}}(|\uparrow\uparrow\uparrow\rangle - |\downarrow\uparrow\uparrow\rangle) \quad E_B = 0 \tag{80}$$

$$\left|\frac{1}{2}, \frac{1}{2}\right\rangle_0 = \frac{1}{\sqrt{6}}(2|\uparrow\downarrow\uparrow\rangle - |\downarrow\uparrow\uparrow\rangle - |\uparrow\uparrow\downarrow\rangle) \quad E_B = -J \tag{81}$$

Hence for $\Delta = 0$ we could choose the spin 1/2 irrep. with basis vectors $\left|\frac{1}{2}, \frac{1}{2}\right\rangle_0$ and $\left|\frac{1}{2}, -\frac{1}{2}\right\rangle_0$ in order to define the intertwiner operator T_0.

However, if $\Delta \neq 0$ the states (78) -(81) are not eigenstates of (76). The full rotation group is broken down to the rotation around the z-axis. The states $\left|\frac{3}{2}, \frac{1}{2}\right\rangle$ and $\left|\frac{1}{2}, \frac{1}{2}\right\rangle_1$ are mixed in the new ground state which is given by:

$$\left|+\frac{1}{2}\right\rangle = \frac{1}{\sqrt{1 + 2x^2}}\left(2\left|\frac{1}{2}, \frac{1}{2}\right\rangle_1 + \sqrt{2}x\left|\frac{3}{2}, \frac{1}{2}\right\rangle\right) \tag{82}$$

where

$$x = \frac{2(\Delta - 1)}{8 + \Delta + 3\sqrt{\Delta^2 + 8}} \tag{83}$$

and its energy is,

$$E_B = -\frac{J}{4}[\Delta + \sqrt{\Delta^2 + 8}] \tag{84}$$

along with its $\left|-\frac{1}{2}\right\rangle$ partner. This are now the two states retained in the RG method. To be more explicit, we have

$$\left|+\frac{1}{2}\right\rangle = \frac{1}{\sqrt{6(1 + 2x^2)}}[(2x + 2)|\uparrow\downarrow\uparrow\rangle + (2x - 1)|\uparrow\uparrow\downarrow\rangle + (2x - 1)|\downarrow\uparrow\uparrow\rangle] \tag{85}$$

$$|-\frac{1}{2}\rangle = -\frac{1}{\sqrt{6(1+2x^2)}}[(2x+2)|\downarrow\uparrow\downarrow\rangle+$$

$$(2x-1)|\downarrow\downarrow\uparrow\rangle + (2x-1)|\uparrow\downarrow\downarrow\rangle] \tag{86}$$

The intertwiner operator T_0 reads then,

$$T_0 = |+\frac{1}{2}\rangle\langle'\uparrow| + |-\frac{1}{2}\rangle\langle'\downarrow| \tag{87}$$

where $|\uparrow\rangle'$ and $|\downarrow\rangle'$ form a basis for the space $V' = \mathbf{C}^2$. The RG-equations for the spin operators $\vec{S}_i$ $(i = 1, 3)$ are then given by

$$T_0^\dagger \vec{S}_i^x T_0 = \xi^x \vec{S'}_i^x \quad i = 1, 3. \tag{88}$$

$$T_0^\dagger \vec{S}_i^y T_0 = \xi^y \vec{S'}_i^y \quad i = 1, 3. \tag{89}$$

$$T_0^\dagger \vec{S}_i^z T_0 = \xi^z \vec{S'}_i^z \quad i = 1, 3. \tag{90}$$

where ξ^x, etc are the renormalization factors which depend upon the anisotropy parameter by,

$$\xi^x = \xi^y := \frac{2(1+x)(1-2x)}{3(1+2x^2)} \tag{91}$$

$$\xi^z := \frac{2(1+x)^2}{3(1+2x^2)} \tag{92}$$

Observe the symmetry between the sites $i = 1$ and 3 which is a consequence of the even parity of the states (85) -(86).

The renormalized Hamiltonian can be easily obtained using Eqs.(88)-(92) and (74), and apart from and additive constant it has the same form as H, namely,

$$T_0^\dagger H_N(J, \Delta) T_0 = \frac{N}{3} e_B(J, \Delta) + H_{N/3}(J', \Delta') \tag{93}$$

where

$$J' = (\xi^x)^2 J \tag{94}$$

$$\Delta' = (\frac{\xi^z}{\xi^x})^2 \Delta \tag{95}$$

Iterating these equations we generate a family of Hamiltonians $H_{N/3^m}^{(m)}(J^{(m)}, \Delta^{(m)})$. The energy density of the ground state of H_N in the limit $N \to \infty$ is then given by,

$$\lim_{N \to \infty} \frac{E_0}{N} = e_\infty^{BRG} = \sum_{m=0}^{\infty} \frac{1}{3^{m+1}} e_B(J^{(m)}, \Delta^{(m)}) \qquad (96)$$

where initially $J^{(0)} = J$, $\Delta^{(0)} = \Delta$ and Eqs.(94)-(95) provide the flow of the coupling constants.

The analysis of Eq.(95) shows that there are 3 fixed points corresponding to the values $\Delta = 0$ (isotropic XX-model), $\Delta = 1$ (isotropic Heisenberg model) and $\Delta = \infty$ (Ising model). The properties of these fixed points are given in Table 6. The computation of e_∞^{BRG} in this case is facilitated by the fact that (96) becomes a geometric series at the fixed point. The exact results concerning the models $\Delta = 0$ and $\Delta = 1$ are extracted from references [28] and [29]. The case with $\Delta \to \infty$ is exact because the states $| \pm \frac{1}{2} \rangle$ given in (85) - (86) tend in that limit to the exact ground state $| \uparrow\downarrow\uparrow \rangle$ and $| \downarrow\uparrow\downarrow \rangle$ of the Ising model. As a matter of fact,

$$| + \frac{1}{2} \rangle \simeq_{\Delta \to \infty} | \uparrow\downarrow\uparrow \rangle - \frac{1}{\Delta} | \uparrow\uparrow\downarrow \rangle - \frac{1}{\Delta} | \downarrow\uparrow\uparrow \rangle$$

$$| - \frac{1}{2} \rangle \simeq_{\Delta \to \infty} -| \downarrow\uparrow\downarrow \rangle - \frac{1}{\Delta} | \downarrow\downarrow\uparrow \rangle - \frac{1}{\Delta} | \uparrow\downarrow\downarrow \rangle$$

The region $0 < \Delta < 1$ which flows under the RG-transformation to the XX-model is massless since both $J^{(m)}$ and $\Delta^{(m)}$ go to zero. We showed at the begining of this section that all this region is critical (a line of fixed points) and therefore massless. The RG-equations (94)-(95) are not able to detect this criticality except at the point $\Delta = 0$. Only the masslessness property is detected.

The region $\Delta > 1$ which flows to the Ising model is massive and this follows from the fact that the product $J^{(m)}\Delta^{(m)}$ goes in the limit $m \to \infty$ to a constant quantity $J^{(\infty)}\Delta^{(\infty)}$ which can be computed from Eqs. (94)-(95) and (87),

$$J^{(\infty)}\Delta^{(\infty)} = \prod_{m=0}^{\infty} \frac{4}{9} \frac{(1 + x_m)^4}{(1 + 2x_m^2)^2} \qquad (97)$$

where x_m is given by (83) with Δ replaced $\Delta^{(m)}$. This quantity gives essentially the mass gap above the ground state and also the end-to-end or LRO order (Long Range Order) given by the expectation value $|\langle \vec{S}(1) \cdot \vec{S}(N) \rangle|$ in the limit $N \to \infty$.

In summary, the properties of the Heisenberg-Ising model are qualitatively and quantitatively well described in the massive region $\Delta > 1$ while in the massless region $0 < \Delta < 1$ one predicts the massless spectrum *but no criticality at each value of* Δ. This latter fact is rather subtle and elusive. One would like to construct a RG-formalism such that the Hamiltonian $H_N(\Delta)$ would be a fixed point Hamiltonian for every value of Δ in the range from -1 to 1.

The phase transition between the two regimes is correctly predicted to happen at the value $\Delta = 1$. This is a consequence of the rotational symmetry, namely at $\Delta = 1$ the system is $SU(2)$ invariant and the RG transformation has been defined as to preserve this symmetry. When $\Delta \neq 1$ the $SU(2)$ symmetry is broken and this is reflected later on in the RG-flow of the coupling constant Δ.

IX. RG FOR HEISENBERG SPIN LADDERS

Much of the El Escorial Summer School has been devoted to the nowdays very active field known as ladders systems (spin, t-J, Hubbard ...), see [32], [33], [2]. What does the BRG method have to say on these systems? We again emphasize that this is a technically simple method which produces qualitative correct results when properly applied. Later it is possible to look for numerical accuracy using DMRG, second order RG (see appendix) or some other means.

The Hamiltonian of a Heisenberg spin ladder with n_l legs, each of length N is given by,

$$\mathcal{H}_{ladder} = \mathcal{H}_{leg} + \mathcal{H}_{rung}$$
$$\mathcal{H}_{leg} = \sum_{a=1}^{n_l} \sum_{n=1}^{N} J\mathbf{S}_a(n) \cdot \mathbf{S}_a(n+1)$$
$$\mathcal{H}_{rung} = \sum_{a=1}^{n_l-1} \sum_{n=1}^{N} J' \, \mathbf{S}_a(n) \cdot \mathbf{S}_a(n+1) \tag{98}$$

where $\mathbf{S}_a(n)$ are spin-1/2 matrices acting on the a-th leg at the position $n = 1, \ldots, N$, and J is the intraleg coupling constant while J' is the interleg exchange coupling constants, both beign positive to guarantee AF spin ladders.

We shall concentrate on the uniform Heisenberg ladders with no staggering. There are many examples that can be worked out but to be concrete we shall pick up the 3-leg ladder system [30]. In advance, what we are going to obtain is the RG-flow towards the strong coupling limit of spin ladders. This is an alternative to the determination of the RG-fow using bosonization as performed by H.J. Schulz [31], [32]. To apply the BRG we need to set up what is the block Hamiltonian H_B which we do by forming blocks of 3 sites each along every leg and located one block on top of another as in fig.7 In this fashion we are selecting a subset of couplings from the whole spin ladder Hamiltonian in (98). The remaining terms involving links with neighbouring blocks make up for the interblock Hamiltonian H_{BB}. We shall not write down explicitly the analytical expressions for H_B and H_{BB} as it is quite clear what is meant simply by looking at Fig.7.

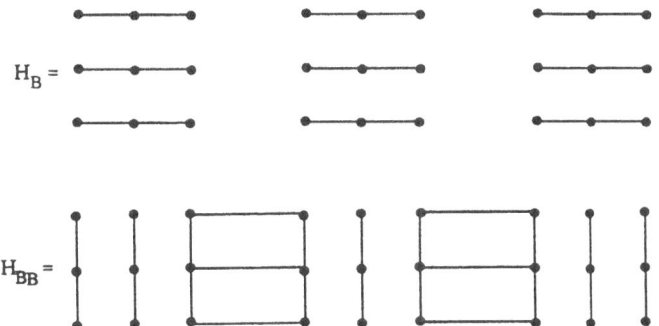

FIG. 7 Block decomposition associated to the standard BRG method applied to the uniform 3-leg Heisenberg ladder.

Now it is aparent that the standard results of the previous section are at work for spin ladders. Notice that H_B is made up of small block Hamiltonians of 3 sites as in (76) whose eigenstates and energies are already computed in (78)-(81). The renormalization process goes through all the way by truncating the block states to the lowest eigenstates, i.e., the spin doublet $|\frac{1}{2}, \pm\frac{1}{2}\rangle_0$, with energy $e_0 = -J$. In this case the embedding operator $T^{(\alpha)}$ for each block is nothing but the projector $P_0^{(\alpha)}$ onto these states; denoting

$$P_B = \prod_{\alpha=1}^{N/3} P_0^{(\alpha)} \tag{99}$$

then, to first order in J the renormalized or effective Hamiltonian acting on the states left out after the truncation is simply,

$$H_{\text{eff}} = P_B (H_B + H_{BB}) P_B \tag{100}$$

Using the embedding operator (87) we arrive at,

$$P_B H_B P_B = +\frac{N}{3} e_0 = -\frac{N}{3} J \tag{101}$$

In this case the renormalization of the block Hamiltonian gives the identity because the two states retained within each block are degenerate by rotational invariance. The renormalization of the interblock Hamiltonian is also simple if we observe that H_{BB} contains products of spin operators belonging to different blocks, say $\mathbf{S}_i^\alpha \cdot \mathbf{S}_j^\beta$ where $\alpha \neq \beta$ denote neighbouring blocks and $i, j = 1, 2, 3$ are the intrablock labels used in Eq. (98). Then, according to Eq. (99) we have,

$$P_B \mathbf{S}_i^\alpha \cdot \mathbf{S}_j^\beta P_B = P_B (P_0^{(\alpha)} \mathbf{S}_i^\alpha P_0^{(\alpha)}) (P_0^{(\beta)} \mathbf{S}_i^\beta P_0^{(\beta)}) P_B \tag{102}$$

Hence we only need to know how the spin operators renormalize within each block onto the new spin operators. By symmetry arguments, the renormalization spin factor denoted by ξ_i must be the same for the 3 components of the spin operators, i.e.,

$$(P_0^{(\alpha)} \mathbf{S}_i^\alpha P_0^{(\alpha)}) = \xi_i \mathbf{S}'_\alpha \tag{103}$$

where $\mathbf{S}'_\alpha$ denotes the spin $1/2$ operator acting on the effective spin $1/2$ subspace of the αth-block. The renormalization spin factors are known from (91)-(92) to be given by,

$$\xi_1 = \xi_3 = \frac{2}{3}$$
$$\xi_2 = -\frac{1}{3} \tag{104}$$

Now we are ready to compute the renormalization of the interblock Hamiltonian H_{BB}. According to Eqs. (102) and (103), the renormalization of the horizontal couplings between blocks of H_{BB} (see Fig. 7) is given by,

$$J \, \mathbf{S}_3^{(\alpha)} \cdot \mathbf{S}_1^{(\beta)} \longrightarrow J \, \xi_1 \xi_3 \, \mathbf{S}_a(n') \mathbf{S}_a(n'+1) = \frac{4}{9} J \, \mathbf{S}_a(n') \mathbf{S}_a(n'+1) \qquad (105)$$

while the 3 vertical couplings between two blocks are renormalized to

$$J' \, (\mathbf{S}_1^{(\alpha)} \cdot \mathbf{S}_1^{(\beta)} + \mathbf{S}_2^{(\alpha)} \cdot \mathbf{S}_2^{(\beta)} + \mathbf{S}_3^{(\alpha)} \cdot \mathbf{S}_3^{(\beta)}) \longrightarrow$$

$$J' \, (\xi_1^2 + \xi_2^2 + \xi_3^2) \, \mathbf{S}^{(\alpha)} \cdot \mathbf{S}^{(\beta)}$$
$$= J' \mathbf{S}_a(n') \cdot \mathbf{S}_{a+1}(n'+1) \qquad (106)$$

We have then obtained that the renormalized Hamiltonian (100), apart from the constant term (101), is the as the original ladder Hamiltonian, but with length $N/3$ and the following renormalization of the coupling constants,

$$J \longrightarrow \frac{4}{9} J$$
$$J' \longrightarrow J' \qquad (107)$$

Hence the ratio J'/J increases as,

$$\frac{J'}{J} \longrightarrow \frac{9}{4} \frac{J'}{J} \qquad (108)$$

after each step of the RG showing that $J'/J = \infty$ is a stable fixed point which controls the behaviour for all values of J and J', while $J'/J = 0$ is an unstable fixed point. Had we chosen blocks made up of more than 3 sites we would have obtained essentially the same result. The RG method for the simple spin chain (i.e. $n_l = 1$) where first obtained in reference [20]. According to Eq. (107) if we start in the weak coupling regime $J'/J \ll 1$, after sufficient iterations of the RG we would get an effective Hamiltonian in the strong coupling regime where we can apply the arguments of section II to derive the nature of the low lying spectrum of the theory.

X. REAL-SPACE RG APPROACH TO THE QUANTUM 2D-AF HEISENBERG MODEL

In this section we present a real-space RG treatment of the quantum two-dimensional Heisenberg antiferromagnetic model with arbitrary spin S. Most of the work using real-space RG methods has been devoted to one-dimensional problems. This is very useful because it is crucial to have an aproximate method which gives good results for both 1D and 2D problems, for it is known that mean field theory methods fail in low dimensional problems. In this regard we have recently shown that the use of quantum groups in combination of real-space methods in 1D captures the essential features exhibited by the exact solutions of models such as Heisenberg and ITF [15], [16]. Nevertheless, the main reason which has prevented the applications of the real-space RG in 2D quantum lattice Hamiltonians is the rapid growth of the number of states to be kept in a reasonable scheme of truncation of states in dimensions higher than one.

We shall be using the Block RG method in our study of the 2D Heisenberg model. This version of the RG method is suitable to achieve fully analytical treatments of interacting many-body problems. The reason for searching for complete analytical approaches as opposed to purely numerical studies relies on the necessity of having a qualitative understanding of the mechanisms responsible for the different behaviors exhibited by the Heisenberg model itself and for its connections to more complicated related Hamiltonians such as t-J and Hubbard where the understanding of the doping effects is a big issue at stake. In order for there to be a completely analytical RG treatment in 2D we need a juidicious choice of the states to be kept as we shall see [34].

Despite of some initial controversies there is by now sufficient theoretical and experimental evidence for the existence of antiferromagnetic long range order (AF LRO) in the 2d spin 1/2 Heisenberg antiferromagnet [35] (and references therein). This property has been observed in parent compounds of hight-T_c materials such as La_2CuO_4 [35].

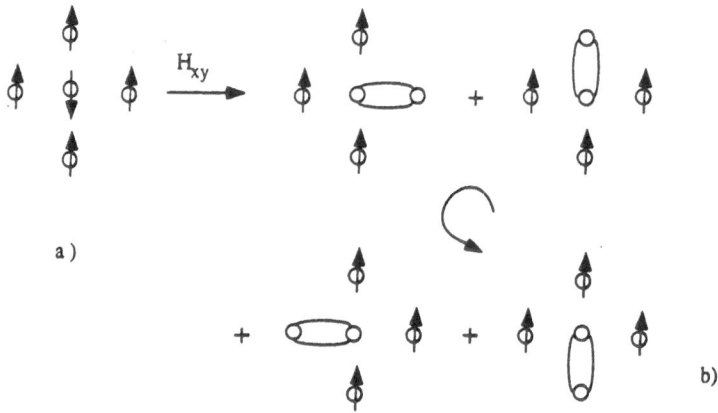

FIG. 8 a) The antiferromagnetic 5-block state. b) Formation of a rotating-valence-bond state upon applying the H_{xy} part of the Hamiltonian to the AF 5-block state. .

From a theoretical point of view this means that the strong quantum fluctuations implied by the low dimensionality and low spin do not destroy completely the Neel order, as it happens [36] in 1d. Though there is no a satisfactory physical explanation of this fact, which may be important regarding the interplay between antiferromagnetism and superconductivity upon doping. The RVB scenario originally proposed by Anderson [37,38], while yielding an appealing picture of the ground state, does not explain the presence of AF LRO. This type of order may however be incorporated a posteriori in long range RVB ansatzs of factorized form [39], with predictions similar to the ones obtained using Quantum Monte Carlo methods [40] and variational plus Lanczos techniques [41]. A class of physical systems where the RVB approach may be actually realized is in spin ladders with an even number of chains [42,43]. The previous works leave still room to investigate in more depth the interplay between the RVB scenario, or more generally "valence bond scenarios", and the AF order present in the 2d AF-magnets, described by the AF Hamiltonian $H = J \sum_{\langle i,j \rangle} \mathbf{S}_i \cdot \mathbf{S}_j$.

We have proposed a new scenario where the valence bonds, instead of resonating as in the RVB scenario, rotate around their ends under the influence of the AF background. To test this idea we propose a variational ground state in which the bonds rotate but do not resonate among themselves. We shall start by considering how the quantum fluctuations affect the classical Neel state. This is also the starting point of the spin wave theory (SW), which we would like to use for comparison of our theory. An important ingredient of our construction is the use of real-space RG techniques, which allows us to obtain exact analytical results for any value of the spin S of the model (S is integer or half-integer and in the discussion above S=1/2.) The advantage of using a real-space RG method is that one can treat in an exact manner the local quantum fluctuations of the classical Neel state. By this we mean that if we divide the square lattice into blocks of 5 sites each, as in Figs. 8,9 and 10, then the Heisenberg Hamiltonian restricted to the blocks can be solved exactly. The ground state for every block is a spin 3/2 irrep. (if S=1/2) which is obtained by forming a singlet (bond) between the spin at the center and the ones surrounding the center (Fig. 8)

FIG. 9 Artistic tesselation of the square lattice with 5-site blocks. .

According to the RG method, the spin 3/2 can be chosen as an effective spin for the renormalized lattice which now has N/5 sites. We shall show later on that the interaction between those effective spins 3/2 (or 3S more generally) is also governed by an AF-Heisenberg model with a renormalized coupling constant. Hence the RG procedure can be iterated yielding a series of effective spins which ultimately goes to its classical value, i.e. infinity! $(S \rightarrow 3S \rightarrow 3^2 S \rightarrow \infty.)$ We thus obtain in an economical and simple way the important result that the 2D AF-Heisenberg models belongs to the universality class of the 2D classical Heisenberg model. For a sigma model derivation of this result see ref. [45]. The rotating bond picture puts in correspondence various approaches to the 2D AF-Heisenberg model.

Let us begin our approach by considering the cluster of 5 spins 1/2 of Fig.8 a). The configuration showed in Fig.8 a) is the exact ground state of the Ising piece of the Heisenberg Hamiltonian, given by $H_z = J\sum_{i=1,\ldots,4} S_0^z S_i^z$, where S_0^z and S_i^z are the third component of the spin operators at the center and the ith position off the center respectively. As soon as the "transverse" Hamiltonian $H_{xy} = J\sum_{i=1,\ldots,4}(S_0^x S_i^x + S_0^y S_i^y)$ is switched on, the down-spin in the middle starts to move around the cluster, and a valence bond between the center and the remaining sites is formed in a s-wave ($l = 0$) symmetric state as shown in Fig.8 b). Other rotational states with $l \neq 0$ may appear corresponding to excitations (l being the orbital angula momentum of the bond). An alternative description of this state is given by first combining the 4 spins sourrounding the center into a spin 2 irrep, which in turn is combined with the spin 1/2 at the center yielding a spin 3/2 irrep with energy $e_0 = -3J/2$. If instead of the spin 1/2 at each site there is a spin S the previous analysis can be easily generalized as follows: the ground state of the AFH Hamiltonian of the 5-cluster has total spin 3S and is obtained by first combining all the surrounding spins into a spin 4S, which in turn becomes a spin 3S after multiplication with the spin S at the center. In a certain sense this state can be viewed as the formation of bonds between the center and its four neighbours. After applying several steps of the real-space RG, as we shall see below, new bonds are generated between sites at longer distances apart. Thus our valence-bond scenario is a type of long range valence bond state.

To study the AFH model in the entire square lattice we begin by first tesselating this plane using the cluster of Fig.8 as the fundamental cell (see Fig. 9). Notice that the centers of the 5-cluster form a new square lattice with lattice spacing $a' = \sqrt{5}a$. Given this tesselation we can apply the standard RG method of replacing clusters of spins by an effective spin [13,14]. This method has been applied for the 1d AFH model by Rabin [20] for clusters or blocks with 3 sites, obtaining a ground state energy with an error of 12%. The effective spin of every 3-block in 1d has spin 1/2. In our case, as we have discussed above, the effective spin of the 5-blocks have spin 3S and the energy per block equal to $e_0 = -JS(4S + 1)$. The effective spins $S' = 3S$ interact by means of an effective Hamitonian which to first order in perturbation theory can be derived if we know the renormalization of the spin operators $\mathbf{S}_\alpha \to \xi_\alpha \mathbf{S}', \alpha = 0, 1, \ldots, 4$.

The *renormalization spin factor* ξ_α can be shown to be given by the sum $\xi_\alpha = \frac{1}{3S} \sum_{m_0,m_1,\ldots,m_4} m_\alpha (C^{3S}_{m_0,m_1,\ldots,m_4})^2$ subject to the constraint $\sum_{\alpha=0}^4 m_\alpha = 3S$. $C^{3S}_{m_0,m_1,\ldots}$ is the CG coefficient which describes the ground state of spin 3S in terms of the 5 original spins S, whose expression is a product of 4 standard CG coefficients. The ξ_α satisfy the *sum rule* $\sum_{\alpha=0}^4 \xi_\alpha = 1$. We arrive at the following result,

$$\xi_\alpha(S) = \frac{1}{3S}\frac{6S+1}{8S+1}\frac{[(2S)!]^5}{[(8S)!]^2}$$
$$\times \sum_{m_1,\ldots,m_4} m_\alpha \frac{(4S - \sum_1^4 m_i)! \, [(4S + \sum_1^4 m_i)!]^2}{\prod_1^4 (S - m_i)! \, (S + m_i)! \, [-2S + \sum_1^4 m_i]!} \tag{109}$$

where if $\alpha = 0$ then $m_0 = 3S - \sum_1^4 m_i$. It follows that the renormalization factors for the four external spins in the 5-block are all equal $\xi_1 = \xi_2 = \xi_3 = \xi_4 \equiv \xi(S)$, while that of the central spin ξ_0 is determined by the sum rule. Amazingly enough the sum (109) can be performed in a close manner yielding,

$$\xi(S) = \frac{1}{3}\frac{S+\frac{1}{4}}{S+\frac{1}{3}} \tag{110}$$

For spin $S = \frac{1}{2}$ one obtains $\xi(\frac{1}{2}) = \frac{3}{10}$. Moreover, Eq. (110) correctly reproduces the classical limit $\lim_{S\to\infty}\xi(S) = \frac{1}{3}$ (recall $S = S^{old} = \frac{1}{3}S' = \xi_{cl}S'$). Notice also that the value for $S = \frac{1}{2}$ is already close to the classical value.

The RG-equations for the spin operators $\mathbf{S_i}$ $i = 1,2,3,4$ allows us to compute the renormalized Hamiltonian H' which turns out to be of the same form as the original AFH Hamiltonian. In fact, we arrive at the following RG-equations,

$$H'(N, S, J) = -JS(4S+1)\frac{N}{5}$$
$$+H(\tfrac{N}{5}, 3S, 3\xi^2(S)J) \tag{111a}$$

$$N' = \frac{N}{5}, \quad S' = 3S, \quad J' = 3\xi^2(S)J \tag{111b}$$

where the first contribution in Eq. (111a) comes from the energy of the blocks. As $3\xi^2(S) < 1$, the flow equation (111b) implies that the coupling constant flows to zero $J^{(n)} \overset{n\to\infty}{\to} 0$ which means that the AFH model remains *massless* for arbitrary value of the spin S. This fact allows us to compute the density of energy $e_\infty(S)$ (per site) as the following series,

$$e_\infty(S) = -\frac{1}{5}\sum_{n=0}^{\infty}\frac{1}{5^n}J^{(n)}S^{(n)}(4\times 3^n S + 1) \tag{112a}$$

$$S^{(n+1)} = 3S^{(n)}, \quad J^{(n+1)} = 3\xi^2(S^{(n)})J^{(n)} \tag{112b}$$

Using eqs. (112a) and (112b) we can compute the ground state energy of our variational RG state for any value of the spin S. In particular for S=1/2 we get the value $e_\infty = -0.5464$. This value has to be compared with the "exact" numerical result -0.6692, which is obtained using Green-function Monte Carlo methods [40], and the spin wave value which is -0.6703. The difference between our result and the Green Function MC or SW is quite big and around 0.12. To clarify the origin of this departure we have considered the semiclassical expansion of Eq. (112a) and compare it with the standard formula of Anderson and Kubo [49],

$$e_\infty^{RG} = -2S(S + \frac{0.0223}{S} + \cdots) \tag{113a}$$

$$e_\infty^{sw} = -2S(S + 0.158 + \frac{0.0062}{S} + \cdots) \tag{113b}$$

The important observation is that the term linear in S is absent in our formula (113a). The reason for this is that we are using a first order RG method for which the ground state energy follows from the formula $E_{GS} = \langle \Psi_0|H|\Psi_0 \rangle$, where $|\Psi_0\rangle$ is the variational ground state constructed by the RG method. Now it is easy to see that taking $|\Psi_0\rangle$ to be simply the Neel state one has to go to second order perturbation theory (PT) to get a linear term in S, which turns out to be given by $S/4 + 1/32 + O(1/S)$. It is clear that the "missing energy" 0.12 is due to this peculiarity of the first order PT. To remedy this one should implement the RG method with second order PT. In 1D and for S=1/2 this can be done, obtaining for the ground state energy density $e_\infty = -0.4530$ which is comparable in precision with the spin wave result -0.4647 (recall that the exact value is -0.4431.) The latter computation in 2D is much more involved but it is expected to yield a result close to the spin wave result.

In order to have a better insight into the physics of the model it is convenient to compute the staggered magnetization $M \equiv \langle \frac{1}{N} \sum_j (-1)^j S_j^z \rangle$. We have been able to obtain a closed formula for arbitrary spin S which is capable of analytical study. To this purpose, we use the RG-equality for V.E.V. $\langle \psi_0|\mathcal{O}|\psi_0 \rangle = \langle \psi_0'|\mathcal{O}'|\psi_0' \rangle$ for renormalized observables $\mathcal{O}'$ in the ground state and divide the sum in M into 5-block contributions. With the help of the renormalization spin factors we arrive at the RG-equation for the staggered magnetization,

$$M_N(S) = \frac{8\xi(S) - 1}{5} M_{N/5}(3S) \tag{114}$$

The explicit knowledge of $\xi(S)$ (109) allows us to solve this RG-equation for the staggered magnetization in the thermodynamic limit $N \to \infty$. In fact, as we know by now that the Hamiltonian renormalizes to its classical limit, we have $\lim_{S\to\infty} M(S) = S$. Defining $M(S) \equiv Sf(S)$, Eq. (114) amounts to solving the equation $f(S) = \frac{S+1/5}{S+1/3} f(3S)$ subject to the boundary condition $f(\infty) = 1$. Thus, we obtain the following formula for the staggered magnetization for arbitrary spin,

$$M(S) = S \prod_{n=0}^{\infty} \frac{S + \frac{1}{5}3^{-n}}{S + \frac{1}{3}3^{-n}} \tag{115}$$

This is a nice formula in several regards. For spin $S = \frac{1}{2}$ we get $M(\frac{1}{2}) = 0.373$ to be compared with the most accurate Quantum Monte Carlo reslult [46] which is 0.3074 (earlier numerical results were obtained with Green function QMC methods [40] and Variational Monte Carlo plus Lanczos algorithm [41]). Other approximate methods employed so far lead to values of $M(\frac{1}{2})$ such as, e.g., spin wave theory plus $1/S$-expansion to order S^{-2} gives [47,48] 0.3069 (earlier SW results were provided by Anderson and Kubo [49]), spin wave theory plus perturbation theory gives [50] 0.313, etc. Our value is close to the one found [51] with pertubation theory around the Ising model to order 4 which is 0.371. We can equally get values of the staggered magnetization for arbitrary spin. For spin S=1, our formula (115) gives 0.8454 to be compared with 0.8043 using SW to order [47] $1/S^2$ and 0.8039 obtained by Wheihong et al. [52] using series expansions.

Another interesting feature of our formula (115) is that it allows us to make a $1/S$-expansion yielding the result,

$$M_\infty^{\rm RG}(S) = S - 0.2 + 0.06\frac{1}{S} + O(1/S^2) \tag{116a}$$

$$M_\infty^{sw}(S) = S - 0.198 + O(1/S^2) \tag{116b}$$

Observe the excellent agreement between the order S^0 term in both formulas. Recall that equation (116a) is derived using first order PT. We expect that a second order RG would further lower the value of $M(S)$, in agreement with the numerical result.

In summary we can claim that the rotating-valence-bond scenario gives a consistent and suggestive picture of the ground state of the 2D AF-Heisenberg model: the quantum fluctuations of the Neel state consist of rotating bonds which appear at all scales corresponding to effective spins which renormalize towards the classical value.

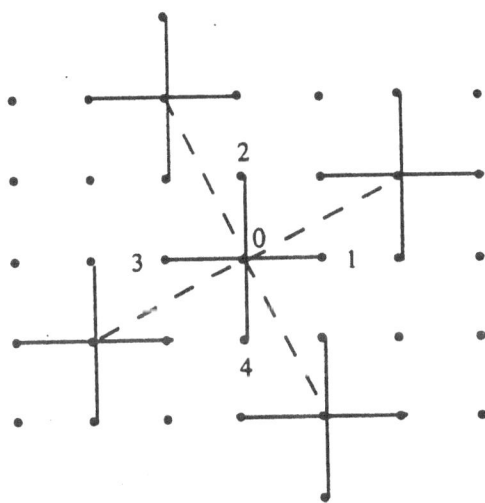

FIG. 10 The two-dimensional square lattice tesselated by the 5-block. Dahed lines are nearest-neighbours in the renormalized lattice. .

XI. APPENDIX: SECOND ORDER FORMALISM FOR THE STANDARD BRG METHOD

The modern fashion to include correlations between blocks in the real-space RG is the DMRG [1]. Nevertheless, there is an old way to include those correlations which we believ has been overlooked in the past. It amounts to include a second order contribution to the BRG. Recall that the standard BRG in previous sections is a first order method from the point of view of Perturbation Theory (P.T.), i.e., the interblock Hamiltonian H_{BB} is treated perturbatively in first order. We can extend this treatment [30] to second order P.T. in the usual fashion and thus arrive to an effective Hamiltonian given by:

$$H_{eff} = P_B[H_B + H_{BB} + H_{BB}(1 - P_B)\frac{1}{E_B - H_B}(1 - P_B)H_{BB}]P_B \qquad (117)$$

where E_B is the ground state energy of the block and P_B denotes the projector onto te ground state of the block. This is nothing but the intertwiner operator of section 2.

As a matter of illustration, we shall work out this formalism for the isotropic AF Heisenberg model in 1D using the 3-site block BRG explained in section 8. Moreover, we restrict to spin 1/2. Denote each block with an index α. Thus, the Hilbert space for each block $\mathcal{H}^\alpha$ is decomposed into 1/2, 1/2 and 3/2-spin subspaces,

$$\mathcal{H}^{(\alpha)} = \mathcal{H}_0^{(\alpha)}(1/2) \oplus \mathcal{H}_1^{(\alpha)}(1/2) \oplus \mathcal{H}_2^{(\alpha)}(3/2) \qquad (118)$$

Correspondingly, we introduce 3 projectors onto those subspaces,

$$P_0^{(\alpha)} + P_1^{(\alpha)} + P_2^{(\alpha)} = \mathbf{1}^{(\alpha)} \qquad (119)$$

They satisfy the following properties that we will be useful,

$$P_B = \prod_{\alpha=1}^{N'} P_0^{(\alpha)} \qquad (120a)$$

$$1 - P_B = 1 - \prod_{\alpha=1}^{N'} P_0^{(\alpha)} \neq \prod_{\alpha=1}^{N'}(1 - P_0^{(\alpha)}) \qquad (120b)$$

$$(P_m^{(\alpha)})^2 = P_m^{(\alpha)}, \quad P_m^{(\alpha)} P_{m'}^{(\alpha)} = 0, \quad m \neq m' \qquad (120c)$$

$$P_m^{(\alpha)} P_{m'}^{(\beta)} = P_{m'}^{(\beta)} P_m^{(\alpha)} \qquad (120d)$$

where $N' = N/3$ and m indicates the site in each block.
With the help of these properties, the second order contribution to H_{eff} in (117), denoted by $H_{eff}^{(2)}$, can be given the following form containing 3 types of terms:

$$H_{eff}^{(2)} = P_B \sum_{\alpha=1}^{N'} \sum_{m_\alpha \neq 0} \frac{1}{e_0 - e_{m_\alpha}}$$

$$\{[(P_0^{(\alpha-1)}\mathbf{S}_3^{(\alpha-1)}P_0^{(\alpha-1)}) \cdot (P_0^{(\alpha)}\mathbf{S}_1^{(\alpha)}P_{m_\alpha}^{(\alpha)})]$$

$$[(P_{m_\alpha}^{(\alpha)} \mathbf{S}_3^{(\alpha)} P_0^{(\alpha)}) \cdot (P_0^{(\alpha+1)} \mathbf{S}_1^{(\alpha+1)} P_0^{(\alpha+1)}) +$$

$$[(P_0^{(\alpha)} \mathbf{S}_3^{(\alpha)} P_{m_\alpha}^{(\alpha)}) \cdot (P_0^{(\alpha+1)} \mathbf{S}_1^{(\alpha+1)} P_0^{(\alpha+1)})]$$

$$[(P_0^{(\alpha-1)} \mathbf{S}_3^{(\alpha-1)} P_0^{(\alpha-1)}) \cdot (P_{m_\alpha}^{(\alpha)} \mathbf{S}_3^{(\alpha)} P_0^{(\alpha)})]\} P_B +$$

$$P_B \sum_{\alpha=1}^{N'} \sum_{(m_\alpha, m_{\alpha+1}) \neq (0,0)} \frac{1}{2e_0 - e_{m_\alpha} - e_{m_{\alpha+1}}}$$

$$[(P_0^{(\alpha)} \mathbf{S}_3^{(\alpha)} P_{m_\alpha}^{(\alpha)}) \cdot (P_0^{(\alpha+1)} \mathbf{S}_1^{(\alpha+1)} P_{m_{\alpha+1}}^{(\alpha+1)})]$$

$$[(P_{m_\alpha}^{(\alpha)} \mathbf{S}_3^{(\alpha)} P_0^{(\alpha)}) \cdot (P_{m_{\alpha+1}}^{(\alpha+1)} \mathbf{S}_1^{(\alpha+1)} P_0^{(\alpha+1)})] P_B \tag{121}$$

In order to work out this expression (121) towards a manageable result, we need to perform a renormalization of the spin operators both in both subspaces of spin-1/2 (recall that in sect.8 we did it only for the lowest energy spin 1/2.)
Let us introduce the following notation for the 4 states of spin 1/2:

$$|m, \beta\rangle \quad \text{with} \quad m = \pm 1/2, \beta = 0, 1 \tag{122}$$

Denote by $\mathbf{S}'$ the effective spin-1/2 coming out of the block renormalization. Then,

$$\langle m, \beta | \mathbf{S}_i | m', \beta' \rangle = \langle m | \mathbf{S}' | m' \rangle (\rho_i)_{\beta,\beta'} \tag{123}$$

with the ρ-matrices given by,

$$\rho_1 = \begin{pmatrix} 2/3 & -1/\sqrt{3} \\ -1/\sqrt{3} & 0 \end{pmatrix}, \quad \rho_2 = \begin{pmatrix} -1/3 & 0 \\ 0 & 1 \end{pmatrix}, \quad \rho_3 = \begin{pmatrix} 2/3 & 1/\sqrt{3} \\ 1/\sqrt{3} & 0 \end{pmatrix} \tag{124}$$

Thus, the spin renormalization that we were searching for is summarized in

$$P_\beta \mathbf{S}_i P_{\beta'} = \mathbf{S}'(\rho_i)_{\beta,\beta'} \tag{125}$$

Namely,

$$P_0 \mathbf{S}_1 P_0 = P_0 \mathbf{S}_3 P_0 \equiv (\xi^{(0)} = \frac{2}{3}) \mathbf{S}' \tag{126a}$$

$$P_0 \mathbf{S}_1 P_1 = P_1 \mathbf{S}_1 P_0 \equiv (\xi_1^{(0)} = \frac{-1}{\sqrt{3}}) \mathbf{S}' \tag{126b}$$

$$P_0 \mathbf{S}_3 P_1 = P_1 \mathbf{S}_3 P_0 \equiv (\xi_3^{(0)} = \frac{1}{\sqrt{3}}) \mathbf{S}' \tag{126c}$$

Upon substitution of these expressions in (121) we are led to the renormalization of the Hamiltonian:

$$H_{eff}^{(2)} = \sum_{\alpha=1}^{N'} [d^{(2)} + J_1^{(2)} \mathbf{S}_\alpha \cdot \mathbf{S}_{\alpha+1} + J_2 \mathbf{S}_\alpha \cdot \mathbf{S}_{\alpha+2}] \tag{127}$$

$$H_{eff}^{(0+1)} = \sum_{\alpha=1}^{N'} [d^{(0)} + J_1^{(1)} \mathbf{S}_\alpha \cdot \mathbf{S}_{\alpha+1}] \tag{128}$$

with the following numerical values,

$$\begin{aligned} d^{(0)} &= -1, & J_1^{(1)} &= \tfrac{4}{9} = 0.44 \\ d^{(2)} &= -0.104861, & J_1^{(2)} &= \tfrac{211}{1620} = 0.130247, & J_2 &= \tfrac{10}{243} = 0.0411523 \end{aligned} \tag{129}$$

Altogether, we end up with the following effective Hamiltonian in which the second order formalism employed shows up as a nearest-neighbour coupling J_2,

$$H_{eff} = \sum_{\alpha=1}^{N'} [d + J_1 \mathbf{S}_\alpha \cdot \mathbf{S}_{\alpha+1} + J_2 \mathbf{S}_\alpha \cdot \mathbf{S}_{\alpha+2}] \tag{130}$$

with,

$$d = -1.104861, \quad J_1 = 0.574691, \quad J_2 = 0.0411523 \tag{131}$$

We can now iterate this RG procedure as usual to obtain the RG-flow equations for the two coupling constants J_1 and J_2:

$$\begin{aligned} J_1^{(m+1)} &= a J_1^{(m)} - b J_2^{(m)} \\ J_2^{(m+1)} &= c J_2^{(m)} \end{aligned} \tag{132}$$

$$a = 0.57491 \quad b = 0.44444 \quad c = 0.041152$$

The fixed points of these RG-eqs. are simply,

$$\left(\frac{J_2}{J_1}\right)_c = \frac{1}{2}[\frac{a}{b} \pm \sqrt{(\frac{a}{b})^2 - 4(\frac{c}{b})}] \tag{133}$$

Upon iteration the system flows towards the smallest fixed point,

$$\left(\frac{J_2}{J_1}\right)_c = 0.076084 \tag{134}$$

This happens to be an understimation of the numerical value.

Finally, we get a series expressing the ground state energy to second order in RG,

$$e_\infty = \sum_{m=0}^{\infty} \frac{1}{3^{m+1}} [-\gamma_1 J_1^{(m)} + \gamma_2 J_2^{(m)}] \tag{135}$$

with $\gamma_1 = -1.104861$, $\gamma_2 = 0.25$.

The above sum can be computed exactly by introducing generating functions $J_i(x) \equiv \sum_{m=0}^{\infty} x^m J_i^{(m)}$, $i = 1, 2$ and using the RG-eqs. (132),

$$J_1(x) = \frac{1}{1 - ax + bcx^2}, \quad J_2(x) = \frac{cx}{1 - ax + bcx^2} \tag{136}$$

Thus,

$$e_\infty = -\frac{1}{3} \frac{\gamma_1 - \gamma_2 \frac{c}{3}}{1 - \frac{a}{3} + \frac{bc}{9}} \tag{137}$$

and substituting the values of γ_1 and γ_2, we get

$$e_\infty^{(2RG)} = -0.453002 \tag{138}$$

This is to be compared with the exact value $e_\infty^{(exact)} = -0.4431$ which amounts to a 2.2 % error. This results improves even the spin wave result $e_\infty^{(sw)} = -S^2 - 0.36338s - 0.033011 = -0.4647$ which is a 4 % off the exact value. Recall that $e_\infty^{(1RG)} = -0.391304$ (11.6 % .)

Acknowledgements The work presented in these notes has been done in collaboration with Germán Sierra and I have enjoyed many conversations on real-space RG methods and related topics with him. Sections on the CBRG method are also in collaboration with J. Rodriguez-Laguna. I also thank R. Shankar for many comments on renormalization group during his visit at CSIC (Madrid). I acknowledge many useful discussions with the lecturers at the El Escorial Summer School, specially Steve White, T. Nishino, A. Sandvik.
This work has been partially supported in part by CICYT under contract AEN93-0776.

---

[1] S.R. White, *Phys. Rev. Lett.* **69**, 2863 (1992); *Phys. Rev. B* **48**, 10345 (1993).

[2] S.R. White, Proceedings of El Escorial Summer School 1996, and references therein.

[3] S.R. White, Preprint cond-mat/9604129.

[4] T. Nishino, "Density Matrix and Renormalization for Classical Lattice Models", proceedings in this volume.

[5] J. Pérez-Conde, Proceedings of El Escorial Summer School 1996, and references therein.

[6] K.R. Wilson, *Rev. Mod. Phys.* **47**, 773 (1975).

[7] P.W. Anderson, *J. Phys. C* **3**, 2436 (1970).

[8] Hirsch, J. *1983, Phys. Rev. B* **28**, *4059, 1985, Phys. Rev. B* **31**, *4403*; Hirsch, J. and Lin, H.Q., *1988, Phys. Rev. B* **37**, *5070*.

[9] S.D. Drell, M. Weinstein, S. Yankielowicz, *Phys. Rev. D* **16**, 1769 (1977).

[10] R. Jullien, P. Pfeuty, J.N. Fields, S. Doniach, *Phys. Rev. B* **18**, 3568 (1978).

[11] Hirsch, J. *1980, Phys. Rev. B* **22**, *5259*.

[12] M.A. Martín-Delgado and G. Sierra, "Analytic Formulations of the Density Matrix Renormalization Group", Int. J. Mod. Phys. **A11** 3145, (1996).

[13] "Real-Space Renormalization", editors Burkhardt, T.W. and van Leeuwen, J.M.J., series topics in Current Physics **30**, Springer-Verlag 1982.

[14] J. González, M.A. Martín-Delgado, G. Sierra, A.H. Vozmediano, *Quantum Electron Liquids and High-T_c Superconductivity*, Lecture Notes in Physics, Monographs vol. **38**, Springer-Verlag 1995.

[15] M.A. Martín-Delgado and G. Sierra, "Real Space Renormalization Group Methods and Quantum Groups". Phys. Rev. Lett. **76**,1146 (1996).

[16] M.A. Martín-Delgado and G. Sierra, in "From Field Theory to Quantum Groups". World Scientific Publishers 1996.

[17] S.R. White, R.M. Noack, *Phys. Rev. Lett.* **68**, 3487 (1992).

[18] Wilson, K.G., *1986, unpublished informal talk*.

[19] V. Karimipour and A. Langari, "A Modified Quantum Renormalization Group for the XXZ Spin Chain", preprint 1996, private communication.

[20] J.M. Rabin, *Phys. Rev. B* **21**, 2027 (1980).

[21] M.A. Martín-Delgado and G. Sierra, "The Role of Boundary Conditions in the Real-Space Renormalization Group". Phys. Lett. **B364** 41, (1995).

[22] M.A. Martín-Delgado, J. Rodriguez-Laguna and G. Sierra, "The Correlated Block Renormalization Group". Nucl. Phys. **B473** 685, (1996).

[23] A. Fernández-Pacheco, *Phys. Rev. D* **19**, 3173 (1979).

[24] T. Nishino, "Density Matrix Renormalization Group Method for 2D Classical Models", J. Phys. Soc. Jpn. **64** (1995) 3958-3961.

[25] T. Nishino and K. Okunishi, "Corner Transfer Matrix Renormalization Group Method", J. Phys. Soc. Jpn. **65** (1996) 891-894. Cond-mat/9507087.

[26] S. Ostlund and S. Rommer, "Thermodynamic limit of the density matrix renormalization for the spin-1 Hesisenberg chain". Phys. Rev. Lett. **75**, 3537 (1996).

[27] T. Nishino and K. Okunishi, "Product Wave Function Renormalization Group". J. Phys. Soc. Jpn. **64** (1995) 4084-4087. Cond-mat/9510004.

[28] Lieb, E., Schultz, T. and Mattis, D. *1961, Ann. Phys.* **16**, *407*

[29] Orbach, R., *1958, Phys. Rev.* **112**, *309*

[30] M.A. Martín-Delgado and G. Sierra, unpublished work 1996.

[31] H. J. Schulz, Phys. Rev. B **34**, 6372 (1986).

[32] H.J. Schulz, Proceedings of El Escorial Summer School 1996, and references therein.

[33] G. Sierra, Proceedings of El Escorial Summer School 1996, and references therein.

[34] G. Sierra and M.A. Martín-Delgado, "Real Space Renormalization Group Approach to the 2D Antiferromagnetic Heisenberg Model", preprint (1996).

[35] E. Manousakis, Rev. Mod. Phys. **63**, 1 (1991).

[36] H.A. Bethe, Z. Phys. **71**, 205 (1931).

[37] P.W. Anderson, Science **235**, 1196 (1987).

[38] S.A. Kivelson, D.S. Rokhsar and J.P. Sethna, *Phys. Rev. B* **35**, 8865 (1987).

[39] S. Liang, B. Doucot and P.W. Anderson, *Phys. Rev. Lett.* **61**, 365 (1988).

[40] J. Carlson, *Phys. Rev. B* **40**, 846 (1989).

[41] Hebb and T. M. Rice, *Z. Phys. B* **90**, 73 (1993).

[42] S. White, R. Noack and D. Scalapino, *Phys. Rev. Lett.* **73**, 886 (1994).

[43] E. Dagotto and T.M. Rice, *Science* **271**, 618 (1996).

[44] P. Fazekas and P.W. Anderson, *Phil. Mag.* **30**, 432 (1974)

[45] S. Chakravarty, B.I. Halperin and D.R. Nelson, *Phys. Rev. Lett.* **60**, 1057 (1988).

[46] Wise and Ying, *Z. Phys. B* **93**, 147 (1994).

[47] C.J. Hamer, Z. Weihong and P. Arndt, *Phys. Rev. B* **46**, 6276 (1992).

[48] C.M. Canali and M. Vallin, *Phys. Rev. B* **48**, 3264 (1992).

[49] P.W. Anderson, *Phys. Rev. B* **6**, 694 (1952); R. Kubo, *Phys. Rev.* **87**, 568 (1952).

[50] D. Huse, *Phys. Rev. B* **37**, 2380 (1988).

[51] M. Parrinello and T. Arai, *Phys. Rev. B* **10**, 265 (1974).

[52] Z. Weihong, J. Oitmaa and C.J. Hamer, *Phys. Rev. B* **43**, 8321 (1991).

m	N=12 2^m	$E_1^{(exact)}(N)$	$E_1^{(CBRG)}(N)$	$E_1^{(DMRG)}(N)$
0	12	6.8148×10^{-2}	6.8148×10^{-2}	6.8148×10^{-2}
1	24	1.7110×10^{-2}	1.7375×10^{-2}	1.7110×10^{-2}
2	48	4.2826×10^{-3}	4.4694×10^{-3}	4.2826×10^{-3}
3	96	1.0708×10^{-3}	1.1515×10^{-3}	1.0708×10^{-3}
4	192	2.6772×10^{-4}	2.9681×10^{-4}	2.6772×10^{-4}
5	384	6.6932×10^{-5}	7.6552×10^{-5}	6.6932×10^{-5}
6	768	1.6733×10^{-5}	1.9752×10^{-5}	1.6733×10^{-5}
	$\gg 1$	π^2/N^2	$9.8080/N^2$	$9.8696/N^2$

TABLE I. Exact and new CBRG values of the first excited state for the 1D Tight-Binding Model with Free-Free BC's. DMRG values are also given.

Energies	Exact	CBRG	DMRG
E_0	0	1.1340×10^{-14}	1.0×10^{-6}
E_1	1.6733×10^{-5}	1.9752×10^{-5}	1.6733×10^{-5}
E_2	6.6932×10^{-5}	7.6552×10^{-5}	6.6932×10^{-5}
E_3	1.5060×10^{-4}	1.8041×10^{-5}	1.5060×10^{-4}
E_4	2.6772×10^{-4}	2.9681×10^{-4}	2.6772×10^{-4}
E_5	4.1831×10^{-4}	5.1078×10^{-4}	4.1831×10^{-4}

TABLE II. Exact and CBRG Values of Low Lying States for the 1D Tight-Binding Model for a chain of $N = 12 \times 2^6 = 768$ sites with Free-Free BC's. DMRG values are also given.

Energies	Exact	Standard BRG	CBRG
E_0	1.7754×10^{-5}	1.5771×10^{-2}	1.8409×10^{-5}
E_1	1.5043×10^{-4}	4.2679×10^{-2}	1.6655×10^{-4}
E_2	4.1761×10^{-4}	4.2794×10^{-2}	4.6408×10^{-4}
E_3	8.1831×10^{-4}	4.3053×10^{-2}	9.1450×10^{-4}
E_4	1.3520×10^{-3}	4.3173×10^{-2}	1.5179×10^{-3}
E_5	2.0196×10^{-3}	4.4288×10^{-2}	2.2852×10^{-3}

TABLE III. Exact, Standard RG and CBRG Values of Low Lying States for the 1D Tight-Binding Model for a chain of $N = 12 \times 2^5 = 384$ sites with Free-Fixed BC's.

Energies	Exact	Standard BRG	CBRG	DMRG
E_0	6.6585×10^{-5}	5.8116×10^{-2}	7.0843×10^{-5}	6.7×10^{-5}
E_1	2.6633×10^{-4}	5.8155×10^{-2}	2.9403×10^{-4}	2.66×10^{-4}
E_2	5.9924×10^{-4}	5.8268×10^{-2}	6.3690×10^{-4}	5.99×10^{-4}
E_3	1.0653×10^{-3}	5.8470×10^{-2}	1.2289×10^{-3}	1.065×10^{-3}
E_4	1.6644×10^{-3}	5.8717×10^{-2}	1.7707×10^{-3}	1.664×10^{-3}
E_5	2.3966×10^{-3}	5.9106×10^{-2}	2.7311×10^{-3}	2.397×10^{-3}

TABLE IV. Exact, Standard RG and CBRG Values of Low Lying States for the 1D Tight-Binding Model for a chain of $N = 12 \times 2^5 = 384$ sites with Fixed-Fixed BC's. DMRG values are also given.

Energies	Exact	CBRG
E_0	0	9.6114×10^{-35}
E_1	1.5056×10^{-4}	1.9390×10^{-4}
E_2	1.5056×10^{-4}	1.9390×10^{-4}
E_3	3.0012×10^{-4}	3.8781×10^{-4}

TABLE V. Exact and CBRG Values of Low Lying States for the 2D Tight-Binding Model for a lattice of $N = 4 \times 4 \times 4^6 = 65536$ sites with Free BC's.

TABLE VI. Fixed Points of the Anisotropic AF-Heisenberg Model

Δ	0	1	∞
R_∞^{BRG}	-0.2828	-0.3913	$-\frac{1}{4}\Delta$
e_∞^{exact}	-0.3183	-0.4431	$-\frac{1}{4}\Delta$
$\frac{e^B - e^{exact}}{e^{exact}} \times 100$	11%	12%	0

A Critical View of the Real-Space Renormalization Group Method Applied to the Hubbard Model

J. Pérez-Conde
Department of Physics
Universidad Pública de Navarra
E-31006 Pamplona, Spain

ABSTRACT

We review the real-space renormalization group method applied to the Hubbard model from one to two dimensions. It is also shown how to avoid the proliferation of new terms during the process if the shape of the blocks and the symmetries of the kept states are chosen conveniently. We characterize the ground state by its energy per site, the gap of charge excitations and the double occupancy. The system shows an insulating behavior at half-filling for all positive values of the interaction parameter, U, in one dimension as well as in two dimensions if the hopping is isotropic ($t_y = t_x$). If the hopping anisotropy, $\alpha = t_y/t_x$, is different from the unity and zero the ground state is a conductor up to a critical value $(U/t)_{c_1}$, which depends on α. The Insulating one-dimensional behavior is recovered at $\alpha = 0$.

1 Introduction

The Hubbard model is thought to be appropriate to describe phenomena such as itinerant magnetism [1], metal-insulator transitions [2] and high-T_c superconductors [3] It consists of two parts: a tight-binding band H_t, and an interaction term H_u,

$$H = - \sum_{<i,j>,\sigma} t_{ij} c_{i\sigma}^+ c_{j\sigma} + U \sum_i c_{i\downarrow}^+ c_{i\downarrow} c_{i\uparrow}^+ c_{i\uparrow} = H_t + H_u \tag{1}$$

where the $<>$ means hopping restricted to nearest neighbors and $c_{i,\sigma}^+$ represents a $s-$like Wannier orbital. The model shows several symmetries if the lattice belongs to the bipartite type [4]. The Schrödinger equation corresponding to (1) is in general very difficult to solve even if we are only interested in the ground state description. The exact solution is known, however, in the one dimensional case [5] as well as some of its ground state properties [6, 8].

It seems then reasonable to develop a method with a minimum of constraints able to describe the different phases of the Hubbard Hamiltonian in any dimension. This is the aim of the real-space renormalization group method which we present here. The renormalization group techniques have had their main success in the analysis of the Kondo problem by Wilson [9], who applied the renormalization procedure in the k-space. Afterwards, S. Jafarey, R. Pearson, B. Stockley and D.J. Scalapino [10] developed the same idea in the real-space. The first published paper was, however, by Drell et al. [11] about the iterative construction of the ground state of the $\lambda\varphi^4$ theory. Later, the same ideas were applied to one and two dimensional spin systems and finally to the one dimensional Hubbard model [12, 13]. A pedagogical history of the real space renormalization can be found in [14] and [15]. Recently, another approach which is very accurate for finite systems has been proposed by White [16]. The present paper is devoted to the renormalization technique applied to a fermion system in two dimensions. In Sec. 2 we briefly discuss the symmetries of the model which will be used later. Sec. 3 describes the block method and the renormalization of the more important quantities: Hamiltonian, double occupancy, etc.. Finally, Sect. 4 contains an account of the main results.

2   Symmetries

There are three global symmetries of H independent of the lattice: the total number operator $N_e = \sum_{i,\sigma} c_{i\sigma}^+ c_{i\sigma}$, the projection of the total spin $S_z = \sum_{i,\sigma} \sigma c_{i\sigma}^+ c_{i\sigma}$ and the total spin $S^2 = S_z^2 + 1/2(S^+ S^- + S^- S^+)$, where $S^+ = \sum_i c_{i\uparrow}^+ c_{i\downarrow}$ and $S^- = (S^+)^-$. The rest of symmetries are lattice-dependent and even if most of them remain in three dimensions we shall focus our attention on the two dimensional square lattice.

The first symmetry pointed out in the literature is the invariance of the energy spectra under the transformation $t \longrightarrow -t$ [5], which works for many other quantities as well, like the double occupancy, the local moment, the effective hopping, etc. Another interesting symmetry of the Hubbard model is the pseudospin Z. It was first discovered by Castellani et al. [17], analyzed later by Nowak [18] and applied by others [19, 20] in different situations. It consists in a generalization of the spin algebra $SU(2)$ to the charge space [21]. This generalization is in principle not possible for an extended system. For the Hubbard model one can write however,

$$Z^+ = \sum_{k,k'} d_{k\downarrow}^+ d_{k'\uparrow}^+ \delta(\varepsilon(k) + \varepsilon(k') - y) \tag{2}$$

where the $d_{k\sigma}$ are the operators which diagonalize the band term

$$H_t = \sum_\sigma \varepsilon(\mathbf{k}) d_{k\sigma}^+ d_{k\sigma} \quad , \quad \varepsilon(\mathbf{k}) = -2(t_x \cos k_x + t_y \cos k_y) \tag{3}$$

¿From (2) and (3) if we choose $g = 0$ and k' so that

$$k' = k + \pi \tag{4}$$

then we obtain

$$\begin{aligned}
[H_t, Z^+] &= 0 \\
[H_u, Z^+] &= U Z^+ \\
[H, Z^+ Z^-] &= 0
\end{aligned}$$

$$\tag{5}$$
$$\tag{6}$$

It is also convenient to write down the explicit form of Z^+ in terms of the Wannier operators $c_{i\sigma}$ in order to get a real-space insight into this pseudospin operator,

$$Z^+ = \sum_i (-1)^i c_{i\downarrow}^+ c_{i\uparrow}^+$$

The Casimir Z^+Z^- is therefore another symmetry of the Hamiltonian (1). It should be noted that (4) is only possible in an $N_s = L \times L$ lattice if L is even(odd) with periodic(free) boundary conditions. Finally, we shall use later the spatial symmetry to construct explicitly the Hamiltonian. In the case of the square lattice and an anisotropic hopping $t_y \neq t_x$, H is invariant under the C_{2v} point group [22]. This additional symmetry enables us to reduce by four the dimension of the greatest subspace defined by N_e and S_z (see also Sect. 3).

3   The block method

Condensed matter physicists are mostly interested in the low energy properties of the system. It is therefore reasonable to integrate out the short distance or high energy degrees of freedom. The real-space renormalization group has been developed following this idea [11, 12, 13]. The actual procedure is qualitatively based on Kadanoff's decimation. We present below an outline of the standard procedure as well as some improvements we have performed in the understanding of this technique. We can then, following the Kadanoff's idea, divide the lattice into blocks of size $(as)^d$, where a is the lattice parameter, d is the dimension of the space and $s > 1$ is the size factor. Afterwards, we average over this block the quantities we are dealing with: the Hamiltonian, the double occupation, etc. In doing so we have neglected the fluctuations down to a size sa. This procedure can be repeated until we reach convergence, the so-called fixed point in the parameter space (t, U).

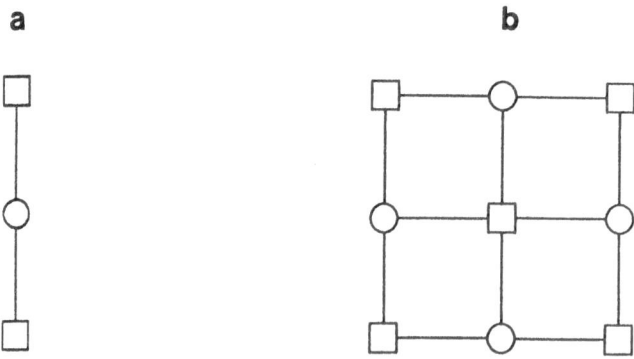

Figure 1 Adequate blocks to carry out the real-space renormalization group procedure in $1d$ (a) and $2d$ (b).

In order to clarify this process we can visualize it in one and two dimensions where we have chosen blocks with $s = 3$ (see Fig. 1). We assimilate then each block to a point belonging to a new lattice with a parameter equal to $3a$. So far, we don't know what the Hamiltonian becomes. To have a first insight we label the parameters of the Hamiltonian with a for the original lattice and with $3a$ for the transformed one, there will be then some relations between the initial (t_a, U_a) and the renormalized, (t_{3a}, U_{3a}, J_{3a}),

$$t_{3a} = \varphi_1(t_a, U_a), \quad U_{3a} = \varphi_2(t_a, U_a), \quad J_{3a} = \varphi_3(t_a, U_a), \tag{7}$$

where J_{3a} represents the possible couplings generated by the process. Furthermore, we are looking for a transformation that conserves the fermionic character of the second quantization operators $c_{i\sigma}$ and, in addition, avoids the proliferation of couplings, i.e. $J_{3a} = 0$.

We shall focus our attention on the transformation of the Hamiltonian. The expression (1) can be separated into two terms,

$$H = \sum_\alpha H_{b\alpha} + \sum_{<\alpha,\beta>} V_{\alpha\beta} = H_o + V \tag{8}$$

where α is a block label, $H_{b\alpha}$ is the intra-block Hamiltonian and V stands for the hopping part between the sites on the border of contiguous blocks. Let us assume that we know how to solve the $H_{b\alpha}$'s and obtain all its eigenvectors $|\phi_{\{i\}\alpha}\rangle$ exactly. We can then formally rewrite H in this basis

$$H = \sum_{\{\psi,\psi'\}} |\psi\rangle\langle\psi|H|\psi'\rangle\langle\psi'| \tag{9}$$

where $|\psi\rangle = \prod_\alpha |\phi_{\{i\}\alpha}\rangle$.

Despite the tautological appearance of (9) it is useful to indicate the first approach of the method: we shall keep four of the eigenvectors in each block from the 4^{N_s} ($N_s = s^d$) which compose the total Hilbert space. This choice allows us to reduce drastically the number of configurations as well as to conserve the form of the Hamiltonian in the truncated basis. We shall restrict ourselves to the situations where the intra-block term reproduces the structure of the operator $n_\downarrow n_\uparrow$. Each block will be then represented by four states $|\phi_{\{o\}\alpha}\rangle$ that we shall call $|0'\rangle$, $|\downarrow'\rangle$, $|\uparrow'\rangle$ and $|\downarrow\uparrow'\rangle$ by analogy with the original sites. The explicit choice of these states will be done below.

We want now to project H onto the subspace spanned by $|\phi_{\{o\}\alpha}\rangle$'s. If we call P_o the projector on to this subspace,

$$P_o = \sum_{\{\psi_o\}} |\psi_o\rangle\langle\psi_o| \quad , \quad |\psi_o\rangle = \prod_\alpha |\phi_{o\alpha}\rangle \tag{10}$$

We are interested in the explicit expression of the effective Hamiltonian,

$$H' = P_o P H P P_o, \tag{11}$$

where P is the perturbed projector of the truncated basis (the perturbation is V). It is now straightforward to get [23]

$$H' = \sum_{\psi_o, \psi'_o} |\psi_o\rangle \left\{ E_{\psi_o} \delta_{\psi_o \psi'_o} + \langle \psi_o | V | \psi'_o \rangle + \langle \psi_o | \frac{V(1 - P_o)V}{E - H_o} | \psi'_o \rangle + \cdots \right\} \langle \psi'_o | \quad (12)$$

We have already mentioned that we want to avoid the proliferation of couplings. An inspection of (12) indicates that the two first terms in the right hand side will reproduce "roughly" the original model, and the third one will introduce some new couplings like spin-spin interactions and the hopping between next-nearest neighbors blocks. Hence, we keep the perturbative expression of H' up to the first order in V.

We want now to express the effective Hamiltonian, H', in terms of some second quantization operators $c'_{\alpha,\sigma}$ in order to obtain a Hamiltonian as similar as possible to the original one. We choose then the two first eigenstates,

$$|0'\rangle \equiv |N_e = N_s - 1, s = s_1, S_z = -s_1, z_1, \Sigma_1, E_1\rangle$$
$$|\downarrow'\rangle \equiv |N_e = N_s, s = s_1 + \tfrac{1}{2}, S_z = -s_1 - \tfrac{1}{2}, z_1 - \tfrac{1}{2}, \Sigma_2, E_2\rangle, \quad (13)$$

where N_e is the number of electrons, s and z the total spin pseudospin, and Σ_i stands for the irreducible representations of the spatial group C_{2v} [22]. The symmetries which are not explicitly given in (13) will be provided by the conditions that we shall impose later. We complete the four states by applying the raising spin and pseudospin operators on $|0'\rangle$ and $|\downarrow'\rangle$,

$$|\uparrow'\rangle \equiv \frac{1}{\sqrt{2(s_1 + 1/2)}} S_\alpha^+ |\downarrow'\rangle, \quad E = E_2$$
$$|\downarrow\uparrow'\rangle \equiv \frac{1}{\sqrt{2(z_1 + 1/2)}} Z_\alpha^+ |0'\rangle, \quad E = E_1 + U, \quad (14)$$

where α indicates that S^+ and Z^+ are defined on the α block.

The states $|\uparrow'\rangle$ and $|\downarrow\uparrow'\rangle$ are again eigenfunctions of the intra-block Hamiltonian because S^2 and Z^2 are symmetries of the system. The energies E_1 and E_2 will be the lowest compatible with all other symmetries.

We need now to establish a correspondence like,

$$P_o H_b P_o \longrightarrow H'_u + H'_\mu + K'$$
$$P_o V P_o \longrightarrow H'_t.$$

Here H'_μ represents a chemical potential term and K' is a constant which will give later the ground state energy. H'_t stands for the hopping term between nearest neighbor blocks. Hence, we can obtain explicit expressions for φ_1 and φ_2 in (7).

¿From the intra-block part it is not very difficult to obtain

$$P_o H_{b\alpha} P_o = U' n'_\downarrow n'_\uparrow - \mu'_\downarrow n'_\downarrow - \mu'_\uparrow n'_\uparrow + K'_\alpha$$

where

$$
\begin{aligned}
U' &= 2(E_1 - E_2) + U \\
\mu'_\sigma &= (E_1 - E_2) + \mu_\sigma \\
K'_\alpha &= E_1 - \mu_\downarrow N_{1\downarrow} - \mu_\uparrow N_{1\uparrow}
\end{aligned} \tag{15}
$$

and $N_{1\downarrow} + N_{1\uparrow} = N_o - 1$ and $N_{1\uparrow} - N_{1\downarrow} = 2S_{z1}$. It is now clear why we have to choose only four states to conserve the repulsion term structure: had we chosen eight states for example, we would have created two kind of electrons, and therefore some new couplings.

The projection of the hopping term , $P_o V P_o$, is in general more difficult to obtain. First, it is interesting to write the $c_{i\sigma}$ operators on the border of the block in terms of the $|0\rangle_i, ..., | \downarrow\uparrow\rangle_i$ states,

$$
\begin{aligned}
c_{i\sigma} &= c_{i\sigma}(1 - n_{i,-\sigma}) + c_{i\sigma} n_{i,-\sigma} \\
&= |0\rangle_i \langle \sigma|_i + sgn(-\sigma)\, |\sigma\rangle_i \langle \uparrow\downarrow |_i
\end{aligned} \tag{16}
$$

and then, assuming a natural expansion of the operators on the border of the block α in the subspace spanned by $|0'\rangle, | \downarrow'\rangle, ...$

$$
\begin{aligned}
c_{i\sigma}(\alpha) &= r_{i_\alpha,\sigma} c'_{\alpha\sigma}(1 - n'_{\alpha,-\sigma}) + v_{i_\alpha,\sigma} c'_{\alpha\sigma} n'_{\alpha,-\sigma} \\
&= r_{i_\alpha,\sigma} |0'\rangle_i \langle \sigma'|_i + sgn(-\sigma)\, v_{i_\alpha,\sigma} |\sigma'\rangle_i \langle \uparrow\downarrow' |_i,
\end{aligned} \tag{17}
$$

where the $r_{i_\alpha,\sigma}$'s and the $v_{i_\alpha,\sigma}$'s are real numbers and the other matrix elements, such as $|0'\rangle_i \langle 0'|_i$ vanish. It has been demonstrated [24] that if we choose the $|\prime\rangle$'s so that

$$
Z_\alpha^- | \downarrow'\rangle = 0,
$$

with the spatial symmetry given by $\Sigma_1(\Sigma_2) = A_1(A_2)$ the completely symmetric (antisymmetric) irreducible representation and, say

$$
| \uparrow'\rangle = \frac{1}{\sqrt{2(s_1 + \frac{1}{2})}} S_\alpha^+ | \downarrow'\rangle
$$

and, in addition, a convenient shape for the blocks (we restrict ourselves to the blocks shown in Fig. 1), we get a spin-dependent renormalized hopping,

$$
t'_\downarrow = 2(s_1 + \tfrac{1}{2}) t'_\uparrow \tag{18}
$$

In particular for the half-filled band we are looking for the ground state of minimum spin [19], $s_1 = 0$, so that there is no spin dependence for the renormalized band,

$$
t'_{x,y} = -2\lambda^2 t_{x,y}, \tag{19}
$$

where $\lambda = \langle 0'|c_{i_\alpha\sigma}| \downarrow'\rangle$.

¿From (19) we notice that the renormalization process preserves the spatial anisotropy and, also, that the hopping parameter changes its sign. This last feature happens only for $s = 3$ and is not important as long as we are interested in obtaining quantities which are invariant under the change $t \rightarrow -t$ [5].

We can extract directly the gap of the charge excitations and the ground state energy per site from the analysis of (15). The insulating gap is defined for a given N by [5, 25],

$$\Delta(N) = [E_o(N+1) - E_o(N)] - [E_o(N) - E_o(N-1)] \tag{20}$$

For the half-filled system, in the frame of the renormalization group, we identify the ground state for N electrons to be $|\sigma^{(\infty)}\rangle$ and the $N \pm 1$ electron states to be $|0^{(\infty)}\rangle$ and $|\downarrow\uparrow^{(\infty)}\rangle$. Here the symbol ∞ represents the convergence limit and will be clarified later. The gap (20) is then given by

$$\Delta = U^{(\infty)} \tag{21}$$

If the system is not half-filled $U^{(\infty)}$ does not represent the gap any more because the eigenstate $|\downarrow\uparrow'\rangle$ defined by (14) is no longer the ground state in the $N+1$ subspace.

¿From the constant K' in (15) we obtain the ground state energy per site

$$\varepsilon_o = \lim_{n \to \infty} \frac{K^{(n)}}{9^n}, \tag{22}$$

where

$$K^{(n+1)} = E_1^{(n)} - (N_o - 1)\mu^{(n)} + 9K^{(n)} \tag{23}$$

In the free electron limit it is possible to compare the analytical results from the standard method with those from the real space renormalization group. We can compare for example the exact value of the ground state energy in two dimensions, $\varepsilon_o^{ex} = -(4/\pi)^2 t = -1.62t$ to that obtained by applying (23), $\varepsilon_o^{rg} = -1.3t$ $(t_y < t_x)$. That is, the ε_o^{rg} is about a 80% of the exact one. Another interesting property is the dynamical exponent z defined by

$$t' = s^{-z}t$$

which tells us how the energy is rescaled depending on the length scaling. In Table 1 we show the renormalization factors t'/t for two fillings and different hopping anisotropies. ¿From these values we obtain a dynamical exponent which varies at the half-filling from $z = 0.89$ in the isotropic case $(\alpha = 1)$ to $z = 0.63$ if $\alpha \neq 1$. These results have been obtained as the limiting case of $U \ll 1$ and they already show the odd behavior of the isotropic system at the half-filling (see discussion about this anomaly in Sect. 5).

When we study highly correlated systems it is also interesting to analyze the qualitative behavior of some quantities such as the double occupancy. We need, therefore, to construct an algorithm to obtain these values in the renormalization group method.

We give here, as an example, how the double occupancy can be obtained. The double occupancy is defined like,

$$D = \frac{1}{N} \left\langle \sum_i n_{i\downarrow} n_{i\uparrow} \right\rangle. \tag{24}$$

We use, in fact, a more general expression

$$C(r,s) = \frac{1}{N} \left\langle r \sum_i n_{i\downarrow} n_{i\uparrow} - s \sum_i n_i \right\rangle \tag{25}$$

After the first iteration, for the 3×3 block, we obtain

$$C(r,s) = r\frac{d_1}{9} - s\frac{N_o - 1}{9} + \frac{1}{9}C'(r',s') $$

where

$$r' = r + 2r(d_1 - d_2), \quad s' = s + r(d_1 - d_2) \tag{26}$$

and

$$d_1 = \langle 0' | \sum_{i=1}^{9} n_{i\downarrow} n_{i\uparrow} | 0' \rangle, \quad d_2 = \langle \downarrow' | \sum_{i=1}^{9} n_{i\downarrow} n_{i\uparrow} | \downarrow' \rangle.$$

¿From (26) it is clear that if we choose $s = r/2$ then this ratio keeps for the following iterations: $s^{(n)} = r^{(n)}/2$. We obtain thus a closed expression for D

$$D = C(1, \frac{1}{2}) + \frac{1}{2}(1 - \delta) \tag{27}$$

There are some points where the double occupancy is easy to evaluate, these are the fixed points in the (U,t) space so that

$$\frac{U^{(n+1)}}{t^{(n+1)}} = \frac{U^{(n)}}{t^{(n)}}$$

and then

$$d_1^{(n+1)} = d_1^{(n)} \quad , \quad d_2^{(n+1)} = d_2^{(n)} \tag{28}$$

In these cases C takes a simple form

$$C_{(fixed)}(r, \frac{r}{2}) = \frac{r}{9}(d_1 - \frac{1}{2}(N_o - 1))(1 + \nu + \nu^2 + \ldots)$$

where

$$\nu = 1 + 2(d_1 - d_2)$$

and the double occupancy has an analytical expression

$$D_{(fixed)} = \frac{1}{2}\frac{d_2}{4 - d_1 + d_2}. \tag{29}$$

Two of the trivially fixed points are $U/t = 0$ and $U/t = \infty$, where we know that, in the half-filling, $D(0) = 1/4$ and $D(\infty) = 0$. The results using the expression (29) are shown in the Table 2 for $U = 0$ and two different values of the anisotropy ratio, $\alpha = t_y/t_x$ (the limit $U/t = \infty$ does not present any interest because $D = 0$ anyway). It should be noted that for the isotropic system, $\alpha = 1$, the method does not reproduce the expected value, but, when $\alpha \neq 1$ we recover the a priori normal result.

4   Results and discussion

We are interested in the macroscopic properties of the Hubbard Hamiltonian. In the frame of the real-space renormalization the thermodynamic limit would correspond in principle to nearly 16 iterations, so that each site of the renormalized lattice represents close to $7.13 \cdot 10^{15}$ original sites (equivalent to Avogadro's number in two dimensions). One can, however, wonder about the necessity to reach this number of iterations to get reliable results. In addition, we have not yet defined (numerically) the concept of fixed point which is crucial to interpret the results. To clarify these two questions we report in Tables 3 and 4 the results for $U/t = 0.5$, $\alpha = 1, 0.9$ and different numbers of iterations (n_{it}). From the analysis of the data we can conclude that:

- The amount of the quantities per site like energy, double occupancy (extensive ones) does not depend on the iteration number if $n_{it} > 4$.

- On the other hand, the quantities such as the gap are more sensitive to the $U/t(n_{it})$ value, that is, to the number of iterations n_{it} and to the lattice size represented. Therefore, to assure that we have reached a fixed point, the value of $U/t(n_{it})$ must be either *zero* or *infinity*. We have chosen our *zero* in determining the number of iterations needed to get $t^{(n_{it})} \sim 10^{-14}$ for $U = 0$. This corresponds to 40 iterations approximately. The *infinity* has been arbitrarily chosen as $U/t = 1000$ because of technical reasons: the CPU-time allowed was finite so we have restricted the diagonalizations to the values of U/t up to 1000.

If we would carry out strictly the renormalization procedure for each group of values (U, t, α) we should diagonalize the Hamiltonian and then, from (15) and (19), we would obtain the input values U', t' of the renormalized Hamiltonian (α does not change in the renormalization process as it is indicated in (19)) and so on. This can be done in one dimension and $s = 3$ but in two dimensions the smallest block is the 3×3 block and the procedure of diagonalizing at each time is, in practice, inoperative because of the size of matrix and the huge number of iterations we need to realize. We make use then of the fact that U' and t' as functions of U/t are smooth and monotonous (as well as all other properties of a finite system [31]). We can therefore exactly compute these *functions* at some points and use afterwards an interpolation algorithm to obtain other values. We restrict the U/t interval to [0, 1000]. Furthermore, to avoid the errors due to

α	1	7/9
1	3/8	1/2
0.9	1/2	1/2
0.0001	1/2	1/2
0	1/2	1/2

Table 1: Renormalizing factor of the hopping parameter t'/t for different values of $\alpha = t_y/t$ in two dimensions and two filling cases $n = 1, 7/9$.

$U = 0$	$\alpha = 0.9$	$\alpha = 1$
d_1	1.81246	1.67155
d_2	2.18746	1.98434
D	0.2499	0.230

Table 2: Total amount of the double occupancy on the 3×3 block ground state in the $N_e = 8$ (d_1) and $N_e = 9$ (d_2) subspaces. It should be noted the difference between the isotropic case, $\alpha = 1$, and the anisotropic one, $\alpha = 0.9$. We report as well the double occupancy per site in the macroscopic limit, D.

n_{it}	$-\varepsilon_0/t$	Δ	D	$(U/t)(n_{it})$
2	1.21069	$7 \cdot 10^{-2}$	0.16270	0.548
4	1.21238	$1 \cdot 10^{-2}$	0.16269	0.610
10	$=$	$4 \cdot 10^{-5}$	$=$	0.936
15	$=$	$4 \cdot 10^{-7}$	$=$	1.819
20	$=$	$11 \cdot 10^{-9}$	$=$	27.45
22	$=$	$9 \cdot 10^{-9}$	$=$	114495

Table 3: Renormalization results at $U/t = 0.5, \alpha = 1$ and different number of iterations. The fixed point in this case is $U/t = \infty$.

n_{it}	$-\varepsilon_0/t$	Δ	D	$(U/t)(n_{it})$
2	1.18780	$4 \cdot 10^{-2}$	0.13954	0.189
4	1.19075	$3 \cdot 10^{-3}$	0.13955	0.054
10	1.19076	$7 \cdot 10^{-7}$	$=$	0.00087
15	$=$	$7 \cdot 10^{-10}$	$=$	0.00003
40	$=$	0	$=$	0

Table 4: The same analysis than in the preceding table in the anisotropic case, $\alpha = 0.9$. The fixed point is here $U/t = 0$

the interpolation algorithm the first iteration is done at the points where we have performed the diagonalization.

A. Half-filled system

One of the first tests of a method, not necessarily conclusive, consist in obtaining the ground state energy per site. In Fig. 2 we present the result obtained with the renormalization group method for the isotropic system ($\alpha = 1$) to be compared with the lower bound [32] and a trivial upper bound (Hartree-Fock approximation). The renormalization group energy lies for moderate values of U/t above the limiting interval. This feature is probably due to the neglected terms when we obtained the effective Hamiltonian (12). These terms could enhance the kinetic part and so lower the total energy of the system.

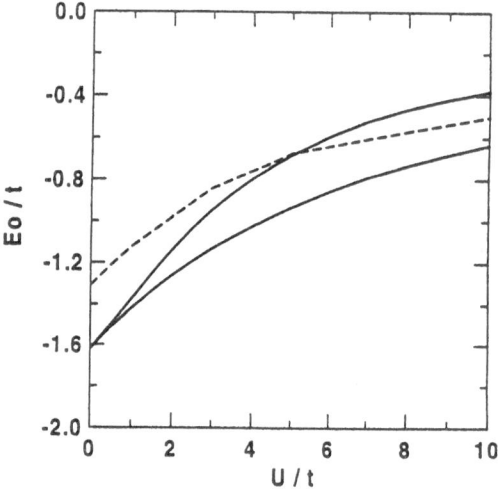

Figure 2 The ground-state energy per site: renormalization group result eq. (22) (broken line) and upper (Hartree-Fock) and lower bound results from ref. [32] (full lines). For the values in the interval $U/t \, \epsilon(5, 10]$ only three iterations are possible because off the cut-off $U/t = 1000$ discussed in the text. The actual value of the renormalization group energy should be therefore even slightly small than the one presented here.

The insulating nature of the ground state in the two dimensional Hubbard model has not yet been completely understood. There is the general belief that an insulating gap is open for all positives values of U/t. This gap would be induced by the presence of spin density waves and be roughly proportional to the staggered magnetization [26]. Another kind of approach consists in obtaining the moment distribution $n(k)$ and analyze its behavior around the Fermi moment k_F [27]. The continuity of $n(k)$ at this point would imply the existence of no Fermi surface. Unfortunately all these methods have been applied to the isotropic hopping system ($t_x = t_y$) where, as it is well known, the Fermi level coincides (at $U = 0$) with a

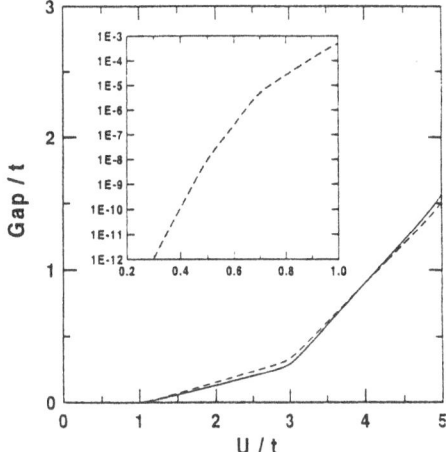

Figure 3 Variation of the insulating gap with U/t for two values of the hopping, $\alpha = 0.9$ (full line) and $\alpha = 1$ (broken line). The upper couple of lines corresponds to the mean-field results (see for example refs. [32, 26]. The two lower lines represent the renormalization gap. The results for the RG isotropic case for the small values of U/t are shown in the inset.

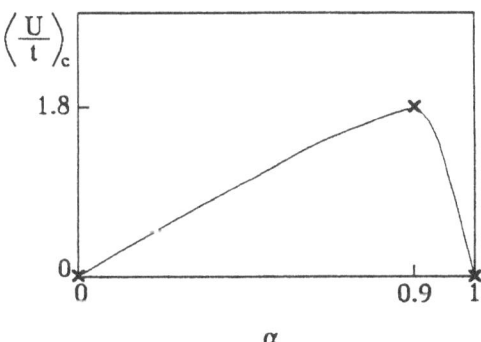

Figure 4 Phase diagram for the Hubbard model at half-filling depending on the anisotropy parameter $\alpha = t_y/t_x$. In fact only three critical values of the interaction strength have been computed, for $\alpha = 0, 0.9$ and $\alpha = 1$.

still holds if $t_x \neq t_y$ and then if a Mott-Hubbard transition occurs for a finite U/t in this case.

We have obtained the insulating gap from the expression (21) and for two values of α. The results are shown in Fig. 3. It should be noted that for the isotropic case, $\alpha = 1$, the gap opens for all positive U and is exponentially small if $U/t < 1$, in agreement with the results mentioned above. Otherwise, if $t_y = 0.9t_x$ the ground state is found conductor up to $(U/t)_c = 1.8$. The system behaves like

α	1	7/9	5/9
1	1	3/4	0
0.9	1/2	3/4	0
0.0001	1/2	3/4	0
0	1	3/4	0

Table 5: Renormalizing factor of the $(U/t)'/(U/t)$ ratio in the $U \ll t$ limit and for three different electronic densities. At the half-filling there are two marginal cases $\alpha = 1, 0$ which corresponds to an insulating ground state.

an insulator in both cases when U/t is large, where we know that the Hubbard model is equivalent to the Heisenberg model (see [33] for example) and then there would be no electron motion. Even if we have not computed the gap numerically for the whole anisotropy interval $\alpha \in [0, 1]$, we can still guess a qualitative phase diagram in the half-filling case (see Fig. 4) where we plot the critical $(U/t)_c$ as a function of α. In fact, we know only three points: $\alpha = 0$, 0.9, 1, but perturbative calculations show that $(U/t)_c$ is, although small, non zero for all $\alpha \neq 0$, 1 (see Table 5). It should be noted here that the results at $U = 0$ are taken as the limit $U/t \to 0$. That is to say, we first obtain the $U \ll t$ result and we assume the same structure for the ground state at $U = 0$. This procedure avoids the ambiguity in defining the ground state in the degenerate case. In fact this ambiguity could be the signature of a non analytical behavior of the thermodynamic limit at $U = 0$, similar to the one of one dimensional case [5]. Recently Zlatić and Horvatić have pointed out that this non-analyticity could occur in two dimension as well in the $n = 1$, $U = 0$ case [34].

B. Non half-filled band, $n = 7/9$

Away from half-filling the expected behavior of the Hubbard Hamiltonian is more rich, in particular concerning the magnetic configurations and at least in the mean-field approach [29]. Unfortunately, there are no exact results about the nature of the ground state like the two Lieb theorems [19] except in the limiting case considered in the Nagaoka theorem [30]. We cannot impose therefore a minimum spin in diagonalizing the block of size s^d as in the half-filled case. On the other hand, if we want to apply the same doublet scheme we cannot choose the states with an even number of electrons as a spin doublet: the electronic density should be then 1/9, 3/9, 5/9 or 7/9 in the two dimensional case, etc. Furthermore, the results when $U \to 0$ have to be consistent with a free electron system. All these conditions and results reported in the Table 5 allows us to state that the real-space renormalization method, restricted to the choice of four states in each iteration, remains valid only in the $n = 7/9$ case. For example if $n = 5/9$ the $|\downarrow'\rangle$ and $|0'\rangle$ states has to be chosen so that $s = 3/2$ and $s = 0$ respectively. The renormalized hopping would be then zero, that is to say, the method would predict an insulating ground state for a free electron system.

The reasons of this behavior have to be sought in an implicit hypothesis on the first approximation: when we had chosen only four states, we implicitly assumed no charge fluctuations. This circumstance is surely correct near the half-filling but, far away of this limit it is too naive to represent a block by just four configurations and an effective repulsion . Instead, it would be more natural to imagine the existence of the different exchange terms produced by keeping more than four states. This is however, an ambitious program and for now we will show and interpret the results obtained at $n = 7/9$ by the "standard" procedure explained in Sect. 2. It should be noted however, that $|\downarrow\uparrow'\rangle$ does not correspond to the true ground state in the $N_e = 8$ subspace. The actual ground state will have an energy of, say, E_3 smaller than $E_2 + U$ for the state defined in (14) [18]; but our choice permits us to avoid, again, the proliferation of couplings in the band term. These couplings are, in addition, energetically unfavorable. We must therefore no longer interpret $U^{(\infty)}$ like a gap but just like a fixed point in the parameter space. The behavior of the ground state energy is similar to the half-filled case (see [28]).

We have also estimated the total spin of the system. We have first to reinterpret the renormalization flow of the U/t parameter. Here the $U = 0$ fixed point refers to a paramagnetic state and the $U/t = \infty$ limit corresponds to a weakly ferromagnetic state. For example, if we begin the process already at the infinity limit ($U > 1000t$), the spin per site will be $(3/2)(1/9) = 3/18$. On the other hand the $U = 0$ case has zero spin per site value. The intermediate values are separated in two zones by a critical point $(U/t)_c \sim 2$. We have carried out the renormalization procedure as if the ground state were paramagnetic. Additional work is needed to take into account the spin symmetry-breaking explicitly produced by the spin-dependent hopping (18). The lower bound of the total spin is then estimated as the ratio $s^{(n_{it})}/9^{n_{it}}$, where $s^{(n_{it})}$ is the spin of block at $(U/t)(n_{it})$ and $9^{n_{it}}$ represents the size of the lattice at the n_{it} iteration. This estimate is computed from an expression for the total spin per site, which is obtained by the same kind of arguments as those we used to get the total energy and the doubly occupancy per site : $\langle S_z \rangle/N = \sum_{i=1}^{n} s_0^{(i)}/9^i + \langle S_z'^{(n)}/9^n \rangle$.

Conclusions

Macroscopic description of the ground state is, in general, difficult to infer from microscopical model Hamiltonians by means of analytical or numerical techniques. The real-space renormalization group analysis of the Hubbard model that we have presented here is an attempt to fill this lacking. We have first improved the understanding of the method itself: we keep the fermionic character of the second quantization operators and, at the same time, we are able to avoid the proliferation of couplings, which is highly desirable in order to keep the computing work up to reasonable limits. Besides, we have outlined the way in which, away from half-filling, the spin and charge symmetries can be "naturally" broken in the frame of the renormalization group.

In what concerns the description of the ground state we have studied two electronic density cases: $n = 1$ and $n = 7/9$. At half-filling $(n = 1)$ we have analyzed the dependence on the hopping anisotropy $\alpha = t_y/t_x$. We find that the isotropic system $\alpha = 1$ behaves like an insulator for all positives values of the interaction parameter U. On the other hand if $\alpha < 1$ the ground state becomes conductor up to a critical value of the ratio $(U/t)_c$. In particular, at $\alpha = 0.9$ we find $(U/t)_c = 1.8$. The one dimensional insulating behavior is recovered at $\alpha = 0$. One should note the remarkable behavior of the system in the α interval. One could naively think that the half-filled system is insulating in the entire α interval because of the two insulating limits. However, the physical picture is different in the one dimensional case $(\alpha = 0)$ compared to the isotropic two dimensional system $(\alpha = 1)$. The former presents a Mott-Hubbard transition at $U = 0$ which can be considered as an one dimensional feature without spin broken symmetries. The latter, on the other hand, presents a degeneracy at $U = 0$ which induces, even in finite clusters, a different behavior for $U/t \to 0$ and $U/t = 0$ cases. The standard treatment of this degeneracy is done by an explicit spin broken symmetry (Hartree-Fock) which can be related, in the frame of the one particle view, with the insulating gap. For the rest of the α values, where no results are known, the system could show a genuine Mott-Hubbard transition, where a broken symmetry is not necessarily implied.

Away from half-filling the situation is less clear. We have to interpret in a slightly different manner the two trivial fixed points. At $U/t = 0$ the system represents a conducting paramagnet and if $U/t = \infty$ we still assume the conducting behavior but in a weak ferromagnetic background. The transition between the two regimes is estimated at $(U/t)_{c2} \sim 2$ and the intermediate values have to be viewed as a paramagnet with renormalized parameters if $U < 2t$ or as a weak ferromagnet if $U > 2t$. More work is needed in this region of the density of electrons in order to improve the renormalization procedure. This improvement must be done in two ways. Firstly, the spin broken symmetry has to be taken into account when the small cluster is diagonalized. Secondly, one should consider the possibility of keeping more than four states in each iteration in order to describe correctly the low density limits, where a block cannot be assimilated to a s-like atomic orbital.

References

[1] Hubbard J., Proc. Roy. Soc. A **276**, 238 (1963). Gutzwiller M. C. , Phys. Rev. **10**, 159 (1963). Kanamori J., Proc. Theor. Phys. **30**, 275 (1963).

[2] Mott N. F., *Metal-Insulator Transitions* (Taylor & Francis, London 1991).

[3] Anderson P. W., Science **235**, 1196 (1987)

[4] From the reference [19]: A lattice is said to be bipartite if it can be divided into two sublattices A et B so that $t_{ij} = 0$ whenever $i \epsilon A$ and $j \epsilon A$ or $i \epsilon B$ and $j \epsilon B$. See also the review by G. Morandi in this issue.

[5] Lieb E. H., Wu F. Y., Phys. Rev. Lett. **20**, 1445 (1968)

[6] Ovchinnikov A. A., Sov. Phys. JETP **30**, 1160 (1970)

[7] Shiba H., Phys. Rev. B **6**, 930 (1972)

[8] Coll C. F., III, Phys. Rev. B **9**, 2150 (1974). Schulz H. J., Int. J. Mod. Phys. **5**, 57 (1991).

[9] Wilson K. G., Rev. Mod. Phys. **47**, 773 (1975).

[10] Jafarey S., Pearson R., Stockley B. and D.J. Scalapino (unpublished)

[11] Drell D., Weinstein M., Yankielowicz S., Phys. Rev. D **14**, 487 (1976)

[12] Hirsch J. E., Phys. Rev. B **22**, 5259 (1980).

[13] Dasgupta C., Pfeuty P., J. Phys. C **14**, 717 (1981).

[14] Jullien R., Can. J. Phys. **59**, 605 (1980)

[15] Pfeuty P., Jullien R., Penson K. A., *Renormalization for Quantum Systems* in "Topics in Currents Physics" vol. 30 (Springer-Verlag 1982).

[16] S. R. White, Phys. Rev. Lett. **69**, 2863 (1992) and review in this issue.

[17] Castellani C., Di Castro C., Feinberg D., Ranninger J., Phys. Rev. Lett. **43**, 1957 (1979)

[18] Nowak, E., Z. Phys. **B45**, 173 (1981)

[19] Lieb E. H., Phys. Rev. Lett. **62**, 1201 (1989)

[20] Zhang S., Phys. Rev. Lett. **65**, 120 (1990)

[21] Lipkin H. J., *Lie groups for pedestrians* (North-Holland 1965).

[22] Hammermesh M., *Group Theory* (Addison-Wesley 1964)

[23] Hirsch J. E., Mazenko G. F., Phys. Rev. B **19**, 2656 (1979).

[24] Pérez-Conde J., J. Phys. A **24**, 2691 (1991).

[25] Mattis D. C., Landovitz L. F., J. of Non-crystalline Solids **2**, 454 (1970)

[26] Hirsch J. E., Phys. Rev. B **31**, 4403 (1985).

[27] Baeriswyl D., von der Linden W., Int. J. Mod. Phys. **5**, 999 (1991).

[28] Pérez-Conde and Pfeuty P., Phys. Rev. B **47**, 856 (1993).

[29] Penn D. R., Phys. Rev. **142**, 350 (1966).

[30] Nagaoka Y., Phys. Rev. **147**, 393 (1966).

[31] Sewell G. L., *Quantum Theory of Collective Phenomena* (Clarendon Press Oxford 1986).

[32] Langer W. D., Mattis D. C., Phys. Let. **36**, 139 (1971)

[33] Harris A. B., Lange R. V., Phys. Rev. **157**, 2 (1967)

[34] Zlatić V., Horvatić B., J. Magn. Magn. Mater. **104**, 593 (1992)

Quantum Dissipative Systems

Francisco Guinea

Instituto de Ciencia de Materiales.
Consejo Superior de Investigaciones Científicas. Cantoblanco. 28049 Madrid. Spain.

1 Introduction

The problem of the interaction of a simple quantum system, described by one or a few degrees of freedom, interacting with a medium characterized by a continuum of excitations arises often in condensed matter and statistical physics. Sometimes the subsystem is a microscopic impurity embedded in a macroscopic system, like in the Kondo model, or in the study of X-ray photoemission from core levels in metals. In other cases, the isolated mode represents a macroscopic, collective, degree of freedom, and the environment is built up by the microscopic excitations which are also part of the spectrum of the whole system. Such situations arise in the study of mesoscopic systems, like small Josephson junctions or SQUIDs.

Systems which can be separated in this fashion into a subsystem described by a few degrees of freedom and an environment with a continuous spectrum share a number of common features. In this work, I will review some of the best known models studied so far. Special emphasis will be made in describing how the knowledge gained in the simplest of them can be used to elucidate more complicated, seemingly unrelated, cases. In order to keep the discussion as simple as possible, I will not follow the historical order in which they were studied.

In the next section, I will introduce the simplest situation in which the interaction with a continuum changes drastically the physics: the orthogonality catastrophe[1]. Directly related to the orthogonality catastrophe lies the so called X-ray edge singularity problem[2], which will be discussed in the same section. Then, the simplest system in quantum dynamics, the two level system, TLS, coupled to a continuous environment, will be studied. This system is perhaps the simplest example of a quantum phase transition of the Berezinskii-Kosterlitz-Thouless type. This phase transition will be analyzed in some detail. This model will provide many of the physical concepts which are used in the discussion of more complicated systems.

The following section is devoted to the Kondo problem, which, cronologically, was studied much earlier than the dissipative two level system. The TLS is, however, simpler to understand, and, on the other hand, more generic than the Kondo model itself. Some interesting variations of the Kondo problem are briefly discussed.

The next models to be studied arise in the context of macroscopic quantum mechanics[3], and the physics of mesoscopic electronic devices, which show the phenomenon of Coulomb blockade[4]. In particular, the model for a dissipative quantum Josephson junction[5], which has attracted recently a great deal of attention in relation to junctions in one dimensional systems[6].

The final section is devoted to a generalization of the previous model: the dissipative quantum rotor. This model exemplifies the variety of applications found in dissipative quantum mechanics as it can either describe a normal tunnel junction in the so called Coulomb blockade regime[7, 8], or the process of monopole induced baryon decay[9]. This system also undergoes a quantum phase transition, although of a different type: the scaling properties are very similar to the Heisenberg ferromagnet near two dimensions, or quantum chromodynamics.

2   The orthogonality catastrophe

The first hint of the existence of non trivial features in systems coupled to an infinite reservoir with a continuum spectrum arise in the study of the sudden switching on of a local potential in a metal. The only degrees of freedom taken into account in the description of the metal are the electronic excitations. At low energies, an independent electron model leads to a continuum of electron-hole excitations around the Fermi level. Anderson[1] studied the changes induced in the metal by the sudden appearance of an external, local potential acting on the electrons. Formally, the starting point is the hamiltonian:

$$\mathcal{H}_0 = \sum \epsilon_k c_k^\dagger c_k \tag{1}$$

whose ground state wavefunction is $|\Psi_0\rangle$. In the presence of an external potential, the hamiltonian has to be replaced by:

$$\mathcal{H}_V = \mathcal{H}_0 + V \sum_{k,k'} c_k^\dagger c_{k'} \tag{2}$$

whose ground state wavefunction is $|\Psi_V\rangle$. A perturbative treatment of the effects of V on $|\Psi_0\rangle$ is only possible if $\langle \Psi_V | \Psi_0 \rangle \neq 0$. Otherwise, there is no continuity arguments which allow us to go from the non interacting to the interacting problem. This quantity of interest is the overlap of two Slater determinants, which can be written as the determinant of the matrix built up by the overlaps of pairs of one electron wavefunctions. As the number of electrons N goes towards infinity, Anderson proved that $\langle \Psi_V | \Psi_0 \rangle \to N^{-\alpha}$, with $\alpha > 0$.

The previous result not only implies the failure of a perturbation expansion. It also gives a hint on the way the expansion fails. As the size of the system grows, the lowest order corrections diverge as $\log(N)$, where N is a measure of the total number of electrons. When the low energy part of the spectrum is decoupled from V, either by thermal fluctuations, by V being time dependent

..., $\log(N)$ is replaced by $\log[\epsilon_c/\max(\omega, T)]$. ϵ_c is the maximum energy available for the formation of electron-hole pairs, and ω and T are the driving frequency and the temperature at which the system is probed. This type of divergence has been extensively studied in quantum field theories. It leads to renormalizable models, in which scale dependent couplings are defined, which depend on the ratio between ϵ_c and the scale of interest for the particular experimental situation. This "renormalizability" is a common feature of the models studied here.

****

Exercise.
The electronic degrees of freedom can be replaced by harmonic oscillators, in which case 1 and 2 change into:

$$\mathcal{H}_0 = \sum_n \frac{p_n^2}{2m} + \frac{m\omega_0^2(x_n - x_{n+1})^2}{2}$$
$$\mathcal{H}_V = \mathcal{H}_0 + \Delta(x_0 - x_1) \tag{3}$$

Assuming that there are N sites in the previous hamiltonian, and imposing periodic boundary conditions, calculate the overlap $\langle \Psi_V | \Psi_0 \rangle$. Hint. There is a canonical transformation, U, such that $U^{-1}\mathcal{H}_0 U = \mathcal{H}_V$.

****

The simplest application of the preceding ideas is found in the so called X-ray edge singularity problem[2]. In an X-ray photoemission experiment, an X-ray photon ejects an electron from shells deep into the core of the ions in a metal. A straightforward analysis leads us to expect that the energy distribution of the emitted electrons should show a sharp peak at an energy related to the position of the level. The core hole left behind, however, interacts with the conduction electrons. Its effect is equivalent to an external potential, which is created when the electron is emitted. Because of the orthogonality catastrophe, the probability that the conduction electrons are left in the ground state which corrsponds to this new situation tends to zero with the size of the system. The conduction electrons are left in a superposition of excited states, whose energy should be compensated by a reduction in the energy of the emitted electron. Thus, the energy spectrum of the electron is an incoherent continuum. It can be shown that this continuum has a power law divergence, with a power related to the exponent α found in the orthogonality catastrophe.

It is interesting to note that similar power law spectra are expected in the conduction band photoemission experiments of one dimensional Luttinger liquids. Roughly speaking, the ejection of a conduction electron in a Luttinger liquid also gives rise to an orthogonality catastrophe.

3 The dissipative two level system

The next level of complexity is given by a quantum two level system (TLS) interacting with a continuum. In the preceding situation, the external potential which
was switched on had no internal dynamics of its own. A TLS can undergo quantum fluctuations between its two states. This problem arises in a variety of guises:
an atom in a double well potential interacting with electron-hole pairs[10], a spin
1/2 magnetic impurity in a metal (the Kondo problem, see next section)[11], an
appropiately biased dissipative SQUID, an 'electron box " in the presence of an
external potential[12] ...

The environment coupled to the TLS can be described in terms of electronic
degrees of freedom, or in terms of conveniently chosen quantum oscillators, which
mimick the response of the electron hole pairs. For convenience, we will use the
second description. Then, the hamiltonian can be written as:

$$\mathcal{H}_{TLS} = \Delta\sigma_x + \sum_k k b_k^\dagger b_k + \lambda\sigma_z \sum \sqrt{k}(b_k^\dagger + b_k) \tag{4}$$

The first term describes the dynamics of the TLS, the second term gives the
environment, and the third is the coupling. The energy of the oscillators is bound
by an upper cutoff, k_c. The type of coupling depends on the problem at hand.
Here, we have chosen the coupling expected if the oscillators represented the electron hole pairs of a metal. This is the coupling relevant for all the cases mentioned
earlier.

The TLS has two positions, to be denoted $|R\rangle$ and $|L\rangle$. When it is frozen
into one of these positions, the oscillators experience a potential proportional to
$\pm\lambda$. The two positions of the TLS define two corresponding relaxed states of
the oscillators, $|bath_R\rangle$ and $|bath_L\rangle$. It can be shown, by using the techniques
needed to solve the exercise in the preceding section, that these two states are
orthogonal. Thus, the fluctuations of the TLS induce an orthogonality catastrophe
in the environment. A perturbation expansion in powers of λ is impossible, as the
corrections are logarithmically divergent.

The nature of the ground state of (4) can be inferred from a scaling procedure.
Let us consider the effect of the high energy oscillators *only*. If they are much
faster than the TLS, that is, if $k_c \gg \Delta$, they will follow quasi instantaneously the
fluctuations of the TLS. We consider those oscillators such that $k_c - dk < k < k_c$.
We can restrict our Hilbert space to those states in which the oscillators are fully
relaxed to the position of the TLS. Let us define:

$$\begin{aligned}
|R'\rangle &= |R\rangle|0\rangle + \sum_{k_c-dk<k<k_c} \frac{\lambda\sqrt{k}}{k}|R\rangle|k\rangle \\
|L'\rangle &= |L\rangle|0\rangle - \sum_{k_c-dk<k<k_c} \frac{\lambda\sqrt{k}}{k}|L\rangle|k\rangle
\end{aligned} \tag{5}$$

The dressing of the TLS by the oscillators leads to an effective reduction in the tunneling amplitude of the TLS:

$$\langle L'|\mathcal{H}_{TLS}|R'\rangle = \Delta' \approx \Delta\left(1 - \lambda^2 \frac{dk}{k_c}\right) \tag{6}$$

Thus, we have an effective problem with a reduced cutoff, $k_c' = k_c - dk$, and a new hopping, Δ'.

The manipulations performed above can be iterated, provided that $k_c' \gg \Delta'$. The ratio between these scales evolves as:

$$\frac{\Delta'}{k_c'} = \frac{\Delta}{k_c}(1 - \lambda^2)\frac{dk}{k_c} \tag{7}$$

Thus, if $\lambda^2 > 1$, we can proceed scaling down k_c all the way to $k_c = 0$. At this point, we have removed all the oscillators. The effective tunneling of the TLS is zero. The ground state is degenerate. The wavefunctions are those of a localized TLS, with the oscillator degrees of freedom fully relaxed to the position of the TLS.

On the other hand, if $\lambda^2 < 1$, the scaling procedure breaks down at some scale, where the effective cutoff becomes comparable to the effective tunneling rate. In renormalization group terms, the system flows towards a strong coupling fixed point. Let us assume that this fixed point corresponds to the situation $\Delta/k_c \to \infty$. Then, the TLS is fluctuating much faster than the typical frequencies of the oscillators. On the time scale at which the oscillators can respond, $\langle \sigma_z \rangle \approx 0$, and the coupling can be neglected. Thus, the picture which emerges is that of a TLS oscillating at a renormalized tunneling rate. The oscillators which are much faster than this rate follow instantaneously the TLS. The low energy oscillators are decoupled from the TLS. There is a crossover between a high energy, high temperature regime, and a low energy or low temperature one . Note that, at low temperatures, the dynamics of the low energy modes are governed by thermal fluctuations. Hence, they are effectively decoupled from the two level system, and play no role in the response of the system.

The picture outlined above is, indeed, correct, and can be justified by more rigorous treatments[13, 14, 15]. The strong coupling fixed point $\Delta'/k_c' \to \infty$ can be used as the starting point in an expansion in powers of Δ_{ren}, the scale at which the weak coupling scaling scheme breaks down[16].

****

Exercise.

Calculate Δ_{ren}, the scale for which $\Delta'/k_c' \sim 1$ when $\lambda^2 < 1$. Hint. It has a power law dependence on ω_c and Δ.

****

4 The Kondo model

One of the major puzzles in condensed matter physics in the 60's was the influence of magnetic impurities on the low temperature resistivity of metals. Ordinary impurities give rise to a temperature independent contribution, while inelastic processes typically freeze out at low temperatures. Experiments showed that, in the presence of magnetic impurities, the resistence rised at low temperatures. These observations lead to an exhaustive theoretical analysis of the problem, which produced an enormous variety of surprises and important results, useful also in other contexts.

The simplest version of the Kondo model describes a spin 1/2 magnetic impurity in a metal. The hamiltonian is:

$$\mathcal{H}_K = \sum_{k,s} \epsilon_k c_{k,s}^\dagger c_{k,s} + J \sum_{k,k',s,s'} \vec{S} c_{k,s}^\dagger \vec{\sigma}_{s,s'} c_{k',s'} \tag{8}$$

where the coupling J is assumed to be antiferromagnetic ($J > 0$). The first term describes the metal, and the second describes the interactions between the spin of the impurity and the spin of the electrons. The impurity spin has no internal dynamics, but a spin flip can occur if the spins of the electrons also change. These spin flip processes are the counterpart of the tunneling events of the TLS discussed in the preceding section. On the other hand, if only the longitudinal coupling J_z is kept, the system is exactly soluble, and is equivalent to the dissipative TLS in the absence of the TLS dynamics. The spin has two orientations, and the conduction electrons relax to them as if acted upon by an external potential.

The mapping of the Kondo hamiltonian onto the TLS can be made more rigorous by noting that the electronic degrees of freedom can be integrated out, leaving an effective theory, with retarded interactions, for the impurity spin. This theory is best expressed in the path integral formulation. The possible paths for the spin are sucessions of binary variables, which represent the z-component of the spin at a given instant. The interactions induced by the integrated out conduction electrons can be written as effective couplings between the spins at different times. This was the path originally used to tackle, sucessfully, the Kondo hamiltonian[17]. The path integral over spin histories was expressed as an integral over spin flips (that is, defects in the spin histories). The effective interaction between the spin flips depends logarithmically on their distance (in time). This dependence is the signature of the existence of a Berezinskii-Kosterlitz-Thouless transition, as a function of the strength of the interaction. The absence of direct spin dynamics is reflected in the fact that the equivalent TLS hamiltonian has $\lambda^2 \sim 1$. The localized phase, discussed for the TLS, in which the dynamics of the TLS is effectively frozen out, corresponds to a ferromagnetic coupling. The nature of the strong coupling fixed point to which the system flows when the coupling is antiferromagnetic was identified by Nozières[18], after an extensive numerical study of the transition by Wilson[19]. Below a certain scale, the Kondo temperature, the spin fluctuates. This situation corresponds, when the conduction electrons are included back into the picture, to the formation of a singlet, in which the impurity spin and a conduction electron are strongly coupled. The singlet binding energy is proportional to the Kondo temperature. The remaining electrons are effectively decoupled from the impurity.

The formation of the singlet quenches the contribution of the impurity spin to the magnetic susceptibility, which approaches a constant at zero temperature. The residual interaction between the conduction electrons and the singlet leads to an enhancement in the density of states near the impurity, which is reflected in the specific heat. From the low temperature value of the susceptibility and the change in the specific heat it is possible to form a universal constant, the Wilson ratio, in which the only parameter which characterizes the low temperature behavior, the Kondo temperature, drops out.

5 Variations on the Kondo model

Once the physics of the Kondo model were well understood, a number of generalizations were tackled. The study of the Kondo model also provided a powerful and reliable method for the analysis of impurity problems, Wilson's numerical renormalization group scheme. An intuitive formulation of field theoretical renormalization group schemes was put forward by L. Kadanoff, in which he proposed that the running coupling constants used in field theory could be associated to effective hamiltonians valid for the low energy behavior of blocks of different sizes. This idea was formulated in a more precise way by K. G. Wilson. In a quantum system, an effective low energy hamiltonian is given by the projection of the full hamiltonian in a restricted basis which contains the low energy states. In such a basis, the intra- and interblock couplings can be separated, and the latter treated perturbatively. Wilson's idea was to make this procedure, rooted in the basic approximation schemes of quantum mechanics, iterative. In such a way, it is possible to treat larger and larger blocks, at the cost of reducing the range in energies spanned by the states in the basis. While this method does not converge too well in extended systems, it gives excellent results for impurity problems. For instance, the dynamics of the dissipative TLS can be fully characterized in ths way[16].

The generalization of the Kondo problem which has attracted most attention is the so called multichannel Kondo model. In it, a magnetic impurity is coupled to N different types of fermions. Note that, in the usual Kondo model, the impurity can be thought of as coupled to a single semiinfinite electronic chain, built up by the succesive spherical shells with s-wave symmetry which surround the impurity. The N-channel Kondo hamiltonian reads[20]:

$$\mathcal{H}_{MK} = \sum_{k,s,i=1}^{i=N} \epsilon_k c_{k,s,i}^\dagger c_{k,s,i} + J \sum_{k,k',s,s',i=1}^{i=N} \vec{S} c_{k,s,i}^\dagger \vec{\sigma}_{s,s'} c_{k',s'.i} \qquad (9)$$

The high energy physics of the multichannel Kondo model are the same as for the standard Kondo impurity. For antiferromagnetic couplings, spin fluctuations are enhanced, and the value of J in units of the effective cutoff grows. A spin zero complex can only be formed if N is equal to the multiplicity of the impurity spin, n. Otherwise, there are two possibilities: i) If $2N < n$, the conduction electrons cannot screen completely the impurity. A complex with spin degeneracy $n - 2N$ is formed, which, in turn, interacts antiferromagnetically with the next shell of conduction electrons. Ultimately, the full impurity spin is screened, and becomes decoupled from the remaining electrons. ii) When $n < 2N$, as J grows upon renormalization, the N electrons closest to the impurity become tightly bound to it. The resulting complex has a spin multiplicity $2N - n$, which is again less than N. This spin couples to the next shell of N conduction electrons, giving rise to a situation similar to the initial one. The $J \to \infty$ fixed point is unstable, signalling the possibility of a non trivial fixed point at intermediate couplings.

The second case has attracted a great deal of attention. A fixed point at a finite value of J^* also implies the existence of non trivial critical exponents. The only gapless electronic system carefully analyzed in the literature is the Fermi liquid, characterized by the marginal (at most) behavior of the interactions at low energies. The multichannel Kondo model gives an example of a genuine non Fermi liquid fixed point. Magnetic impurities do not give rise to an overscreened Kondo system, but it was proposed that it could be realized in models of non magnetic impurities in metals[21]. For these models, an expansion in powers of N^{-1} can be made, and the existence of a non trivial fixed point can be proven[22]. This non magnetic version of the multichannel Kondo model has been proposed as an explanation for the non Fermi behavior observed in many heavy fermion compounds[23].

6 Macroscopic quantum phenomena

The preceding sections have dealt with models used to describe microscopic processes in condensed matter physics. It is remarkable that many of the results can be generalized to a rather different problem, the observation of quantum behavior at macroscopic scales.

Macroscopic quantum phenomena involve the study of a few variables, which, by definition, describe the collective behavior of a large number of microscopic units. The remaining degrees of freedom in the system are not completely decoupled from the dynamics of the collective variable of interest. They manifest themselves in the form of noise and dissipation. A typical equation of motion for a macroscopic variable has the form:

$$M\ddot{Q} = -\frac{\partial V}{\partial Q} - \eta\dot{Q} \tag{10}$$

where η is the friction coefficient, and Q is the macroscopic variable. The dynamics of Q were quantized[3] in the path integral representation, by using the action (in imaginary time):

$$S = \int d\tau M \left(\frac{\partial Q}{\partial \tau}\right)^2 + \int d\tau V[Q(\tau)] + \frac{\eta}{2} \int d\tau \int d\tau' \frac{[Q(\tau) - Q(\tau')]^2}{(\tau - \tau')^2} \qquad (11)$$

The last term can be traced back to the interaction of Q with the remaining degrees of freedom. If we are only interested in the dynamics of Q, all external systems and couplings which lead to this effective equation are equivalent. In particular, we can model the system by a continuum of oscillators, linearly coupled to Q. In a hamiltonian formalism, it leads to:

$$\mathcal{H} = \frac{P^2}{2M} + V(Q) + \sum_k |k| b_k^\dagger b_k + \lambda Q \sum_k \sqrt{k}(b_k^\dagger + b_k) \qquad (12)$$

where $\lambda \propto \sqrt{\eta}$. This coupling has the same dependence on the oscillators' degrees of freedom as the one used in eq. (4). Hence, if $V(Q)$ is the double well potential, it may be possible to restrict the states which describe the dynamics of Q to those localized at the bottom of either minima. Then eq. (12) can be replaced by eq. (4). This is the case for experiments devised to test the Schrödinger's cat paradox, or the validity of the superposition principle at the macroscopic level. In particular, hamiltonian (4) has been used to describe the tunneling between the two lowest states of a SQUID threaded by half of a flux quantum. These states, which are macroscopically different, correspond to currents flowing clock- and anticlockwise around the ring.

Another system which has attracted a lot of interest is a Josephson junction in the quantum regime. Then, the macroscopic variable in (12) is the phase across the junction, the effective mass is the capacitance, and $V(Q) = E_J \cos(Q)$, where E_J is the Josephson coupling. The source of dissipation can be the impedance of the external circuit, or a normal current flowing in parallel across the junction. The resulting action is:

$$S = \int d\tau C \left(\frac{\partial \phi}{\partial \tau}\right)^2 + \int d\tau E_J \cos[\phi(\tau)] + \frac{\eta}{2} \int d\tau \int d\tau' \frac{[\phi(\tau) - \phi(\tau')]^2}{(\tau - \tau')^2} \qquad (13)$$

This action, first studied in[5] has many remarkable properties. It is self dual, and the $E_J \to 0$ and $E_J \to \infty$ can be interchanged (redefining some physical quantities). When $E_J \to \infty$ the phase tunnels between the infinite minima of the Josephson potential. As in the two level case, a transition to a localized state is expected, as a function of the dissipation strength. This is, indeed, the case[24], although there are some interesting differences from the two level case, like the absence of screening, in a representation in terms of charges. In addition, the study of the leading irrelevant operators has caused some surprises recently[25]. The low energy behavior differs markedly from that of the Kondo model.

The model in (13) also arises in the study of tunneling between Luttinger liquids[6]. A detailed analysis of the problem in this context can be found in the lectures by G. Gómez-Santos, in this volume.

7 The quantum rotor

As a last application of the ideas discussed here, we consider the case of a dissipative planar rotor. The rotor's degree of freedom is a phase, ϕ. In the classical limit, the phase obeys the same equation of motion as any other classical variable (10). The only constraint is that the potential $V(\phi)$ should have the right periodicity. The quantization of the dissipation cannot be given by the action (11), because the last term in this equation is not invariant under the replacement $\phi \to \phi + 2\pi$. A more adequate action, with the right symmetries and classical limit is:

$$S = \int d\tau C \left(\frac{\partial \phi}{\partial \tau}\right)^2 + \int d\tau V[\phi(\tau)] + \frac{\eta}{2} \int d\tau \int d\tau' \frac{1 - \cos[\phi(\tau) - \phi(\tau')]}{(\tau - \tau')^{2+g}} \quad (14)$$

where we have introduced the parameter g for future convenience. This action arises in many, apparently unrelated, fields. For instance, we can let $E_J \to o$ in the model for a dissipative Josephson junction of the preceding section. The dissipative term can be interpreted as arising from a normal current shunting the junction. Hence, we have a model which can describe a normal tunnel junction[8]. The macroscopic variable is a "phase " which is conjugated to the number of electrons. The discreteness of the latter variable implies that the phase should be taken as periodic. Then, the appropiate dissipative term is the one in (14) (there are situations, however, where the charge needs not be considered quantized, as when polarization charges are taken into account). A different situation, more alike the Kondo model, is the interaction of a magnetic monopole with a fermion background[9]. The monopole acts as an impurity, which scatters the fermions, described by Dirac's equation. The level spectrum of the monopole can be approximated by that of the quantum rotor. The crucial ingredient is the existence of a gap between the ground state and the next excited state.

In the absence of the external potential, there are two competing interactions in (14): i) the rotor kinetic energy favors quantum fluctuations in ϕ and ii) the dissipative term, which tends to localize the phase. As in the examples of the two level system and the Josephson junction, we expect a localization transition as a function of the strength of the dissipation.

The model (14) can be studied, using standard renormalization group methods, in the strong dissipation limit[26]. While the two level and Josephson models show a transition of the Kosterlitz-Thouless type, the flow of the present model resembles formally that of the so called asymptotically free systems, like the 2D non linear-σ models and 4D non Abelian gauge theories. The parameter g in (14) is the counterpart of the dimension in a $2 + \epsilon$ or $4 + \epsilon$ expansion. In the strong dissipation limit, the system is in the localized phase when $g < 0$, and in the fluctuating phase if $g \geq 1$. Hence, for $g = 1$, quantum fluctuations in ϕ dominate the low energy behavior, irrespective of the strength of the dissipation. Thus, quantum fluctuations are enhanced with respect to the cases where the dissipation is not gauge invariant. The complete phase diagram depends also on the value of the mass[27], and was obtained by combining numerical and variational methods.

References

[1] P.W. Anderson, Phys. Rev. **164**, 352 (1967).

[2] G. D. Mahan, **Many-Particle Physics**, Plenum (New York) 1991. P. Nozières and C. T. de Dominicis, Phys. Rev. **178**, 1097 (1969).

[3] A. O. Caldeira and A. J. Leggett, Ann. of Phys. (NY) **149**, 374 (1983).

[4] D. V. Averin and K. K. Likharev in **Mesoscopic Phenomena in Solids**, edited by B. Altschuler, P. A. Lee and R. A. Webb (Elsevier, Amsterdam, 1991).

[5] A. Schmid, Phys. Rev. Lett. **51**, 1506 (1983).

[6] C. L. Kane and M. P. A. Fisher, Phys. Rev. Lett. **68** 1220 (1992). Phys. Rev. B **46**, 15233 (1992).

[7] E. Ben-Jacob, E. Mottola and G. Schön, Phys. Rev. Lett. **51**, 2064 (1983).

[8] F. Guinea and G. Schön, Europhys. Lett. **1**, 585 (1986); J. Low Temp. Phys. **69**, 219 (1987).

[9] J. Polchinski, Nuclear Physics **B242**, 345 (1984).

[10] K. Vladár and A. Zawadovsky, Phys. Rev. B **28**, 1564 (1983).

[11] J. Kondo, Prog. Theor. Phys. **32**, 37 (1964).

[12] P. Lefargue *et al*, Zeits. für Physik, **85**, 327 (1991).

[13] V. Hakim, A. Muramatsu and F. Guinea, Phys. Rev. B **30**, 464 (1984).

[14] A. J. Leggett, S. Chakravarty, A. T. Dorsey, M. P. A. Fisher, A. Garg and W. Zwerger, Rev. Mod. Phys. **59**, 1 (1987).

[15] U. Weiss, **Quantum Dissipative Systems**, World Scientific, Singapore (1993).

[16] F. Guinea, Phys. Rev. B **32**, 4486 (1985).

[17] G. Yuval and P. W. Anderson, Phys. Rev. B **1**, 1522 (1970).

[18] Ph. Nozières, J. Low Temp. Phys. **17**, 31 (1974).

[19] K. G. Wilson, Rev. Mod. Phys. **47**, 773 (1975).

[20] Ph. Nozières and A. Blandin, J. Physique **41**, 193 (1980).

[21] K. Vladár, A. Zawadovsky and G. Zimanyi, Phys. Rev. B **37**, 2001 (1988).

[22] A. Muramatsu and F. Guinea, Phys. Rev. Lett. **57**, 2337 (1986).

[23] D. L. Cox, Phys. Rev. Lett. **59**, 1240 (1987).

[24] F Guinea, V Hakim and A Muramatsu, Phys. Rev. Lett. **54**, 263 (1985).

[25] F. Guinea, G. Gómez-Santos, M. Sassetti and M. Ueda, Europhys. Lett. **30**, 561 (1995).

[26] J. M. Kosterlitz, Phys. Rev. Lett. **37**, 1577 (1977).

[27] T. Strohm and F. Guinea, preprint.

Impurity Effects in Quantum Wires

G. Gómez-Santos

Departamento de Física de la Materia Condensada.
Universidad Autónoma de Madrid, 28049 Madrid, Spain.

We study the properties of a one-dimensional system of interacting particles, described as a Luttinger liquid, in the presence of a single source of backscattering (*impurity*). After a brief introduction to the Luttinger liquid concept and bosonization, the problem associated with transport in the presence of a local impurity is considered. An efficient numerical treatment is then described. This method allows an analysis of the size dependence of the low energy spectrum, by means of an effective integration of high degrees of freedom. Presence of irrelevant operators and their transport manifestations are issues treated within the present approach. In particular, we confirm the existence of two irrelevant operators, and show that they have distinctive transport manifestation in the temperature and frequency dependence of the conductance.

I. INTRODUCTION AND OUTLINE

The discovery of new high T_c superconductors [1] has contributed vigorously to the study of highly correlated systems. Part of this effort has been motivated by proposals of unusual metallic behavior [2], beyond the standard Landau's Fermi liquid description of ordinary metals [3]. Although the relevance of these ideas in the original high T_c problem is not clear, an output of this study has been the widespread appreciation of one-dimensional interacting systems as a paradigm of non-Fermi liquid behavior. It is well known theoretically that interactions endow particles in one dimension with properties that are qualitatively different from the quasi-particle Fermi liquid picture [4,5]. A new description of these one-dimensional models is required, generically designed with the name of Luttinger liquids (LL) [6–9]. Besides their relevance as examples of unusual metallic behavior, one-dimensional models might have approximate experimental realizations. In addition to bulk materials with extreme anisotropies [10], the continuous progress in man-made nanostructures is making possible the fabrication of systems close to the ideal one-dimensional limit [11,12]. Furthermore, the realization that edge states of quantum Hall effect liquids are intrinsic one-dimensional LL [13], opens new perspectives to the experimental possibilities of this field. In fact, we might be very close to give experimental reality [14] to theoretical idealizations such as those considered in this work.

We will study interacting one-dimensional LL with emphasis on transport properties, a natural property for experimental probes. The problem of a single impurity in an otherwise perfect LL will occupy a central position, both as a first step towards the real (usually dirty) world, and also as a system with possible experimental relevance in its own right [14]. The paper is organized as follows. In Sect. 2 we introduce the basic aspects of the LL concept and bosonization technique, followed by an analysis of transport in the perfect system. In Sect. 3, transport in the presence of a single impurity is considered. We introduce the peculiarities of this situation in an elementary way, describing later the overall picture from a more detailed analysis. In Sect. 4, the repulsive LL plus impurity problem is treated in depth, following recent work of the author [15]. In particular, a thorough numerical analysis of the sources of transport behavior (and their manifestations in both frequency and temperature dependences of the conductance) is carried out. Finally, Sect. 5 summarizes this lecture.

II. FUNDAMENTALS OF THE LUTTINGER LIQUID CONCEPT. TRANSPORT

In this section we introduce the basic ideas and notation of the LL concept which are necessary for the rest of the paper. For a more detailed treatment and balanced overview, I refer to Schulz lectures in this school [16], in addition to standard sources [6–9]. Interacting, spinless (for simplicity) fermions in one dimension share common features in their low-energy physics, and they are generically described as LL. These properties are summarized by the following LL Hamiltonian (H_{LL}), written in bosonized language:

$$H_{LL} = 3D\frac{v_s}{2} \int dx (\pi g \Pi^2(x) + \frac{1}{\pi g}(\partial_x \theta)^2) \tag{1}$$

$\theta(x)$ and $\Pi(x)$ are canonically conjugated scalar fields obeying $[\theta(x), \Pi(y)] = 3Di\delta(x - y)$, v_s is the sound velocity, and g is a dimensionless parameter whose departure from unity measures the effective strength of interactions ($g < 1$ corresponds to the usual case of particle-particle repulsion). H_{LL} is the Hamiltonian of a quantum vibrating string: the simplest field theory in $(1+1)$ dimension. The physical connection with the underlying particle system, of which H_{LL} is supposed to be the effective low energy limit, is provided by the identification of particle density ($\rho(x)$) and current ($j(x)$) with θ and Π in the following manner:

$$\rho(x) = 3D\partial_x\theta/\pi, \quad j(x) = 3Dv_s g\Pi(x) \tag{2}$$

It is customary to introduce a new bosonic field $\phi(x)$ as $\Pi = 3D\partial_x\phi/\pi$. Bosonization is, then, the name given to the translation from the original particle Hamiltonian to the bosonic form H_{LL}, describing the phonon-like density fluctuations of a quantum string.

A sketchy derivation of bosonization is obtained as follows (see the above mentioned sources for proper accounts). Consider a one-dimensional, non-interacting (spinless) fermion system. The Fermi surface reduces to two points in k-space ($\pm k_f$). If we restric our attention to low energy processes, we can classify particles into right (+) and left (−) movers according to whether they move in the vicinity of $+k_f$ or $-k_f$. Fourier components of the usual density operators can be classified as follows:

$$\rho_q^+ = 3D \sum_{k \sim +k_f} c_{k+q}^\dagger c_k, \quad \rho_q^- = 3D \sum_{k \sim -k_f} c_{k+q}^\dagger c_k, \tag{3}$$

with total density $\rho_q = 3D\rho_q^+ + \rho_q^-$, a faithful decomposition in the small q limit. It is a simple exercise to show that

$$< [\rho_{-q}^+, \rho_{q'}^+] > = 3D < [\rho_q^-, \rho_{-q'}^-] > = 3D(qL)/(2\pi)\delta_{q,q'} \tag{4}$$

where the expectation value is taken with respect to the non-interacting Fermi sea, and L is the system size. With the following rescaling: $a_{\pm|q|}^\dagger = 3D(2\pi/|q|L)^{1/2}\rho_{\pm|q|}^\pm$, we recognize Eq. (4) as the usual commutation relations for bosonic creation-annihilation operators. Although Eq.(4) is valid only for the ground state, we follow Tomonaga [6] and employ a continuity argument to promote it to the status of an operator identity, expected to be asymptotically valid in the low energy limit. Upon linearizing the band dispersion at the Fermi points, the following commutation relation holds between the non interacting Hamiltonian H_o and bosonic operators: $[H_o, a_q^\dagger] = 3Dv_f|q|a_q^\dagger$, v_f being the Fermi velocity. This implies the following expression for H_o:

$$H_o = 3Dv_f|q| \sum_q a_q^\dagger a_q \tag{5}$$

The reader will immediately recognize that the previous expression is nothing but the Hamiltonian of a collection of independent harmonic oscillators. They describe, by construction, the low-energy, phonon-like density fluctuations of the non-interacting particle system. It is now a matter of tradition to write the previous collective description in real space. One introduces two bosonic fields (θ, ϕ) describing charge $(\rho^+ + \rho^- - 3D\partial_x\theta/\pi)$ and current $(\rho^+ \quad \rho^- = 3D\partial_x\phi/\pi)$, respectively, rewriting H_o as:

$$H_o = 3D\frac{v_f}{2} \int dx (\pi\Pi^2(x) + \frac{1}{\pi}(\partial_x\theta)^2) \tag{6}$$

This is nothing but Eq. (1), adapted to the non-interacting case: $g = 3D1$, and $v_s = 3Dv_f$.

A nice aspect of bosonization is that the presence of interactions preserves the harmonic structure of the non-interacting case. To see this, notice that interactions add a density-density term to the non-interacting Hamiltonian. A density-density coupling is a quadratic term on top of an already harmonic system: the total Hamiltonian can be easily diagonalized and the final structure is precisely that of H_{LL} of Eq. (1). Notice that the presence of interactions manifests itself in two ways: v_s is not the Fermi velocity of the non-interacting Fermi sea (just a change in the overall energy scale), and the parameter g differs from unity (this has important spectral consequences).

The previous procedure completes the bosonization program at the Hamiltonian level. In addition, one would like to know how to write the natural objects in the particle picture, fermion creation-annihilation operators $(\Psi^\dagger(x), \Psi(x))$, in terms of the new fields of the collective boson description (θ, ϕ). It can be shown that particle conservation and fermion statistics force the following correspondence (see ref. [9] for a rigourous derivation plus qualifications):

$$\Psi^\dagger_\pm(x) \sim e^{\pm i(k_f x + \theta(x))} e^{i\phi(x)} \tag{7}$$

Eq. (7) amounts to a bosonization dictionary that allows us to translate any object written in particle language to the collective LL language, completing our brief introduction to bosonization. One could, for instance, calculate single-particle spectral properties using the previous dictionary and realize that, in the presence of interactions, the quasi-particle features of the familiar Landau's Fermi liquid theory are absent in one-dimension. This breakdown of the Fermi liquid concept in one dimension, and its replacement by a new universal behavior termed LL, is what has attracted much attention in recent times.

Now we ask ourselves about the current (j) that is established in a perfect LL upon applying a voltage drop (V). The reader is surely familiar with the non-interacting result $j = 3D(e^2/2\pi\hbar)V$, corresponding to perfect transmission in the Landauer [17] formula. We seek to generalize this result to the interacting case, using the bosonized LL framework.

Applying a voltage drop is to add a linear (in boson fields) perturbation to the already quadratic H_{LL}. The current being also linear in boson fields, its calculation amounts to an elementary linear response exercise in a harmonic system, which the reader is encouraged to perform. We will obtain the current response to a voltage with the following alternative route which emphasizes energy conservation. If a current j follows a voltage drop V, a power jV is delivered to the system. Where is this power invested?. Imagine that, in a quiet (no current) LL, we suddenly apply a voltage drop V at a particular point. Following the time development of the perturbation, we will see that a current wavefront of intensity j extends from the perturbing site at the sound velocity v_s. By the continuity equation, this current perturbation is accompanied by a density excess (defect) $\pm\delta\rho = 3D v_s^{-1} j$. This current-density distortion costs energy which has to be continuouly supplied to keep the perturbation extending its size at velocity v_s. The power required to maintain this ever-moving wavefront can be read from the Hamiltonian H_{LL} of Eq. (1), with the result: $Power = 3D(2\pi/g)j^2$ ($e = 3D\hbar = 3D1$). Conservation of energy dictates that: $jV = 3D(2\pi/g)j^2$, implying that the conductance $G = 3Dj/V$ is (restoring physical units):

$$G = 3Dg(e^2/2\pi\hbar). \tag{8}$$

This same result would have been obtained with the traditional linear response scheme suggested as an exercise. Notice that the presence of interactions makes the conductance of a perfect LL non universal, depending on interactions through the parameter g (this is, in fact, the reason for such a notation).

III. LUTTINGER LIQUID WITH A SOURCE OF BACKSCATTERING

Now, we consider the problem of transport with a single impurity in an otherwise perfect LL. Recall that for the non-interacting case, Landauer formula simply replaces the perfect transmission of the clean system by the quantum transmission of particles at the Fermi level. We will see that interactions change this picture drastically [18].

First, we have to decide what is the meaning of an impurity in our bosonized LL language. It would be tempting to say that an impurity at position x_o is nothing but a perturbation that couples to the particle density and, therefore, adds a term to H_{LL} proportional to $(\partial_x \theta)_{x_o}$. This is wrong (or rather incomplete): except for a shift in densities, this perturbation does not change the Hamiltonian dynamics. We *do* know that an impurity is a real perturbation (recall Landauer formula for the non-interacting case), therefore something is missing. To solve the problem remember that, in our decomposition of density into right and left components,we restricted ourselves to small q processes. This excludes the $2k_f =$ component of the density [9] (also a low energy process) in which an electron is backscattered at the impurity site. This backscattering process is what we will understand by impurity perturbation. Using the bosonization dictionary of Eq. (7) we can write this impurity perturbation as

$$H_{imp} \sim u e^{i2\theta_o} + H.c. \sim u \, \cos(2\theta_o),\qquad(9)$$

where θ_o is the charge field at the impurity position, and u stands for the probability amplitude for right-left exchange.

Looking at Eq.(9), it is clear that the addition of an impurity to H_{LL} poses a very difficult problem. We will try to study transport properties in the (nominal) weak coupling limit $u << \omega_c$ (ω_c being the cut-off energy of the LL), following the line of reasoning based on energy conservation. The presence of an impurity is an additional source of dissipation, and has to be included in the energy balance that allowed us to obtain the conductance in the clean LL. The power wasted by the impurity can be evaluated with the following trick. Remembering that θ_o has the meaning of total charge at one side of the impurity, we can mimic the presence of a background current (j) in Eq.(9), with the substitution $\theta_o \rightarrow \theta_o + \pi j t$, where t means time. Now the impurity potential acts as a time dependent perturbation, and the power absorbed can be calculated to lowest order with Fermi's golden roule. This calculation shows that the power absorbed by the impurity at temperature T has the following dependences: $\sim u^2 T^{2g-2} j^2 \omega_c^{-2g}$. Including this contribution in the energy balance, one gets the following (lowest order) impurity correction to the conductance (δG):

$$\delta G \sim u^2 T^{2g-2} \omega_c^{-2g}\qquad(10)$$

This result is rather disturbing. It tells us that the impurity correction to the clean conductance adquires a power law temperature dependence, with exponent determined by g, the parameter characterizing the clean LL. In the case of attractive interactions ($g > 1$), the correction vanishes at $T \to 0$: the system shows perfect conductance, even with impurity!. For the usual case of repulsive interactions ($g < 1$), the $T \to 0$ correction diverges!. Because the whole procedure is nominally perturbative, this means that there is no truly perturbative regime at $T = 3D0$ for $g < 1$. Notice that the above conclusions are not in contradiction with the non-interacting results: for $g = 3D1$ there is a finite temperature-independent correction to the conductance (Eq.(10) being its perturbative value), in agreement with Landauer formula.

These results, though strange, are not an artifact of our simple-minded energy-balance approach to transport (intended to bypass the use of renormalization group (RG) as the first approach). A more detailed analysis [18] combining both perturbative and RG methods, provides the same features which we now summarize:

i) The coupling constant u of the impurity perturbation, Eq.(9), obeys the RG flow:

$$du/dln(l) = 3D(1 - g)u, \qquad (11)$$

where l is the running energy scale. This means that the impurity perturbation is irrelevant (relevant) if $g > 1(g < 1)$, and marginal in the non-interacting case $g = 3D1$. This is precisely the physical meaning of Eq. (10), upon interpreting T as the energy scale of the RG flow.

ii) For $g < 1$, the RG flow takes the system away from the weak coupling limit where Eq. (11) is valid. Nevertheless, there is every reason to expect that the final destination of the flows can be described with the simple physical picture that takes $u \to \infty$. Looking at Eq. (9), this means that the system gets trapped in one of the equivalent minima of the cosine pinning potential. Perfect pinning of θ_o implies that, irrespective of the original coupling strength u, the final $u \to \infty$ fixed point corresponds to a perfect insulator (zero conductance). At low energies, though, θ_o owes its residual dynamics to hopping processes between cosine minima: $\theta_o \to \theta_o + n\pi$, see Eq. (9). This is nothing but particle tunnel across the junction. This particle hopping ($t_{eff} =$) is an irrelevant perturbation ($g < 1$) with RG flow

$$dt_{eff}/dln(l) = 3D(1 - g^{-1})t_{eff}. \qquad (12)$$

Standard analysis [18] would then provide the following temperature dependence to the conductance:

$$G \sim T^{2/g-2}. \qquad (13)$$

Following the usual practice of interpreting frequency (ω) as the running energy scale of the RG flow, we can exchange T and ω in Eq. (13), obtaining the same frequency dependence for the dynamic conductance $G \sim \omega^{2/g-2}$.

The phase diagram associated with the RG flows just described are depicted in Fig. 1. These results constitute a summary of what has been obtained in recent times by Kane and Fisher (KF) [18] in the analysis of LL with backscattering, although the basic physical picture was already clear in the eighties from the study of quantum dissipative systems [19,20]. The rest of this lecture will be devoted to a numerical analysis of the insulating regime ($g < 1$), which confirms but also qualifies some aspects of this problem.

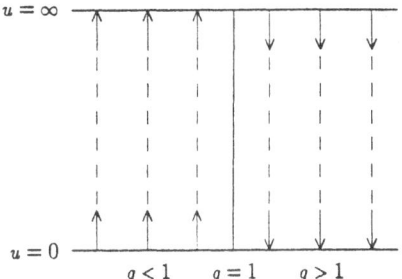

FIG. 1 Phase diagram and RG flows for backscattering intensity u in a LL with perfect conductance g. (See Ref. [18]).

IV. FINITE-SIZE STUDY OF LUTTINGER LIQUIDS WITH BACKSCATTERING: IRRELEVANT OPERATORS AND TRANSPORT

Based on the Self-Consistent Harmonic Approximation (SCHA) [21,22], which substitutes the effect of the cosine in Eq. (9) by a quadratic term $\alpha\,\theta_o^2$, Guinea et al. [23] have qualified the KF picture of transport in the regime $g < 1$. They claim that, in addition to the hopping between minima, the SCHA suggests the existence of another irrelevant perturbation of dimension 2, describing oscillations of θ_o around its pinned value and providing the junction with a capacitance. (A density-density coupling between both sides of the already split system would be an equivalent description.) This new irrelevant operator (dominant for $g < 1/2$) would give rise to a g-independent τ^{-4} contribution to the junction current-current correlation. Scaling then dictates a ω^2 dependence for the conductance, replacing the KF result $\omega^{2/g-2}$ when $g < 1/2$.

The case for this new operator seems strong. Chamon et al. [24] have been forced to invoke the presence of a density-density coupling to understand features of the noise spectrum, and Tsvelik [25] has deduced the existence of a capacitance from Bethe ansatz techniques. Thermal transport also indicates its existence [26]= . On the other hand, the same scaling that gives the ω^2 law for the dynamic conductance would predict a T^2 temperature dependence for the static conductance [23] replacing the KF prediction for $g < 1/2$ =. Unfortunately this last result contradicts the exact solution found for $g = 3D1/3$ from Bethe ansatz techniques [27], which is in agreement with the KF analysis $G \sim T^{2/g-2}$. This confronts us with a puzzling situation.

Naive scaling is not guaranteed to work, and to settle the problem described above one would like to separate the following two issues: (i) presence of irrelevant operators (nature, dimension, and region of dominance), and (ii) their manifestation in physical properties. With this in mind, we have adopted a finite size approach in this paper. The idea is well known: solving $H = 3DH_{LL} + H_{imp}$ in a finite size L amounts to probing the system with a energy scale set by $1/L$. The low energy spectrum of the (scaled) Hamiltonian LH becomes scale invariant in the limit $L = 3D\infty$ (fixed point). The presence of irrelevant operators and their scaling dimensions can be read off from the L dependence of the approach to the fixed point. In fact, this procedure is completely anologous to the numerical description of the Kondo problem carried out by Wilson [28] in a landmark work, where the fixed point vicinity was analyzed by solving the system for increasing sizes (decreasing energies).

Unfortunately, this simple idea faces severe implementation problems. The number of degrees of freedom grows gigantically with size, and there is no real chance of solving exactly a finite system of even modest size. One should resort to approximations that keep a manageable number of states while ensuring that they truly describe the low energy sector of interest. In our case, the use of the SCHA as an intermediate step will help us handle this problem in a transparent and reliable way, as we will see.

It is well known that, under the guise of a variational approach, the SCHA performs a perturbative renormalization [21–23]: replacement of the cosine term in Eq. (9) by a self-consistent $\alpha\theta_0^2$ provides the following dependence: $\alpha \sim u^{1/(1-g)}$. This behavior is precisely what would have been obtained for the running coupling u upon integration of its RG flow Eq. (11), up to the point where u equals the cut-off energy, signalling the end of the perturbative regime. Thus, the low energy states of the SCHA have a built-in effect of the high energy modes. Although the SCHA flows to the same fixed point expected for the original problem (θ_o is also pinned), it loses memory of the discrete (global) translational invariance ($\theta \to \theta + n\pi$) of H. It is clear that the SCHA can pin θ_o at any of the translation-equivalent minima, suggesting the following procedure: replicate the SCHA states for each well, and use these states as a basis for the diagonalization of the original Hamiltonian.

The steps in the precise implementation of this idea are the following. (i) The Hamiltonian H is regularized in a linear chain of L sites with periodic boundary conditions, the cosine perturbation affecting only one lattice site. (ii) The SCHA is performed for every L, providing a basis which is replicated at every equivalent well. (iii) The regularized hamiltonian is scaled, $h(L) = 3DHL/\pi$, and diagonalized in the replicated SCHA basis, keeping a fixed number of states (~ 100) per well and taking advantage of the θ translational symmetry.

It is important to stress that the *true* Hamiltonian is diagonalized in a restricted basis provided by the SCHA as an intermediate step. The effective integration of high modes implied by the SCHA will prove crucial for the success of this approach, (one can consider this step as a RG in disguise). Matrix elements and overlaps pose no severe computational problem owing to the harmonic nature of the basis.

The calculation has all the ingredients of tight-binding band structure problems, and the spectrum adopts the form of enery (ϵ) versus *crystalline* momentum (q), associated with the (discrete) θ-periodicity. θ being charge, q can be identified with flux piercing the ring, enabling transport properties to be read off from $\epsilon(q)$ [29].

We have performed calculations following the above described procedure for values of g in the interval $1/3 \leq g < 1$, with lattice sizes in the range $100 < L < 10^7$, and values of u much smaller than the energy cut-off. The number of states per θ_o well is 100, having checked that doubling this number produces no significant changes. The first important point to remark is the (u, L) dependence of the spectrum of the scaled $h(L)$ through the combination $u^{1/(1-g)}L$, allowing us to measure the effect of the perturbation by the following scaling variable:

$$\tilde{u}(u, L) = 3Du(2/L)^{g-1} \tag{14}$$

Trading $1/L$ for temperature, this is the same scaling variable found by KF [18] to describe their conductance results. This scaling is a consecuence of the RG flow of Eq. (11) for the relevant coupling u. In our scheme, this property comes through the mentioned dependence $\alpha \sim u^{1/(1-g)}$ of the SCHA, emphasizing again its implicit integration of high degrees of freedom.

Representative results for the spectrum are shown in Fig. 2, where the two lower bands $\epsilon_{1,2}(q)$ of $h(L)$ for $g = 3D0.4$ are plotted for increasing values of $\tilde{u}$. (Similar results are obtained for all values of g.) The spectrum evolves smoothly from the fixed point $\tilde{u} = 3D0^+$ [30] to that of $\tilde{u} \to \infty$, where (flat band) energies and degeneracies are, of course, those of a broken chain. The quality of the scaling of Eq. (14) is emphasized by the fact that the central panels of Fig. 2 are superpositions of several (u, L) combinations (see caption). Although our interest is the approach to this last fixed point, the analysis of the departure point $\tilde{u} \sim 0$ provides a critical test of our treatment. In the left panel of Fig. 2 we have superimposed the exact result for the (folded) parabola of current-carrying states of a perfect LL with our results. The agreement is perfect, and one should emphasize that the scheme of our calculation is the same for all values of $\tilde{u}$. That means, in the solid-state language, that we are reproducing the free particle dispersion starting out of a tight-binding basis. Following this parabola upwards in energy, discrepancies between exact and calculated results appear, due to the limited basis employed (this happens around $\epsilon \sim 5$ in the present case). This provides us with a quantitative measure of the validity of our results. All calculations presented here are for energies within this ($\tilde{u} = 3D0$) confidence limit, even though it is clear that this energy window will widens with increasing $\tilde{u}$. A further point to notice in the small $\tilde{u}$ limit is that the *Bragg* gap, $\delta\epsilon_B \equiv \epsilon_2(\pi) - \epsilon_1(\pi)$, opens with a dependence $\delta\epsilon_B \sim \tilde{u} \sim L^{1-g}$. This is the expected perturbative behavior for the $2k_f$ perturbation of a LL, which grows as in the RG flow of Eq. (11). This agreement in the nominally worse situation for a scheme based on a tight-binding approach, gives us confidence in the quality of our approach.

Now we study irrelevant operators. The most prominent effect in the approach to the $\tilde{u} = 3D\infty$ fixed point is the lack of band dispersion. This reflects the vanishing particle hopping across the junction (tunnel between θ_o minima). We characterize this residual hopping by the dispersion of the lowest band, and present

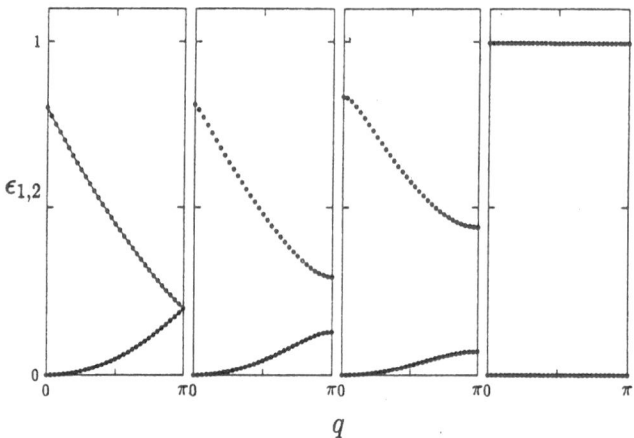

FIG. 2 Two lowest bands $\epsilon_{1,2}(q)$ of the scaled hamiltonian $h(L) = 3DHL/\pi$ with $g = 3D0.4$ for four increasing values of the scaled perturbation [Eq.(14)] $\tilde{u} = 3D0^+, 0.72, 1.63$ and 34.9 (left to right). The continuous line of left panel is the exact result of a perfect LL. The panel for $\tilde{u} = 3D0.72$ is the superposition of results for the following three (u, L) pairs: $(4.5 \times 10^{-2}, 2 \times 10^2)$, $(1.14 \times 10^{-2}, 2 \times 10^3)$, and $(2.87 \times 10^{-3}, 2 \times 10^4)$. The panel for $\tilde{u} = 3D1.63$ is the superposition of results for the following three (u, L) pairs: $(9.3 \times 10^{-3}, 1.1 \times 10^4)$, $(6.14 \times 10^{-3}, 2.2 \times 10^4)$, and $(1.54 \times 10^{-3}, 2.2 \times 10^5)$. (From ref. [15]).

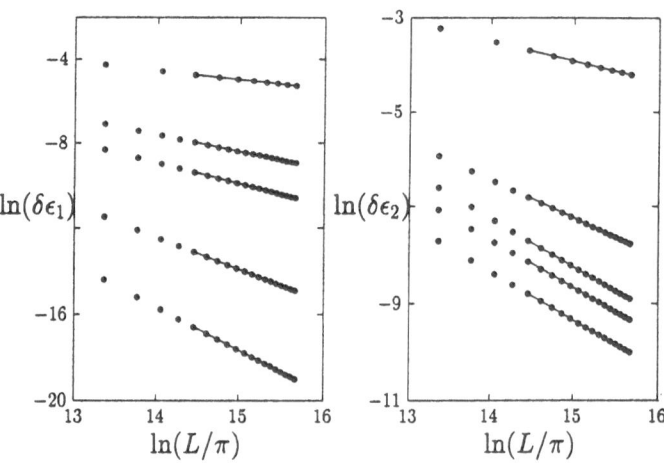

FIG. 3 Left panel: $ln(\delta\epsilon_1)$ [Eq.(15)] versus $ln(L/\pi)$ for $g = 3D0.7, 0.55, 1/2, 0.4$, and $1/3$ (top to bottom). Continuous lines are asymptotic behavior [Eq.(15)]. Right panel: as in left panel for $ln(\delta\epsilon_2)$ [Eq.(16)] with $g = 3D0.7, 0.55, 1/3, 0.4$, and $1/2$ (top to bottom). (From ref. [15].

its L dependence in Fig. 3 (left panel). Notice that this size dependence is not affected by the value of u, which can be chosen arbitrarily (of course, all points for a given value of g share the same value of u). For all values of g, the results show the dependence

$$\delta\epsilon_1 \equiv (\epsilon_1(\pi) - \epsilon_1(0)) \sim L^{1-1/=g}, \tag{15}$$

corresponding to a scaling dimension $1/g$, in agreement with the KF analysis for the hopping t_{eff} across the junction (Eq. (12)).

Let us now consider other clear feature of the spectrum in Fig. 2: the approach of the second band to its asymptotic value at the Brillouin zone center, $\delta\epsilon_2 \equiv (1 - \epsilon_2(0))$. Its L dependence is plotted in Fig. 3 (right panel) and summarized as follows:

$$\delta\epsilon_2 \equiv |1 - \epsilon_2(0)| \sim \begin{cases} L^{1-1/g} & (1/2 < g < 1) \\ L^{-1} & (g < 1/2) \end{cases} \tag{16}$$

If tunnel between wells were the only irrelevant operator, then we would always have the first result of Eq. (16) This is certainly what happens for $g > 1/2$, while for $g < 1/2$ the behavior obtained is precisely that expected from the presence of an operator with scaling dimension 2 describing fluctuations of the pinned phase θ_o, as explained before. These results confirm the picture for two irrelevant operators described at the beginning, with the value $g == 3D1/2$ marking the boundary between regions of dominance.

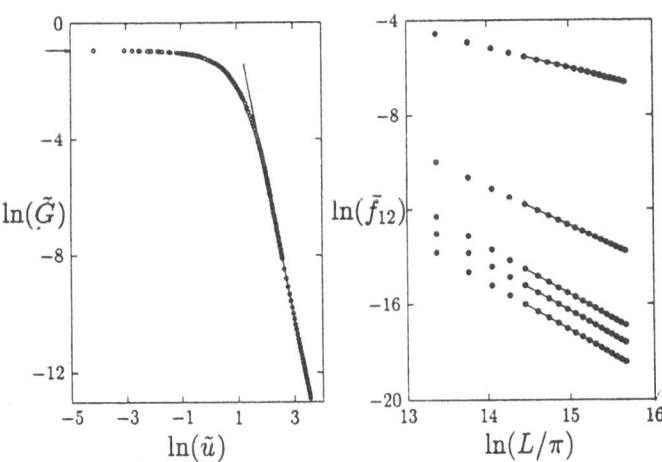

FIG. 4 Left panel: $ln(\tilde{G})$ [Eq.(17)] versus $ln(\tilde{u})$ [Eq. (14)] for $g = 3D0.4$ with results for 200 (u, L) points collapsed onto a single scaling curve. Continuous line is the asymptotic behavior $\sim \tilde{u}^{-2/g}$. Arrow marks the exact limit for $\tilde{u} = 3D0$. Right panel: $ln(\bar{f}_{12})$ [Eq.(19)] versus $ln(L/\pi)$ for $g = 3D0.7, 0.55, 1/2, 1/3$, and 0.4 (top to bottom). Continuous lines are asymptotic behavior [Eqs.(19)]. (From ref. [15]).

Now we study the consequences for transport properties, beginning with the temperature dependence of the static conductance. Although we cannot calculate the conductance of the *infinite* system as a function of T, we can circumvent this problem by defining a running temperature for each size, $T = 3D2/L$, so that the energy scale for a given size is also the temperature. We *define* the conductance for each size-temperature according to

$$\tilde{G}(u, L, T = 3D2/L) = 3D\pi^2 < \partial_q^2 \epsilon >_T, \tag{17}$$

where its well known expression as a thermal average $(<>_T)$ of the second derivative of energy versus flux has been used [29].

Although the $\tilde{G}$ so calculated is not the true (infinite size) conductance as a function of T, it is clear that it will be a universal function of the scaling variable $\tilde{u}$ [Eq. (14)], with the same scaling dependences. Results for $\tilde{G}(\tilde{u})$ are plotted in Fig. 4 (left panel) for $g = 3D0.4$. Notice the correct limit $\tilde{G}(\tilde{u} \rightarrow 0) \rightarrow g$, and a smooth crossover to the asymptotic behavior $\tilde{G}(\tilde{u} \rightarrow \infty) \rightarrow \tilde{u}^{-2/g}$. This limit implies (remenber $T = 3D2/L$) that $\tilde{G} \sim T^{2/g-2}$, in agreement with the KF result. Similar results have been obtained for all values of g, at both sides of $g = 3D1/2$. This should not surprise in view of Eq. (17), where only the operator responsible for band dispersion can appear.

Now we study the dynamic conductance, that is, the ω dependence of the (real) part of G (zero temperature). Placing a time dependent voltage (represented by a vector potential) right at the junction adds a term to the Hamiltonian proportional to the current density Π_o. Calculating matrix elements of Π_o, we can extract the power loss associated with a transition frequency, whereupon $G(\omega)$ can be obtained. We define a (scaled) *oscillator strength* $f_{12}(q)$ for the vertical transition between states of the two lowest bands as

$$f_{12}(q) = 3D(g/2)| < q, 2|\Pi_0|q, 1 > |^2 L^2, \tag{18}$$

and analyze its L dependence for very large sizes. The results are presented in Fig. 4 (right panel), where the (Brillouin-zone averaged) oscillator strength, $\bar{f}_{12} = 3D < f_{12}(q) >_q$, is plotted for several values of g. The following asymptotic behaviors are obtained:

$$\bar{f}_{12} \equiv < f_{12}(q) >_q \sim \begin{cases} L^{2-2/g} & (1/2 < g < 1) \\ L^{-2} & (g < 1/2) \end{cases} \tag{19}$$

It is a simple exercise to show that $\bar{f}_{12} \sim L^{-\eta}$ implies $G(\omega) \sim \omega^\eta$. Thus, our results show that the dynamic conductance changes from $G(\omega) \sim \omega^{2/g-2}$ for $1/2 < g < 1$, to $G(\omega) \sim \omega^2$ for $g < 1/2$.

This behavior is in agreement with the picture presented before, meaning that both irrelevant operators contribute to $G(\omega)$. For $g > 1/2$ hopping dominates and one obtains the KF result. For $g < 1/2$, oscillations of the pinned θ_o dominate, and $G(\omega)$ reflects the capacitance interpretation of this irrelevant peturbation. In physical terms, the oscillations around the pinning center provide the junction with a polarizability that shows up as a capacitor in the dynamic response. It is interesting to notice that a simplified analytical treatment that keeps only the two lowest states per well, shows that the oscillator strength comes from two contributions. One performs the transition for the states of the same well (no particle tunnelling), and is dominated by the *capacitance* operator. The other makes the transition between states of different wells, $\theta_o \to \theta_o \pm \pi$, and only depends on the hopping irrelevant operator. This difference would help explain why the capacitance operator does not show up in the static conductance: its contribution to the oscillator strength is through transitions conserving charge (θ_o).

V. SUMMARY

The core of this lecture has been devoted to the analysis of a LL in the presence of a single impurity as a source of backscattering. After a brief introduction to the fundamentals of a LL and associated bosonization language, transport properties have been studied in both the clean system and system plus impurity, by means of a simplified approach. The results of a more complete analysis have been described, and a finite size approach has been introduced to perform a detailed numerical investigation of the backscattering problem in the $g < 1$ regime. A successful numerical analysis of the spectrum for very large sizes has been made possible thanks to the use of an intermediate step that effectively integrates high degrees of freedom, as in the RG approach. Existence of irrelevant operators and transport consequences are issues treated independently in our treatment. We confirm the existence of two irrelevant operators, representing particle hopping (dimension $1/g$) and charge oscillations (dimension 2), respectively. Temperature dependence of the static conductance has been shown to be controlled by hopping alone, giving $G \sim T^{2/g-2}$. Frequency dependence is affected by both operators, giving $G \sim \omega^\eta$, with $\eta = 3D(2/g - 2)$ for $g > 1/2$ and $\eta = 3D2$ for $g < 1/2$. This provides a clarification of the transport properties of LL with backscattering in the repulsive $(g < 1)$ regime.

Acknowledgements. The author thanks Prof. F. Guinea for very stimulating discussions. This work has been supported by the DGICyT of Spain (grant PB92-0169).

[1] J. G. Bednorz and K. A. Muller, Z. Phys. B **64** , 188 (1986).

[2] P. W. Anderson, Phys. Rev. Lett. **64**, 1839 (1990); Science **235**, 1196 (1987).

[3] See, for instance, A. A. Abrikosov, L. P. Gorkov, and I. E. Dzyaloshinski, *Methods of Quantum Field Theory in Statistical Mechanics* (Dover, New York, 1963).

[4] V. J. Emery, in *Highly Conducting One-Dimensional Solids*, ed. by J. T. Devreese (Plenum Press, New York, 1979).

[5] J. Solyom, Adv. Phys.**28**, 201 (1979).

[6] S. Tomonaga, Prog. Theor. Phys. **5**, 544 (1950).

[7] D. C. Mattis and E. H. Lieb, J. Math. Phys. **6**, 304 (1965). J. M. Luttinger, J. Math. Phys. **4**, 1154 (1963).

[8] M. den Nijs, Phys. Rev. B **23**, 6111 (1981).

[9] F. D. M. Haldane, J. Phys. C **14**, 2585 (1981); Phys. Rev. Lett. **47**, 1840 (1981).

[10] See, for example, *Organic and Inorganic Low Dimensional Crystalline Materials*, ed. by P. Delhaes and M. Drillon, NATO ASI, Ser. B, vol. 168 (Plenum Press, New York, 1987).

[11] G. Timp in *Mesoscopic Phenomena in Solids*, edited by B. L. Altshuler, P. A. Lee, and R. A. Webb (Elsevier, Amsterdam, 1990).

[12] S. Tarucha, T. Honda, and T. Saku, Solid State Commun. **94**, 413 (1995).

[13] X. G. Wen, Phys. Rev. B **43**, 11025 (1991); Phys. Rev. Lett. **64**, 2206 (1990).

[14] F. P. Milliken, C. P. Umbach, and R. A. Webb, Solid State Commun. **97**, 309 (1996).

[15] G. Gómez-Santos, Phys. Rev. Lett. **76**, 4223 (1996).

[16] H. J. Schulz, lectures in this volume.

[17] R. Landauer, IBM J. Res. Dev. **1**, 223 (1957). M. Buttiker, Phys. Rev. Lett. **57**, 1761 (1986).

[18] C. L. Kane and M. P. A. Fisher, Phys. Rev. Lett. **68**, 1220 (1992); Phys. Rev. B **46**, 15233 (1992).

[19] A. Schmid, Phys. Rev. Lett. **51**, 1506 (1983).

[20] F. Guinea, V. Hakim, and A. Muramatsu, Phys. Rev. Lett. **54**, 263 (1985).

[21] M. P. A. Fisher and W. Zwerger, Phys. Rev. B **32**, 6190 (1985).

[22] A. O. Gogolin, Phys. Rev. Lett. **71**, 2995 (1993).

[23] F. Guinea, G. Gómez-Santos, M. Sassetti, and M. Ueda, Europhys. Lett. **30**, 561 (1995).

[24] C. de C. Chamon, D. E. Freed, and X. G. Wen, Phys. Rev. B. **51**, 2363 (1995); preprint cond-mat/9507064.

[25] A. M. Tsvelik, preprint cond-mat/9409027.

[26] C. L. Kane and M. P. A. Fisher, Phys. Rev. Lett. **76**, 3192 (1996).

[27] P. Fendley, A. W. W. Ludwig, and H. Saleur, Phys. Rev. Lett. **74**, 3005 (1995); Phys. Rev. B **52**, 8934 (1995).

[28] K. G. Wilson, Rev. Mod. Phys. **47**, 773 (1975).

[29] W. Kohn, Phys. Rev. **133**, 171 (1964).

[30] The calculation requires a minimum value of $u \ll 1/L$ to be technically possible. For all practical purposes this means $\tilde{u} = 3D0$.

Skyrmions in the Quantum Hall Effect

L. Brey[1]* H.A.Fertig[2], R.Côté[3] and A.H.MacDonald[4].

[1] *Instituto de Ciencia de Materiales (CSIC),*
Universidad Autónoma , 28049, Madrid, Spain.
[2] *Department of Physics, University of Kentucky,*
Lexington, Kentucky 40506-0055.
[3] *Département de Physique,*
Université de Sherbrooke, Sherbrooke,
Québec, Canada J1K 2R1.
[4] *Department of Physics, Indiana University,*
Bloomington, Indiana 47405.

The lowest energy charged excitations of the filling factor $\nu=1$ quantum Hall ferromagnet are Skyrmions. The net spin of the Skyrmion's is always larger than $1/2$, in such a way that adding or removing charge from a Hall ferromagnet rapidly degrades its spin polarization. In the ground state of a two-dimensional electron gas at Landau level filling factor ν near 1 a finite density of Skyrmions could exits. In that case and for Zeeman coupling different from zero, the ground state is a Skyrme crystal with both spontaneous long range order in the charge density and spontaneous long range order in the transverse spin density. The energy of the Skyrme crystal is lowest for a square lattice structure with opposing postures for topological excitations on opposite sublattices. We discuss interpretations of our results in terms of non-linear σ and generalized spin models for quantum Hall ferromagnets. In the zero Zeeman coupling case, we find that the ground state can be interpreted as a meron crystal with two interpenetrating sublattices, each supporting quasiparticles with charge $e/2$.

PACS 73.40.Hm

*Corresponding author. E-mail: brey@marlene.fmc.uam.es;
FAX: 34-1-372-0623

I. INTRODUCTION

The ground state of a two-dimensional electron gas (2DEG) in a strong magnetic field, B, at filling factor $\nu = 1$ is ferromagnetic [1-3] with total spin $S = N/2$. (Here, $\nu=N/N_\phi$ where N is the number of electrons, $N_\phi=AeB/hc \equiv A/(2\pi\ell^2)$ is the orbital degeneracy of the Landau level, ℓ is the magnetic length and A is the sample area.) For weak Zeeman coupling the charged excitations of this system [2,3] are *Skyrmions*, the lowest energy spin texture excitations of the Non-Linear σ (NLσ) model, valid for long length scale excitations in any ferromagnetic system. The Skyrmions carry a charge in quantum Hall ferromagnets because [2] of the commensurability relations between magnetic flux and charge density required by incompressibility. For $N = N_\phi \pm 1$, the ground state contains a single charged

Skyrmion. The Skyrmions can be introduced by changing the total electron number or, in what is the typical experimental situation by changing the magnetic field strength and hence N_ϕ. A Skyrmion is characterized by the sign of its topological charge, the location of its center, its size, and the global orientation of its spin density. In the NLσ model, the energy of an isolated Skyrmion depends *only* on its topological charge. Recently [4], we developed a microscopic approach using the Hartree-Fock approximation (HFA) that can be used for quantitative studies of Skyrmion properties, including the effects of Zeeman coupling and realistic (*i.e.*, Coulomb) electron-electron interaction not captured by the NLσ model. We found that at $\nu = 1$, the Skyrmion energies are always smaller than the excitations energies of localized spin 1/2 quasiparticles. The Zeeman coupling favors small Skyrmions, since the spins near the center of the Skyrmion is oriented in opposition to the Zeeman field. On the contrary the repulsive Coulomb interaction favors long Skyrmions. For Zeeman coupling strengths typical of experimental systems we found that, although quasiparticle energy reductions were modest, the number of reserved spins in a Skyrmion was large. On this basis we predicted that deviations of ν from 1 would lead to large reductions in the spin polarization of the ground state. These predictions were confirmed when Barret *et al* unexpectedly succeeded [5] in using optical pumping techniques to perform NMR Knight shift measurements of the spin-polarization of two-dimensional electron systems in the Quantum Hall regime. The results of this experiment indicate that each extra charge in the system, with respect $\nu = 1$, carries a spin 3.5, in quantitative agreement with the microscopic predictions obtained using the Hartree-Fock approximation [4]. There seems to be little doubt that the elementary charged excitations of the quantum Hall ferromagnets are Skyrmions-like objects that carry large spin quantum numbers. Recent transport [6] and optical [7] experiments add additional support to this conclusion.

In this work we focus on the ground state of a 2DEG for $\nu \neq 1$. In Sect. II we discuss the finite Zeeman coupling case where we have found that the ground state is a square lattice crystal of Skyrmions, with long-range order in both the charge density and components of the spin density perpendicular to the magnetic field. We also present a qualitative interpretation of these results in terms of NLσ model considerations. Section III focuses on the zero Zeeman coupling limit. We have found that the ground state is a crystal which consists of two interpenetrating square lattices supporting quasiparticles with charge $e/2$. Finally in Sect. IV we briefly summarize.

II. SKYRME CRYSTAL

A. Summary of Hartree-Fock Results

At filling factors near $\nu = 1$, a finite density of Skyrmions exists in the 2DEG. In that case, and for finite Zeeman coupling, we have found [8] that the lowest energy state of this system is a Skyrme *crystal*, with lattice parameter that is proportional to $|1 - \nu|^{-1/2}$. Over a wide range of filling factors, the spin textures of the individual Skyrmions are strongly coupled, favoring a state in which nearest neighbor sites have opposing orientations. (For a magnetic field in the $\hat{z}$ direction, we define the orientation of a Skyrmion as the angle formed by transverse $(\hat{x} - \hat{y})$ projection of the spin density at some reference point near the Skyrmion center, and the vector connecting that point to the Skyrmion center.) For an isolated Skyrmion the energy is independent of the orientation. For extremely low Skyrmion densities, a triangular lattice is favored, consistent with the prefered structure in two-dimensions for crystals of point charges. However, for experimentally relevant parameters, the system undergoes a phase transition to the square lattice at very small values of $|\nu - 1|$. Furthermore the spin-polarization of the square lattice Skyrme crystal is much smaller, for a given Zeeman coupling and filling factor, than for the triangular lattice crystal.

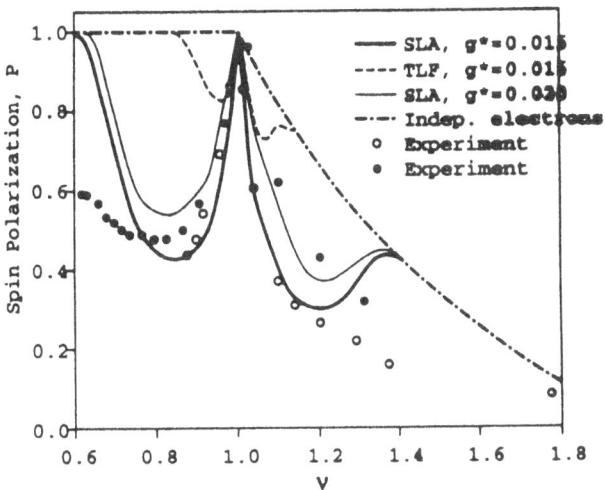

FIG. 1 Variation of the spin polarization, P, as a function of ν, for square (SLA) and triangular (TLF) Skyrme crystal states. The open circles are experimental data obtained for g^* near $0.016(e^2/\ell)$, while the filled circles show data obtained for g^* near $0.021(e^2/\ell)$.

Recently, the Knight shift of the ^{71}Ga nuclei located in n-doped GaAs quantum wells for $0.66 < \nu < 1.76$ has been measured [5], using specialized NMR techniques. This quantity is proportional to the spin polarization, P, of the electron gas confined in the well. It was found that P drops very quickly on either side of ν=1 as we had predicted. Figure 1 compares the experimental spin polarization, with calculated spin polarizations for square and triangular Skyrmion lattices, at different values of the Zeeman coupling. The non-interacting electron spin polarization is also shown, to illustate the dramatic lowering of the spin polarization which occurs because Skyrmion spin texture occur in the ground state of the system. The agreement between experiment and the square lattice results is excellent in the range $0.8 < \nu < 1.2$.

In Fig. 2a we show, for the case of $\nu = 1.1$ and Zeeman energy $0.015e^2/\varepsilon\ell$, the excess of charge density with respect the density of the $\nu = 1$ Hall ferromag-

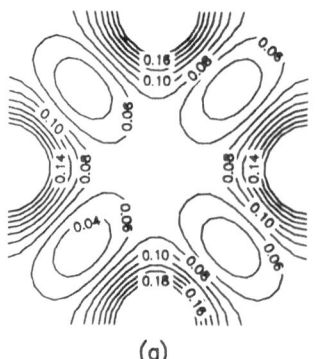 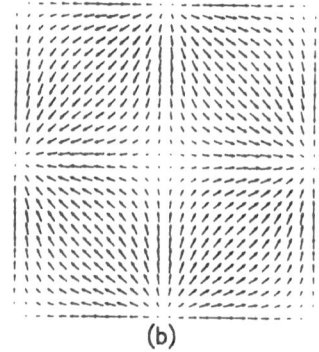

(a) (b)

FIG. 2 (a) Excess charge density and (b)two-dimensional vector representation of the transverse component of the spin density. These results corresponds to the case $\nu = 1.1$ and Zeeman coupling $0.015e^2/\varepsilon\ell$. In both figures the illustrated area is the unit cell used in the calculations. In (a) the numerical values are given in units of $1/2\pi\ell^2$.

net. The figure corresponds to an unit cell of the square lattice of the Skyrme crystal, therefore there are two extra electrons per unit cell. Note there are two maxima of charge density per unit cell. Fig. 2b shows the two-dimensional vector representation of the transverse component of the spin density.

The Hartree-Fock approximation results shown in Figs. 1 and 2 were obtained by using a self-consistently truncated one electron Greens function equation of motion approach [9]. Remarkably simplifications are obtained in this approach by working entirely within the lowest orbital Landau level.

B. Non Linear σ Model Interpretation

Many aspects of the results we have obtained using the microscopic Hartree-Fock approximation can be understood in terms of the NLσ model description of the 2DEG near ν=1. The system near $\nu = 1$ can be described by a unit vector order parameter field, $\mathbf{m}(\mathbf{r})$, directed along the local spin density orientation. In the limit of slowly varying spin textures, macroscopic considerations involving Berry phases and the quantized Hall conductivity, imply the following relation between the excess of charge density compared to that of the uniform of the $\nu = 1$ ferromagnetic ground state, and the order parameter $\mathbf{m}(\mathbf{r})$:

$$q(\mathbf{r}) = -\frac{1}{8\pi}\epsilon_{\nu\mu}\mathbf{m}(\mathbf{r}) \cdot [\partial_\nu\mathbf{m}(\mathbf{r}) \times \partial_\mu\mathbf{m}(\mathbf{r})] \quad . \tag{1}$$

The right hand side of Eq. 1 turns out to be the Pontryagin index density, or topological charge density associated with the vector field $\mathbf{m}$. [2,3] The Pontryagin index for Skyrmion states of the NLσ model is ±1 and these states therefore carry total excess charge $\pm e$.

Using gradient expansions for slowly varying spin densities and the relation between electric and topological charge densities, the Hamiltonian describing low energy excitations of the 2DEG at $\nu=1$ must have the form

$$H = \frac{\rho}{2} \int d\mathbf{r} \, (\nabla m^\mu)^2 + \frac{1}{2} \int d\mathbf{r} d\mathbf{r} \, q(\mathbf{r}) V(\mathbf{r} - \mathbf{r}') q(\mathbf{r}')$$

$$+ \frac{g^*}{\pi \ell^2} \int d\mathbf{r} (m_z(\mathbf{r}) + 1) \ . \tag{2}$$

The first term in Eq. 2 is the only term present in the NLσ model, and the constant ρ is known as the spin stiffnes. The second term describes the Hartree energy of the charge density associated with $\mathbf{m}(\mathbf{r})$. $V(\mathbf{r})$ is the Coulomb interaction. The third term describes the loss in Zeeman energy when electronic spins are not aligned with the magnetic field. Here $g^* = g \mu_B B$ where μ_B the electron Bohr magneton and g is the g-factor of the host semiconductor. It is often convenient to discuss states in terms of the following parametrization for $\mathbf{m}$:

$$m_x + i m_y = \frac{2w}{1 + |w|^2} \ . \tag{3}$$

In terms of w, the charge density is given by :

$$q(z) = \frac{1}{\pi} \frac{|\partial_z w|^2 - |\partial_{z^*} w|^2}{(1 + |w|^2)^2} \ , \tag{4}$$

where $z = x + iy$.

Using this parameterization, the lowest energy non-trivial solutions of the pure NLσ model with topological charge n have the form [10]

$$w(z) = \sum_{i=1}^{n} \frac{\lambda_i}{z - z_i} e^{i\theta_i} \ , \tag{5}$$

where λ_i specifies the size of the Skyrmion centered at z_i, and θ_i specifies the its orientation. In the NLσ model the energy of this solution is $4\pi \rho n$, independent of λ_i, z_i and θ_i.

When the Hartree and Zeeman terms are included, the form of the Skyrmion changes, but many aspects of the positional and orientational order of the solution of the Hamiltonian (2) are still captured by considering the spin-textures defined in Eq. 5. The form of the function w here Eq.(5) ensures that the spin is aligned with the Zeeman field far from the Skyrmion centers. The size of an individual Skyrmion is controlled by the Coulomb repulsion within a single Skyrmion, which favors large size Skyrmions, and the Zeeman coupling and Skyrmion-Skyrmion Coulomb interactions, which favor small sizes. When the Skyrmion size is small compared to the typical inter-Skyrmion distance, the Skyrmion positions at zero temperature, z_i, will be lattice sites. The simplest Skyrme lattices have identical orientations and sizes on all sites, which can be described approximately by a function of the form:

$$w(z) = \lambda e^{i\theta} \sum_{i=1}^{n} \frac{1}{z - z_i} \ . \tag{6}$$

In the model of Eq. 2, the interaction energy between Skyrmions depends on the charge density associated with a spin texture. The charge density is more localized around the center of the quasiparticles in the case of Skyrmions with opposing orientation than in the case of Skyrmions with identical orientation. For this reason the interaction energy between Skyrmions is reduced when they have opposing orientations. This tendency is frustrated on a triangular lattice, hence the Skyrmions prefer to crystallize in a square lattice with two Skyrmions per unit cell (SLA state). This state can be described approximately by the following variational function:

$$w(z) = \lambda e^{i\theta} \sum_{i=1}^{n} \{ \frac{1}{z - z_i} - \frac{1}{z - z_i - z_0} \} \ , \tag{7}$$

with z_i the lattice vectors of the square lattice and z_0 the coordinate of the center of the square lattice unit cell. In the TLF state the repulsion between Skyrmions is stronger than in the SLA state therefore the Skyrmions are smaller in the TLF state. This results in a smaller spin polarization in the TLA state than in the SLF state (Fig.1). In the very dilute limit, when the distance between Skyrmions is much larger than the size of an isolated Skyrmion, we expect than the ground state will be determined by the Madelung energy and the ground state will be a triangular lattice.

At larger densities, we expect that quantum fluctuations will melt the crystal. It is interesting to speculate that the typical coordination number of the charges in the resulting liquid state will be four, in order to minimize the interaction energy of the spin texture.

III. ZERO ZEEMAN COUPLING CASE: THE MERON CRYSTAL

The zero Zeeman coupling case, which is likely to be carefully studied experimentally in the future, is of special interest. Using the Hartree-Fock approximation we have studied the ground state of a 2DEG, with $g^*=0$, as a function of the filling factor. In Fig. 2 we show the excess of charge density and the two-dimensional vector representation of the transverse components of the spin density for the case of $\nu = 1.1$. These figures show the charge density variation over the unit cell used in these calculations, which have two extra electrons compared to the density at $\nu = 1$. The lattice and spin orientation symmetry of the HFA solution in this case can be qualitatively described by a function of the form

$$w(z) = h \prod_{i=1}^{n} \frac{z - a_i}{z - b_i} \ , \tag{8}$$

where h is a parameter and a_i and b_i are the position where the spin texture points in the positive and negative z direction respectively. The ground state of

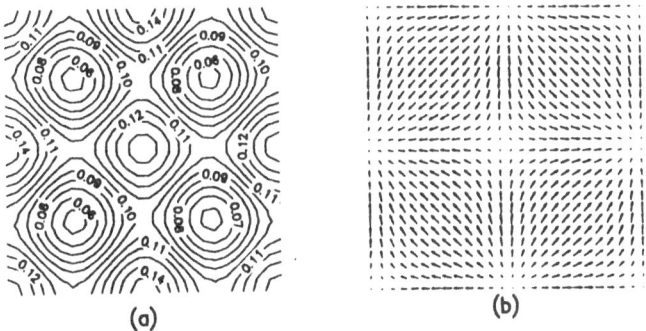

FIG. 3 (a) Excess charge density and (b)two-dimensional vector representation of the transverse component of the spin density. These results corresponds to the case $\nu = 1.1$ and zero Zeeman coupling. In both figures the illustrated area is the unit cell used in the calculations. In (a) the numerical values are given in units of $1/2\pi\ell^2$.

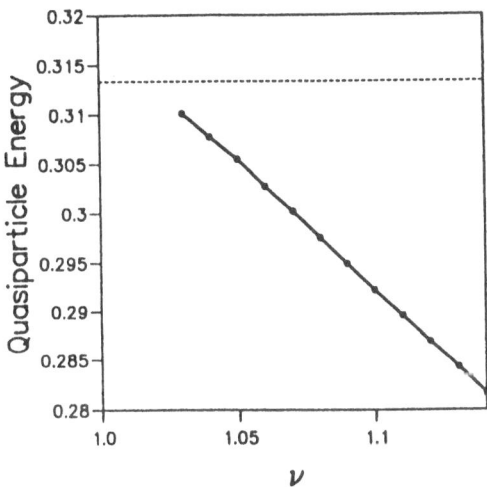

FIG. 4 Variation of the quasiparticle energy as a function of the filling factor, for the case $g^* = 0$. Energies are in units of e^2/ℓ. The dashed line corresponds to the energy of a isolated Skyrmion $\sqrt{\pi/32}$.

the system, shown in Fig.3 is well described by Eq. 8 with the coordinates a_i and b_i on the sites of interpenetrating lattices with lattice parameter $\sqrt{2\pi/|1-\nu|}$. The z-component of the spin density is positive around the nodes of one of the sublattices and negative around the nodes of the other sublattice. In this form the spin polarization of the system is zero for any filling factor.

From the Hartree-Fock results we also obtain the energy (ϵ_{qp}) increase per quasiparticle, when the quasiparticles are created at fixed N by varying N_ϕ:

$$\epsilon_{qp} = \frac{\nu \left[\epsilon + \frac{e^2}{\ell} \sqrt{\pi/8} \right]}{|1 - \nu|}, \tag{9}$$

where $\epsilon = E/N$ is the energy per electron obtained in the Hartree Fock approximation. The term $-e^2/\ell\sqrt{\pi/8}$, is the exchange energy per particle at $\nu=1$. With this convention, we can compare the quasiparticle energy ϵ_{qp} obtained in the HFA at $\nu \neq 1$, with the field theory value for the quasiparticle energy of an isolated infinitely large Skyrmion, $\epsilon_{qp}^{isolated} = \sqrt{\pi/32}e^2/\ell$.

In Fig. 4 we plot the quasiparticle energy, ϵ_{qp} as a function of the filling factor. Although numerical limitations restrict our calculations to the range $|\nu-1| > 0.03$, it is clear from Fig. 4 that in the limit $\nu \to 1$ the quasiparticle energy tends to $\epsilon_{qp}^{isolated}$.

The results in Figs. (3)-(4) demonstrate that to minimize the interaction energy the charge density is modulated and the position of the centers $a_i's$ and $b_i's$ crystallize. The alternanting position of the centers $a_i's$ and $b_i's$ in the crystal, ensures that the spin polarization in the system is zero, as expected in the absence of Zeeman coupling. Since for each Skyrmion there are two centers a_i, b_i, the charge associated with each of these is half an electron [see Fig. (3a)]. In analogy with the charged excitations which occurs in easy-plane ferromagnet states [3,11,12], we call this state a *Meron Crystal*.

Results similar to those shown in Fig.3 are obtained in our dilute limit, ($\nu = 1.03$). From our calculations it seems clear that, for $g^* = 0$, isolated Skyrmion states are not obtained by letting ν tend to 1; the $g^* \to 0$ and $\nu \to 1$ limits do not commute. For $g^* = 0$, the size of the lowest energy isolated Skyrmion is comparable to the the sample size. [3] When $\nu \neq 1$, the size and form of the spin textures are determined by the other spin textures present in the system. That is the reason why even in the very dilute limit the constituent Skyrmions interact strongly and a meron crystal is always energetically favored.

At large Skyrmion densities, we expect that quantum fluctuations will melt the Skyrme crystal even at $T = 0$. For $g^* = 0$ it seems likely that in some regime the ground state will be a liquid state of half electron charged quasiparticles, in agreement with recent theoretical calculations [13], which have shown that at finite temperatures and low Skyrmion density the ground state of the system is described by a classical two-dimensional Coulomb plasma at temperature $1/4\pi$ with a_i and b_i being the coordinates of positive and negative charges.

IV. CONCLUSION

We have computed the energy, spin texture and spin polarization of 2DEG system near $\nu = 1$. For finite Zeeman coupling the lowest energy state is a square lattice of spin-textures with ordered orientations. The spin polarization of this state is in excellent agreement with recent experiments. Here we have presented an interpretation of these results in terms of the NLσ model. We have also studied the case of zero Zeeman coupling. In this case we have found that the ground state is a meron crystal of interpenetrating square lattices supporting quasiparticles with opposite spin orientations and charge $e/2$. This result is in agreement with previous theoretical work.

Acknowledgements. The authors thank S.M. Girvin, C.Tejedor, J.P.Rodriguez, A.Somoza, for helpful discussions. This work was supported in part by NATO CRG No. 930684, by the NSF through Grants Nos. DMR 95-03814 and DMR 94-16906, by the CICyT of Spain under Contract No. MAT 94-0982, by NSERC of Canada, and by FCAR from the Government of Quebec. HAF acknowledges the support of the A.P.Sloan Foundation and the Research Corporation.

[1] E.H. Rezayi, Phys. Rev. B**36**, 5454 (1987); **43**, 5944 (1991).
[2] S.L.Sondhi *et al.*, Phys.Rev.B **47**, 16419 (1993); D.H.Lee and C.L.Kane, Phys.Rev.Lett. **64**, 1313 (1990).
[3] K.Moon*et al.*, Phys.Rev.B **51**, 5138 (1995).
[4] H.A.Fertig, L.Brey, R.Côté and A.H.MacDonald, Phys.Rev.B **50**, 11018 (1994); *ibid* submitted to Phys. Rev. B (1996).
[5] S.E.Barret *et al.* Phys.Rev.Lett. **74**,5112 (1995).
[6] A.Schmeller *et al.* Phys.Rev.Lett.**75**,4290 (1995).
[7] E.H.Aifer *et al.* Phys.Rev.Lett. **76**, 680 (1996).
[8] L.Brey, H.A.Fertig,R.Côté and A.H.MacDonald, Phys.Rev.Lett. **75**, 2562 (1995); Surface Science, *in press*.
[9] R. Côté and A. H. MacDonald, Phys. Rev. B **44**, 8759 (1991) ; R. Côté, L. Brey and A. H. MacDonald, Phys. Rev. B **10**, 10239 (1992).
[10] R.Rajaraman, "Solitons and Instantons" (North-Holland, Amsterdam 1982).
[11] L.Brey, H.A.Fertig, R.Côté and A.H.MacDonald, *umpublished*.
[12] I.Tupitsyn *et al*, Phys.rev.B **53**, R7614 (1996).
[13] A.G.Green, I.I.Kogan and A.M.Tsvelik, Phys.Rev.B, *in press*.

Photoemission Bands in Systems of Strongly Correlated Electrons

Stephan Haas

Theoretische Physik, ETH-Hönggerberg, CH-8093 Zürich, Switzerland

(May 14, 1997)

Recent photoemission experiments on cuprate superconductors in their normal state are analyzed in the context of models of strongly correlated electrons. The phenomenon of antiferromagnetically induced bands close to half-filling is discussed. Also, flat regions in the dispersion of the valence band related to van Hove singularities in the electronic density of states are shown to be present. The possibility of having small, pocket-like Fermi surfaces in the vicinity of half-filling is mentioned. Finally, some analogies are presented between two-dimensional cuprates and related one-dimensional materials.

I. INTRODUCTION

In recent years, there has been much refinement in photoemission experiments. These probes of the electronic structure are of particular interest in the context of high-T_c superconductors because they investigate the nature of the microscopic mechanisms underlying the observed metal-insulator transition, as well as of the formation of a superconducting condensate. Photoemission techniques have by now gained remarkable refinement allowing the observation of weak superstructures in addition to a clearly resolved Fermi surface. For example, the 5×1 structural modulation which has been observed by x-ray spectroscopy for $Bi_2Sr_2CaCu_2O_8$ (Bi2212) is clearly resolved by recent angular resolved photoemission spectroscopy (ARPES) measurements, although its typical amplitude is an order of magnitude smaller than that of the main Fermi surface. [1]

Other groups [2,3] have achieved extremely high energy resolutions, allowing the observation of features in the superconducting gap function, which upon further refinement might give valuable *quantitative* information on the nodal structure of the gap. While the present resolution of features in the energy range of the SC gap (~ 12 meV in the cuprates) is still under debate, in this paper we will concentrate on structures which occur on larger energy scales (i.e. those of $J \simeq 130$meV and $t \simeq 2.5$ J) in the normal state.

The low-energy (valence band) physics of the cuprate materials has been argued to be contained in rather simple, single-band electronic models, such as the t-J and Hubbard models. As reviewed recently by Dagotto, [4] these models can be (approximately) derived from more general, first-principles (multi-band) Hamiltonians, and give a good quantitative description of a number of normal-state observables in the cuprates (including charge transport, NMR, optical conductivity, electronic Raman spectra).

The one-band Hubbard model is given by

$$H_{Hub} = -t \sum_{<ij>,\sigma} (c_{i,\sigma}^\dagger c_{j,\sigma} + h.c.) + U \sum_i (n_{i\uparrow} - \frac{1}{2})(n_{i\downarrow} - \frac{1}{2}), \tag{1}$$

where $c_{i,\sigma}^\dagger$ creates an electron at site i with spin projection σ, $n_{i,\sigma}$ is the number operator, and the sum $\langle ij \rangle$ runs over pairs of nearest neighbor lattice sites. U is the on-site Coulomb repulsion, and t is the hopping amplitude.

On a given site in the Hubbard model, there can thus be either an up-spin or a down-spin electron, a hole, or simultaneously two electrons with opposite spins. For large U, double occupancy is disfavored. In particular, in the limit $U/t \to \infty$, doubly occupied sites will be of infinite energy, and an effective low-energy Hamiltonian, i.e. the t-J Hamiltonian (acting in the restricted Hilbert space of no double-occupancy), is known to be valid. Its Hamiltonian is given by

$$H_{tJ} = -t \sum_{<i,j>\sigma} (\tilde{c}_{i\sigma}^\dagger \tilde{c}_{j\sigma} + h.c.) + J \sum_{<i,j>} (\mathbf{S}_i \cdot \mathbf{S}_j - \frac{1}{4}n_i n_j), \tag{2}$$

where the č-operators are hole operators acting on non-doubly occupied states, J is the exchange integral, and t is the hopping amplitude. The exchange integral is related to the parameters of the one-band Hubbard model in the limit of strong coupling by $J \approx \frac{4t^2}{U}$. At half-filling, both the Hubbard and the t-J model are in an antiferromagnetic phase consistent with experimental findings in the cuprates.

Note that the relevant parameters for the cuprate materials belong to the strong coupling regime, i.e. $U \approx 10t$ ($t \approx 0.4eV$) for the Hubbard model. Unfortunately, this limit cannot be easily treated by perturbative methods because there is no small parameter about which to expand. Then unbiased numerical tools such as Monte-Carlo simulations and exact diagonalizations of small clusters become valuable tools in exploring the properties of these models in the regime of interest.

In this paper, we shall discuss some properties of the low-energy electronic structure of the cuprates close to the half-filled case, which have recently received much attention :

- At half-filling, it is known that both the Hubbard and the t-J model in 2D are in an antiferromagnetic phase, for any finite U and temperature T=0. As a consequence, the corresponding magnetic Brillouin zone has half the volume of the non-interacting one, introducing additional (magnetic) bands into the spectrum. These bands have been observed in photoemission experiments on $Sr_2CuO_2Cl_2$. It has been proposed that these "shadow" features in the photoemission spectra will be present - although of much weaker weight - even away from half-filling, i.e. where only short-range AF fluctuations are present. We discuss some numerical work which attempts to quantify this effect of remnant AF weight in PES.

- The presence of flat regions in the valence band leads to van Hove singularities in the corresponding electronic density of states. In traditional scenarios of superconductivity, the critical temperature is proportional to the DOS at the Fermi level. Thus it has been proposed that the high critical temperature in the cuprates may be related to the occurrence of flat bands which have been observed in PES. We will show that such features are indeed present also in the model Hamiltonians we study.

- It has been proposed that the crossover from AF at half-filling to a Fermi liquid close to zero-filling leads through a non-Fermi-liquid phase between half-filling and optimal hole doping ($\approx 12\%$). A variety of prominent deviations from Fermi liquid behavior is observed in this regime, e.g. a linear temperature dependence of the normal state resistivity. We will briefly comment on the possibility of having small, pocket-like Fermi surfaces in this region as opposed to the large Fermi surface which is observed at larger hole-doping levels. This is a topic of much controversy which is far from being settled.

- Finally, we will comment on some analogies of the 2D cuprates with related, lower-dimensional systems, such as ladder materials and quasi-1D chains. The question will be whether or not some of the strong-coupling paradigms, such as shadow features, occur in other low-dimensional compounds.

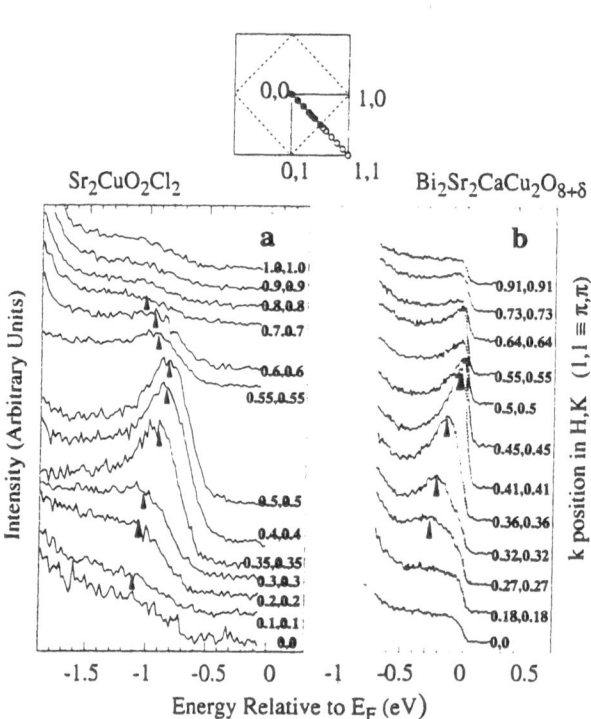

FIG. 1. (a) Angle Resolved Photoemission Spectra in $Sr_2CuO_2Cl_2$. The cut in momentum space shown here is taken along the diagonal of the Brillouin zone, i.e. from $(0,0)$ to (π, π). The spectral weight above the Fermi momentum $(\pi/2, \pi/2)$ is attributed to a folding of the unit cell due to antiferromagnetic long-range order in the system. (b) Same for Bi2212. This system only has short-ranged antiferromagnetic order with a correlation length of 2-3 lattice spacings.

II. ANTIFERROMAGNETICALLY INDUCED PES WEIGHT IN CUPRATE MATERIALS

In a recent letter, Wells et al. [5] reported ARPES measurements on $Sr_2CuO_2Cl_2$. This material is an insulating layered copper oxide which is difficult to dope. The results are interesting because they represent an experimental probe of the well-studied problem of a single hole moving in an antiferromagnetic background. Fig. 1(a) shows the dynamical spectral function for different momenta along the diagonal in momentum space ($k_x = k_y$). One remarkable feature of these spectra is that beyond the naive Fermi momentum of this half-filled system ($\mathbf{p_F} = (\pi/2, \pi/2)$) there is a clearly resolved peak for $p > p_F$. As we will discuss in detail, this "shadow" feature appears to persist even for related materials with a finite amount of doped holes such as $Bi_2Sr_2CaCu2O_{8+\delta}$ (see Fig. 1(b)). It originates from a magnetic superstucture due to strong Coulomb interactions among the electrons in the system which appears in generic models like the t-J model or the one-band Hubbard model with a sufficiently large on-site repulsion U.

There are not many cuprate materials which at finite hole doping levels have a suitably clean surface for PES experiments. An exception is Bi2212. This compound becomes superconducting below a critical temperature of $T_c = 89K$. ARPES experiments in the normal phase (at room temperature) with a fine resolution in k-space have recently been performed by Aebi et al. (Fig. 2).

In addition to a clearly resolved main Fermi surface they observe two weaker superstructures : (i) a 5×1 lattice modulation particular to Bi2212, and (ii) a "shadow Fermi surface" translated by a vector $\mathbf{Q} = (\pi/a, \pi/a)$ from the main Fermi surface. While there is definitely no long-range antiferromagnetic order in this system (as opposed to $Sr_2CuO_2Cl_2$) there is reason to believe that $short - range$ AF correlations are responsible for the appearance of this weak "2×2" structure. Fig. 2 shows a cut from a point near $\mathbf{p} \approx (\pi, \pi)$ (since there is a slight orthorhombic distortion in Bi2212 there are two almost equivalent (π, π) points in this material) to a point near $\mathbf{p} = (\pi, 0)$. This was done by fixing the polar angle at $\theta = 39°$ and scanning the azimuthal angle by rotating the sample in increments of 1°. The spectral weight measured in energy windows of ~20meV is plotted along the cut in k-space starting at the bottom at 0.3 eV above the Fermi energy, and arriving at the last (top) line at E_F. As E_F is approached, the spectral intensity increases, indicating a crossing of the Fermi surface.

Bi2212

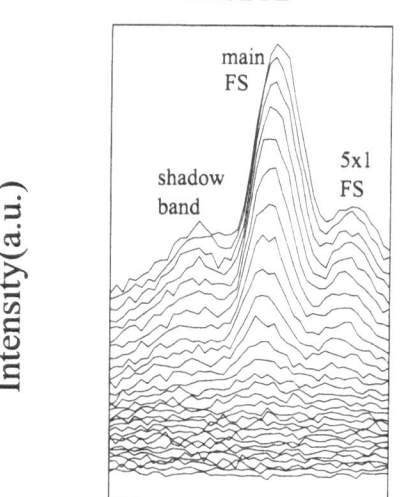

FIG. 2. Angle Resolved Photoemission Spectra in Bi2212. This cut in momentum space shows the main Fermi surface and two weaker superstructures, the 5×1 structural modulation and a 2×2 shadow band due to short-range antiferromagnetic correlations in the material.

In Fig. 3 the Fermi surface is shown along the entire Brillouin zone for Bi2212. The main Fermi surface is open in agreement with predictions by band structure calculations. In addition, a structural 5×1 modulation and antiferromagnetically induced shadow bands can be resolved. However the intensities of these superstructures are about an order of magnitude smaller than those of the main Fermi surface.

In principle, shadow superstructures with a characteristic (antiferromagnetic) wave vector $\mathbf{Q}$ are expected to appear whenever sufficiently strong antiferromagnetic correlations are present. Thus quasi one-dimensional compounds such as the spin-Peierls chain $CuGeO_3$ and the spin ladders $Sr_{n-1}Cu_{n+1}O_{2n}$ are also currently under investigation with respect to possible shadow bands (see section V).

The occurrence of shadow bands was first proposed by Kampf and Schrieffer in the context of the Hubbard model at half-filling. As shown in Fig. 4, a simple spin-density wave mean-field treatment predicts the occurrence of spectral weight obeying the dispersion relations

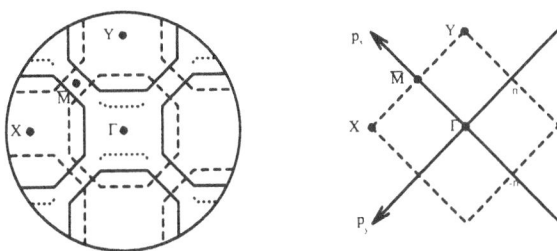

FIG. 3. Fermi surface (schematic) of Bi2212 as seen in ARPES by Aebi et al.: main Fermi surface (solid line), shadow bands (dashed line), and 5×1 structural modulation (dotted line). Note that because of a slight orthorombicity in this compound there are two almost equivalent (π, π) points (X and Y), and the lattice is tilted by $45°$ with respect to the tetragonal convention.

$$E_{\mathbf{p}} = \pm\sqrt{\epsilon_{\mathbf{p}}^2 + \frac{U^2S^2}{4}}, \tag{3}$$

where $\mathbf{p}$ is now restricted to half of the Brillouin zone, S=1/2 is the electronic spin, and $\epsilon_{\mathbf{p}} = -2t(\cos(p_x) + \cos(p_y))$ is the U=0 dispersion in 2D. Here the lattice spacing has been set equal to unity.

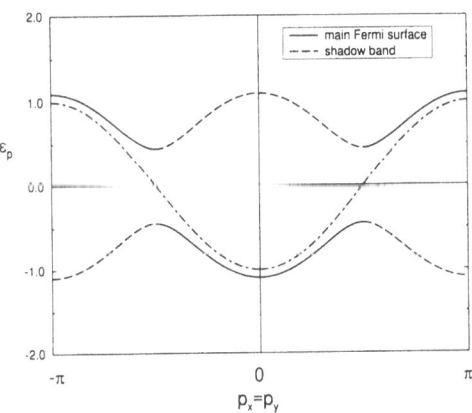

FIG. 4. Energy dispersion in the non-interacting 2D system (dot-dashed), and in the presence of an antiferromagnetic gap.

The corresponding spectral function can be obtained within this treatment in close analogy to the BCS formalism

$$A(\mathbf{p}, \omega) = u_{\mathbf{p}}^2 \delta(\omega - E_{\mathbf{p}}) + v_{\mathbf{p}}^2 \delta(\omega + E_{\mathbf{p}}), \tag{4}$$

where $u_{\mathbf{p}}^2 = \frac{1}{2}(1 + \frac{\epsilon_{\mathbf{p}}}{E_{\mathbf{p}}})$ and $v_{\mathbf{p}}^2 = \frac{1}{2}(1 - \frac{\epsilon_{\mathbf{p}}}{E_{\mathbf{p}}})$.

The two δ-function contributions correspond to the PES (annihilation of an electron) and the inverse PES (annihilation of a hole) process respectively. The latter process is difficult to realize experimentally. In Fig. 5, the PES and IPES results obtained within the SDW treatment are plotted along the line from $\mathbf{p} = (0,0)$ to $(\pi,\pi) \equiv \mathbf{Q}$. Note that if $\mathbf{p} + \mathbf{Q}$ lies exactly on the boundary of the reduced Brillouin zone, then $u_{\mathbf{p}+\mathbf{Q}} = v_{\mathbf{p}+\mathbf{Q}} = \frac{1}{2}$. For $\mathbf{p} \sim \mathbf{0}$, the state $c_{\sim(\pi,\pi)\uparrow}|\phi_0\rangle$ has weight $|u_{\mathbf{p}\sim 0}|^2 \approx 0$, because $E_{\mathbf{p}} \sim \epsilon_{\mathbf{p}}$ if $\mathbf{p}$ is far away from the reduced Brillouin zone boundary. The sizes of the symbols in this figure correspond to the spectral weight of the photoemission peaks, i.e. for $\mathbf{p} \sim \mathbf{0}$ the intensity is of order unity, at $\mathbf{p} \sim (\pi/2,\pi/2)$ it is $\sim \frac{1}{2}$, and at $\mathbf{p} \sim (\pi,\pi)$ it is 0. If $U \to \infty$, then $\frac{\epsilon_{\mathbf{p}}}{|E_{\mathbf{p}}|} \approx 0$, and in this case all peaks have intensity $\frac{1}{2}$. Thus the ratio of intensities at $\mathbf{p} = (\pi/2,\pi/2) + \mathbf{q}$ and $(\pi/2,\pi/2) - \mathbf{q}$ give an indication of the strength of the coupling U.

The result for the limit U=0 is shown in Fig. 5(b). In this case, the PES signal above $(\pi/2,\pi/2)$ is exactly 0, and below it is exactly 1 (opposite for the IPES). Thus the appearence of spectral weight above $(\pi/2,\pi/2)$ in addition to the U=0 IPES weight gives an indication of the strength of the antiferromagnetic correlations. The additional "band" above $(\pi/2,\pi/2)$ generated by the antiferromagnetic correlations is the so-called *shadow band* of Kampf and Schrieffer. [6] The gap and the shadow region are a non-trivial consequence of AF correlations in the ground state which can be verified experimentally.

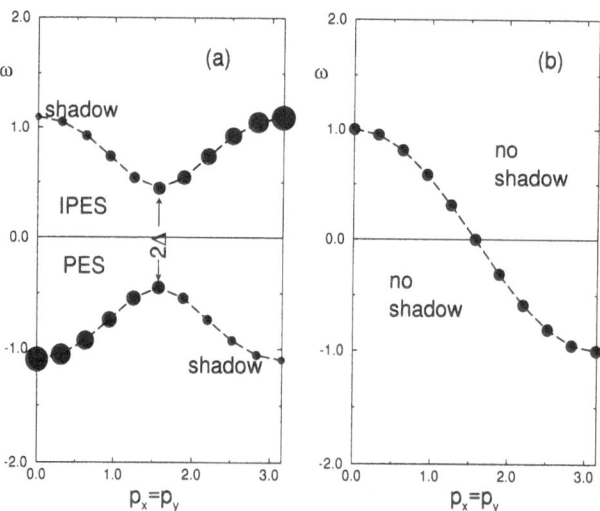

FIG. 5. (a) Photoemission bands of the 2D Hubbard model (U$\neq$0) within the SDW mean-field approximation. The spectral intensities correspond to the size of the symbols. (b) Same as (a) but for U=0. There are no shadow bands in this case.

While at half-filling the SDW treatment gives a sufficient description of the qualitative features observed in PES, it is not extendable into the region close to

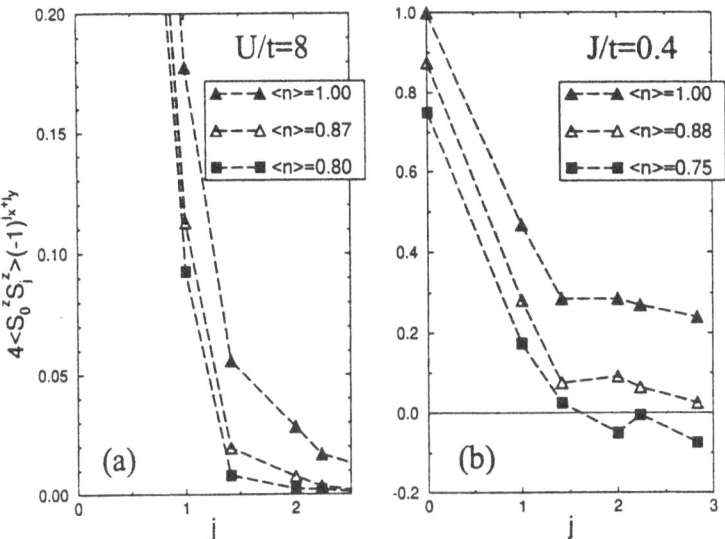

FIG. 6. (a) Spin-spin correlation $4\langle S_0^z S_j^z\rangle(-1)^{j_x+j_y}$ versus distance $j=|\mathbf{j}|$, where $\mathbf{j} = (j_x, j_y)$, for the 2D Hubbard model calculated by QMC at T=t/2, U/t =8, and for certain hole densities on an 8x8 cluster. (b) Spin-spin correlation as defined in (a), for the 2D t-J model calculated using ED techniques on a 4x4 cluster at similar densities.

optimal hole doping. However, it is of interest whether some (certainly weaker) spectral shadow weight induced by remnant short-range AF fluctuations is to be expected in this regime of interest. From NMR measurements it is known that the AF correlation length at optimum doping is between 2 and 3 lattice spacings. [8] This is consistent with exact diagonalizations (ED) we have performed on 2D clusters (shown in Fig. 6). While at half-filling (n=1) the spin-spin correlation function, $\langle S_i^z S_j^z\rangle$, approaches a constant value, indicating long-range AF order, it decreases quite rapidly at finite hole-doping levels. We extract an AF short-range correlation length $\xi_{AF} \approx 1.5$ lattice spacings at n$\approx$0.87, consistent also with Quantum Monte Carlo (QMC) simulations on the one-band Hubbard model.

2D t-J, PES, J/t=0.4

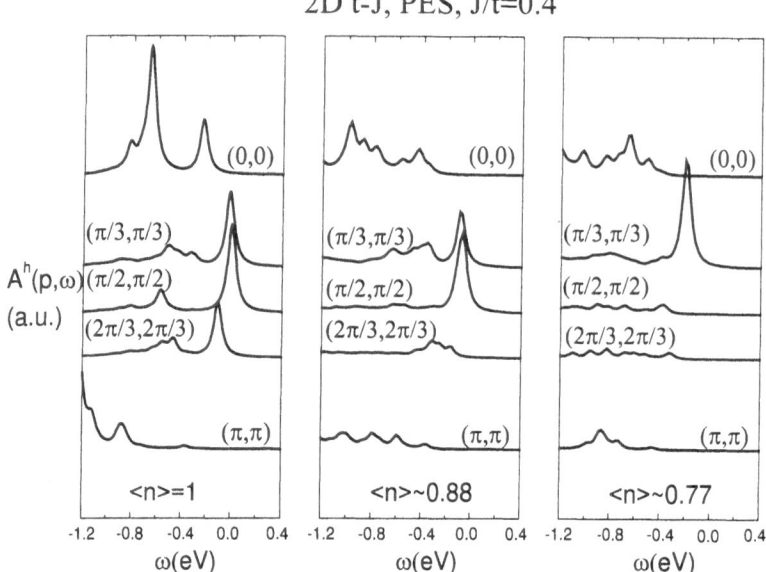

FIG. 7. PES spectral functions evaluated using ED techniques on 4×4 and $\sqrt{18} \times \sqrt{18}$ clusters. The densities are shown in the figure. We assumed t=0.4 eV, and provided a width $\delta = 0.1t$ to the δ-functions. The momenta are indicated on the figure.

In Fig. 7, the PES $A(\mathbf{p}, \omega)$ spectrum is shown for the t-J model. $J/t = 0.4$ was selected as an example, but we have verified that the results are qualitatively similar in the range between $J/t = 0.2$ and $J/t = 0.8$. As expected, at half-filling the largest peak near the chemical potential (i.e., the quasiparticle peak) is obtained at $\mathbf{p} = (\pi/2, \pi/2)$. Increasing the diagonal momentum away from this value, a considerable amount of spectral weight induced by AF correlations is observed. Moving away from half-filling into the subspace of two holes (nominal density $\langle n \rangle \sim 0.88$), the dominant peak remains at $\mathbf{p} = (\pi/2, \pi/2)$ within the momentum resolution. At $(\pi/3, \pi/3)$, the quasiparticle strength is still large and coherent. On the other hand, at $\mathbf{p} = (2\pi/3, 2\pi/3)$, the peak appears broader on the scale used, although its integrated spectral weight remains close to that of $\mathbf{p} = (\pi/3, \pi/3)$. The height of the peak at $\mathbf{p} = (2\pi/3, 2\pi/3)$ as a percentage of the largest peak, located at $\mathbf{p} = (\pi/2, \pi/2)$ (with or without holes), is 15-20% i.e. within the experimentally "observable" region. Finally, at density $\langle n \rangle = 0.77$, the result resembles that of a non-interacting system with a Fermi momentum close to $\mathbf{p} = (\pi/3, \pi/3)$, above which the signal is too weak to be observable in PES experiments. Then, our rough estimations within the t-J model are similar to those of the Hubbard model [9], i.e. weight above $\mathbf{p}_F$ can still be observed at $\langle n \rangle \sim 0.88$ but no longer at density $\langle n \rangle \sim 0.77$.

III. OBSERVATION OF FLAT BANDS IN THE QUASIPARTICLE DISPERSION

In the non-interacting limit (U=0), the electronic dispersion in the square lattice is given by $\epsilon_{\mathbf{p}} = -2t(\cos(p_x) + \cos(p_y))$, i.e. there is a saddle point at momentum $\mathbf{p} = (\pi, 0)$ and its symmetry points, leading to a van Hove singularity in the density of states. The main effect on the valence band dispersion of introducing interactions is a considerable reduction of the bandwidth. This corresponds to the concept of quasiparticles in a Fermi liquid acquiring an effective mass due to electron-electron interactions (although it is far from obvious that close to half-filling the system is a Fermi liquid). In general, in the strongly-correlated models we study, a narrow bandwidth is observed close to half-filling (n=1), leading to an enhanced density of states in this regime. However, as the doping level is increased the bandwidth widens, and an only slightly renormalized non-interacting electron picture emerges in the dilute limit (n → 0).

Here we will concentrate on the effect of strong interactions on the flat region of the quasiparticle dispersion around $\mathbf{p} = (\pi, 0)$. The shape of $\epsilon_{\mathbf{p}}$ in this region is not expected to be changed significantly by the quenching of the band due to strong interaction effects. However, a slight widening of the flat regime can be expected, leading to some further enhancement of the van Hove features in the density of states. This effect is indeed observed in our calculations, and will be discussed in the following.

In Fig.8(a), $A(\mathbf{p}, \omega)$ is shown for the $t - J$ model at $half-filling$ and $J/t = 0.4$ using the ED technique applied to 2D clusters with 16 and 18 sites. The combination of these clusters allows enough resolution in momentum space to analyze quantitatively the dispersion of the main features in the spectral weight. It is clear that near the Fermi energy ($\omega = 0$) there is a robust peak that disperses weakly in the scale of the figure. Remnants of this low-energy peak exist at momenta $(0, 0)$ and (π, π), albeit in the latter barely visible to the eye. (However the intensity and position of this feature can be studied readily with the continued fraction expansion approach [4]). In Fig.8(b), the position of the low-energy peak is shown with full dots, with the convention that the area of the dot is proportional to the intensity of the peak. The bandwidth of this sharp quasiparticle-like peak is $\sim 0.8t = 2J$, in excellent agreement with our expectations based on previous ED [4] and Born approximation calculations. The flat region near $(\pi, 0)$ is clearly visible in the figure. From Fig.8(a) it is clear that additional PES spectral weight in $A(\mathbf{p}, \omega)$ is located at higher energies $|\omega|$. As discussed before in the literature, the strong correlation effects force the hole-like quasiparticle to carry only a fraction of the integrated spectral weight, [7] and thus the presence of considerable incoherent intensity at very low energies is reasonable. An estimate of their position is shown in Fig.8(b) (open squares). This feature is not relevant for the low temperature behavior of the model, which is dominated by the quasiparticle peak at the top of the valence band.

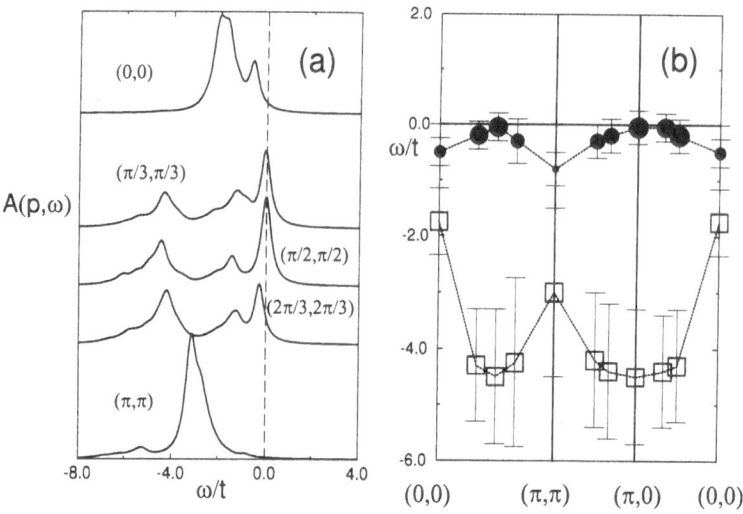

FIG. 8. (a) Spectral weight $A(\mathbf{p}, \omega)$ of the 2D $t - J$ model at $J/t = 0.4$, evaluated using clusters of 16 and 18 sites, for wavevectors along the diagonal in momentum space. The δ-functions have been given a width $\epsilon = 0.25t$ in the plots; (b) position of the two dominant peaks in $A(\mathbf{p}, \omega)$ as a function of momentum. The area of the circles is proportional to the intensity of the quasiparticle peak they represent. The error bars denote the width of the peak as observed in Fig.8(a) (sometimes to a given broad peak several poles contribute appreciably). The full squares at $\omega \sim -4t$ represent the center of the broad valence band weight, and the area of the squares is *not* proportional to their intensity.

Let us now discuss the results away from half-filling. In Fig.9(a) and (b), ED data for density $\langle n \rangle \approx 0.88$ is shown (two holes in the 16 and 18 sites clusters). The PES results in momentum space present a structure very similar to that discussed at half-filling. The low-energy peak is well-defined at all momenta, even those above the naive non-interacting Fermi momentum located near $(\pi/2, \pi/2)$, and still it disperses with a bandwidth of order J. Most of the spectral weight comes from the flat region around $(\pi, 0)$.

The large accumulation of weight at higher energies $|\omega|$ remains localized at $\omega \sim 4t$. Then, to the extent that the one-band models reproduce the physics of the high-T_c cuprates, it is reasonable to expect that PES experiments carried out at half-filling *and* near the optimal doping should produce dispersive features of similar intensity and bandwidth. The clear similarity between the experimental bandwidth of the Bi2212 PES data and the recent results for the *insulating* $Sr_2CuO_2Cl_2$ compound [5] provides more evidence for the validity of strongly correlated one band models for the cuprates. However, it is important to remark that while the concrete prediction of our calculations is that the bandwidth of the hole carriers is of order J, the particular details of the dispersion may *differ* from compound to compound. For example, it has been remarked recently that to reproduce the data for $Sr_2CuO_2Cl_2$, the addition of a small t'-term to the 2D $t-J$ model is necessary. [11] Thus care must be taken when the fine details of different compounds at different dopings are compared.

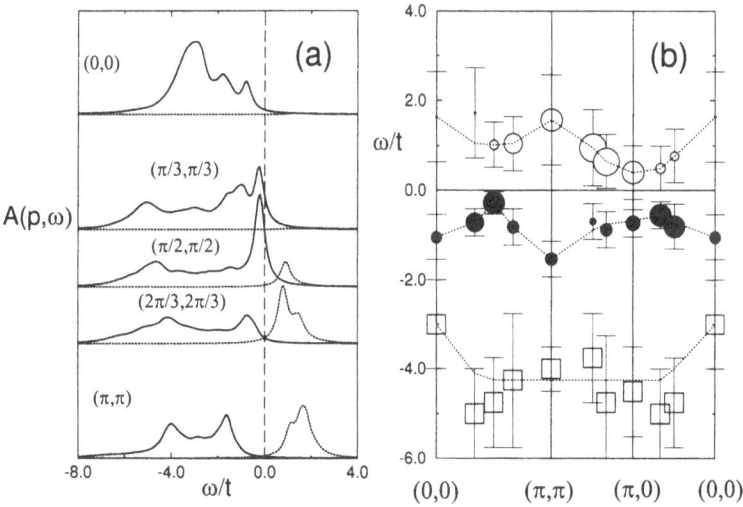

FIG. 9. Same as Fig.8 but for density $\langle n \rangle \approx 0.88$ (i.e. two holes on the 16 and 18 sites clusters). In (a) the PES intensity is shown with a solid line, while the IPES intensity is given by a dotted line. The chemical potential is located at $\omega = 0$. In (b) the full and open circles represent the PES and IPES intensities, respectively, of the peaks the closest to the Fermi energy. Their area is proportional to the intensity.

DOS for the 2D t-J model

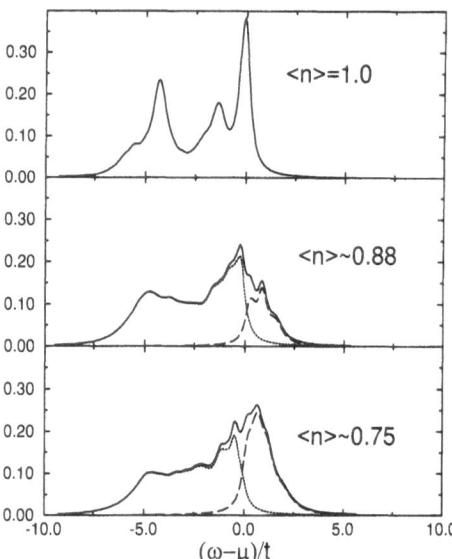

FIG. 10. Electronic density of states in the 2D t-J model at J/t=0.4. In this plot, results for the 16-site and the 18-site square clusters are combined. The dotted lines are the PES, the dashed lines are the IPES, and the solid lines give the total spectral function.

Now consider the inverse photoemission (IPES) ($\omega > 0$) intensity in Fig.3(a),(b). The observed spectral weight in the vicinity of (π, π) somewhat resembles the distribution for a non-interacting Fermi system. At first sight, this effect does not seem to be reproduced by a simple assumption of rigid bands up to half-filling. However, recently Eder and Ohta [12] have shown that if proper *quasiparticle* operators [7] are used in the calculation of the spectral weight (i.e. operators dressed by spin fluctuations, instead of bare electronic operators), then the intensity of the IPES region is much reduced and the quality of the rigid-band description of the t − J model appears more clearly. This is an important point not much emphasized in the literature, namely that the robustness of the rigid-band picture in a given model *cannot* be tested by analyzing the removal of "bare" electrons (sudden approximation) as produced by a PES experiment, but instead "dressed" carriers must be used. Thus, PES and transport experiments may differ in their predictions if holes are heavily renormalized, as in the cuprates.

Finally, let us turn to the density of states defined by $A(\omega) = \int d\mathbf{p} A(\mathbf{p}, \omega)$, shown in Fig. 10. At half-filling, three features are observed in the PES spectrum (whereas there is no IPES at this filling for the t-J model because of the constraint of no double-occupancy). The most prominent low-energy peak is due to the quasiparticle band of width $\sim J$. Note that most of the spectral weight of this feature comes from the region around the $(\pi, 0)$ points. There is another low-energy "string" feature which can be identified with excitations due to strings of flipped spins along the path [10] of a single hole in an antiferromagnetic Neel background. At an energy further away from the chemical potential ($\sim$ -4.8t) there is a broader peak corresponding to the lower Hubbard band.

As the system is doped with holes (n$\sim$0.88), weight appears in the inverse PES spectrum above μ. Also, there remains a strong quasiparticle feature at the bottom of the PES spectrum (close to $\mu = 0$): this peak is clearly distinct from the spectral weight derived from the lower Hubbard band at energies around -5t. Finally, at n$\sim$0.75, most of the quasiparticle weight has moved into the IPES band. This situation corresponds to the "overdoped" regime of the cuprates.

IV. POSSIBILITY OF SMALL FERMI SURFACES CLOSE TO HALF-FILLING

While at the end of the last section we have focussed on the momentum-integrated spectral response, in this section we shall concentrate on the frequency-integrated spectral function, the momentum distribution function $n_\mathbf{p} \equiv \int d\omega A(\mathbf{p}, \omega)$. This quantity is commonly used to determine the shape of the Fermi surface in many-electron systems.

The shape of the Fermi surface in models of strongly correlated electrons has recently been a controversial issue. [13,14] It is known that long-range antiferromagnetic order in two spatial dimensions is only present at half-filling and $T = 0$. However, some theories for the formation of superconducting pairs at finite hole density have been guided by this limit, supplemented by the observation of robust short-range antiferromagnetic correlations in the high-T_c compounds. [15–17] Some unusual normal-state properties, like the linear temperature dependence of the resistivity and the change of sign in the Hall coefficient, can be accounted for in terms of strong antiferromagnetic correlations in these materials. [16,17]

The nature of quasiparticles in these systems is intimately related to the topology of the Fermi surface: do all electrons participate in the response to external fields, which would imply a large Fermi surface ? Or is it possible to understand the low-energy properties in terms of a dilute gas of dressed holes occupying preferred points in momentum space on bands whose particular shape is produced by strong correlations ?

At half-filling, models of strongly correlated electrons are known to be unstable towards the formation of a spin-density wave commensurate with the lattice. They are antiferromagnetic insulators because of strong on-site repulsion, and their Fermi surfaces have perfect nesting properties with nesting vector $\mathbf{Q} = (\pi, \pi)$ in 2D, or $Q = \pi$ in the one-dimensional analogue. How does the shape of the Fermi surface change upon hole doping ?

There are two competing scenarios depicted in Fig. 11. Approximations based on holon-spinon decoupling [18] and high-temperature expansions [19] suggest a large Fermi surface compatible with Luttinger's theorem, and similar to the non-interacting case (Fig. 11(a)). However, mean-field calculations based on spin-density fluctuations [15] and computational techniques [14,20,21] suggest the presence of hole pockets at low doping and low temperatures. The latter scenario does not necessarily contradict the apparent large Fermi surface observed by early photoemission experiments in the cuprates, [22] since thermal effects can easily "wash out" and connect the pockets, producing a "large" surface (Fig. 11(b)). [16,17,20,23]

Recent photoemission data by Aebi et al. [1] (discussed in section II, Figs. 2,3) obtained by a novel technique which permits the mapping of the entire Fermi surface, indicates the existence of hole pockets for Bi2212. The suggestion of a crossover from a small to a large Fermi surface as a function of hole doping emerges also from the measurement of the optical conductivity in $La_{2-x}Sr_xCuO_4$. [24] It is observed that the spectral weight contained in the Drude peak first rises as a function of Sr doping, and then decreases beyond a critical doping level. This observation is consistent with a hole pocket scenario for small dopings, where the number of charge carriers at the Fermi level is proportional to the doping level x, while the volume of a large Fermi surface is expected to decrease approximately with $\sim(1-x)$, as is observed for $La_{2-x}Sr_xCuO_4$ for larger Sr concentrations.

If there are indeed hole pockets constituting a pocket-like Fermi surface, a "dip" in the momentum distribution function should be seen at momenta close to the Fermi surface of the half-filled system. Previous calculations [14–16,20] suggest that these minima are centered around $(\pi/2, \pi/2)$ and its rotational symmetry points as indicated in Fig. 11(b).

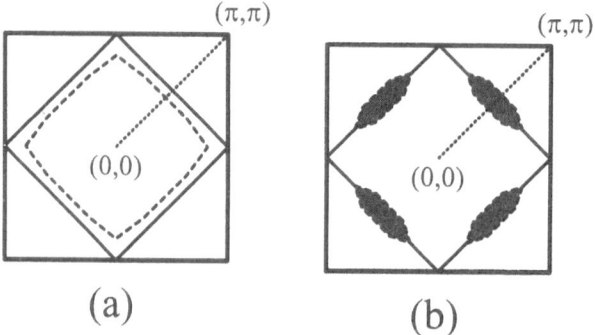

FIG. 11. Schematic plot of the Fermi surface in strongly correlated electronic systems at low hole-doping ($\approx 10\%$). The solid line denotes the non-interacting case. (a) Scenario 1: large Fermi surface; (b) Scenario 2: hole pockets.

Here we will discuss briefly the results of some recent Quantum Monte Carlo simulations in the Hubbard model close to half-filling. [25] In this regime, the fermion sign problem is rather severe, forbidding access to low temperatures. Instead of well-defined hole pockets, the most one could expect from this data is a depletion of n_p in certain areas of the Brillouin zone, indicating the precursor effect of such pockets.

In Fig. 12 is shown the Fermi surface of the one-band Hubbard model at $\langle n \rangle = 0.87$ and U/t=8, as determined from QMC calculations using various criteria. [25] The most common criterion for the position of the Fermi surface is to define it along those points in momentum space where $n_{\mathbf{p}} = 0.5$. Applying this criterion (open squares and solid line in Fig. 12) gives an electron-like Fermi surface closed about $\mathbf{p}=(0,0)$ as previously observed in Ref. [26]. However, because these simulations had to be conducted at high temperatures ($T=t/2$), this criterion is not necessarily reliable because it might miss some low-energy effects which are "washed out" by thermal fluctuations. Thus alternative criterion has been suggested, namely the use of the positions in momentum space where $n_{\mathbf{p}}$ changes most rapidly. Indeed, when applying this criterion (stars and dashed line in Fig. 12) two Fermi surfaces consistent with the picture of pockets are observed. A third way of extracting the Fermi surface is to take the points in momentum space where the quasiparticle peak in $A(\mathbf{p}, \omega)$ crosses the chemical potential (solid circles with error bars and dotted line). Once this data is supplemented with the proper error bars, the results are consistent with those of the criterion of maximum variation in $n_{\mathbf{p}}$. However, due to the large error bars it is still not possible to decide whether the Fermi surface is closed about $\mathbf{p}=(0,0)$ or about $\mathbf{p} = (\pi/2, \pi/2)$.

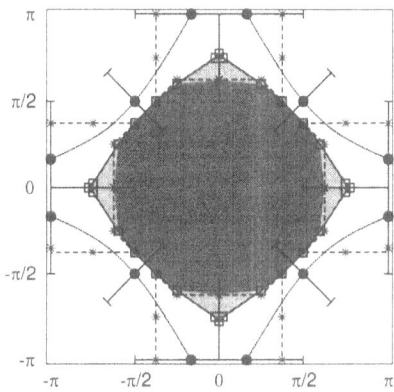

FIG. 12. Fermi surface for the 2D Hubbard model with $U/t =8$, $\langle n \rangle = 0.87$ at $T/t =1/2$ using the $n_{\mathbf{p}}=0.5$ criterion (open squares and solid line), the maximum variation of $n_{\mathbf{p}}$ criterion (stars and dashed line), and from maximum entropy results (solid circles with error bars and dotted line).

V. ANALOGY TO ONE-DIMENSIONAL MATERIALS: CHAINS AND LADDERS

Recently, $CuGeO_3$ was found to be the first inorganic system which shows a spin-Peierls phase. [27] With decreasing temperature, a regime described by uniform S=1/2 Heisenberg chains undergoes a phase transition at $T_{SP} = 14K$ to a dimerized system due to the coupling of the spin-1/2 Cu^{2+}-ions to the three-dimensional lattice phonons. At T_{SP} a structural transition takes place where alternate atoms are displaced in opposite directions, and an energy gap appears for magnetic spin triplet excitations.

$CuGeO_3$ has a c-direction lattice constant of 2.941Å, much smaller than the

a and b lattice parameters. Although the c-axis lattice constant corresponds to the $Cu^{2+} - Cu^{2+}$ distance, the actual links between coppers are provided by the edge sharing of CuO_6 octahedra. The angle defined by each Cu-O-Cu bond is close to $90°$ (Ref. [28]). This induces an exchange coupling J_1 in the effective nearest-neighbor (n.n.) S=1/2 Heisenberg model for the Cu-O chains which is much smaller than that found in the two dimensional (2D) cuprate superconductors. The measured susceptibility, χ, above 14K while showing characteristics of 1D antiferromagnets, is not quantitatively reproduced by calculations corresponding to a n.n. S=1/2 Heisenberg model. Actually, Lorenzano et al. suggested on the basis of a neutron scattering study that a next-nearest-neighbor (n.n.n.) coupling J_2 may arise from the Cu-O-O-Cu path. [29] By fitting the temperature dependence of the magnetic susceptibility and the spin-wave dispersion curve obtained by inelastic neutron scattering techniques, various groups have arrived at parameters similar to: $J_1 = 150K$, $J_2 = 0.24J_1$, and $\delta = 0.03J_1$. [30,31] Here, δ is a n.n. dimerization parameter.

Thus, $CuGeO_3$ can be considered to correspond to a Heisenberg model with n.n. and n.n.n. interactions, and a term which dimerizes the lattice to mimic the effects of phonons. The Hamiltonian is

$$H = J_1 \sum_{\langle ij \rangle} S_i \cdot S_j + J_2 \sum_{\langle \langle im \rangle \rangle} S_i \cdot S_m + \delta \sum_{\langle ij \rangle} (-1)^i S_i \cdot S_j, \qquad (5)$$

where i, j, m denote sites of a 1D chain with N sites, S_i are S=1/2 spins, and $\langle ij \rangle$ ($\langle \langle im \rangle \rangle$) corresponds to n.n. (n.n.n.) spins along the chain.

We have calculated the spectral function $A(q, \omega)$ which corresponds to the creation of holes by the removal of spins on the chains in the sudden approximation. As Hamiltonian for this calculation we use the t − J model generalization of Eq.(5), i.e. we add a hopping term

$$H_t = -t \sum_{\langle ij \rangle \sigma} (\bar{c}_{i,\sigma}^\dagger \bar{c}_{j,\sigma} + h.c.), \qquad (6)$$

where $\bar{c}_{i,\sigma}$ are hole operators, and the other notation is standard. [4] In the absence of experiments testing the properties of carriers in $CuGeO_3$, it is difficult to predict the value of the hopping amplitude t. Thus, in the results quoted below three different amplitudes are used such that more information is available to compare with experiments once ARPES data becomes available. However, if the parameters J_1 and J_2 are caused by an exchange mechanism it is likely that the amplitude t would be larger than both. For simplicity, we have neglected the influence of the dimerization on t which in principle should also be modulated as the exchange J_1. Since the dispersive features observed in our calculations have bandwidths of order J_1 ($\gg \delta$), the influence of δ on the ARPES data is expected to be small. Actually J_2 is more relevant than δ for ARPES, where states deep in energy are tested.

In Fig.13, $A(q, \omega)$ is shown for different momenta, and three couplings t/J_1, with J_1, J_2 and δ defined as before. A substantial width is given to the δ-functions to mimic the effect of the large backgrounds appearing in ARPES data. Note the presence of a dominant peak at low binding energy which moves towards the Fermi energy (at $\omega = 0$ in the figure) as the momentum is changed from q = 0 to q = $\pi/2$ for all couplings. The lowest binding energy is precisely reached at q = $\pi/2$. Based on previous experience with the 2D copper oxide compounds, [4] this large peak may correspond to a hole immersed in a *locally* ordered AF spin background which increases its effective mass. An interesting feature of our results is the presence of substantial weight at large ω for all q and couplings. At q = $\pi/2$, the shape of A(q, ω) at the top and bottom of the band seems similar i.e. a symmetry line appears to run through the middle of the total bandwidth. This feature also appears in the 2D t − J model but only at very small J/t. [32] This is another effect predicted by theoretical calculations that an ARPES experiment could analyze, although we are aware that in such experiments it is difficult to investigate structure at large binding energies due to the presence of core backgrounds in the data that grow with binding energy, as well as the influence of other bands not taken into account in simple one-band models such as the one used in the present paper.

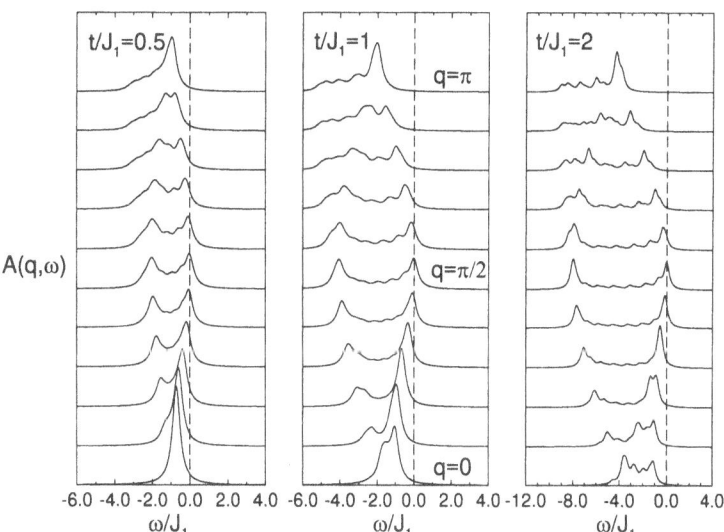

FIG. 13. Spectral function A(q, ω) on a chain of 20 sites, at several values of t/J_1. The broadening is $\epsilon = 0.25 J_1$, and the Fermi energy is arbitrarily located at $\omega = 0$. The other parameters are $J_2 = 0.24 J_1$ and $\delta = 0.03 J_1$.

A curious effect obvious in Fig.13 is the presence of a robust peak in A(q, ω) for momenta larger than q = $\pi/2$. This behavior is quite different from what is observed in normal metals, and here we argue that it is "shadow bands" which are caused by the strong AF correlations in the ground state. This calculation shows that shadow bands can be robust *even* in 1D systems, where we are used to the concept that there are no ordered phases at zero temperature. Our prediction can be extended to materials like Sr_2CuO_3 whose properties are accurately described by the n.n. Heisenberg model. [33] The existence of shadow bands in 1D systems should not be surprising since the hole dispersion tested in ARPES experiments is dominated by *short* distance correlations.

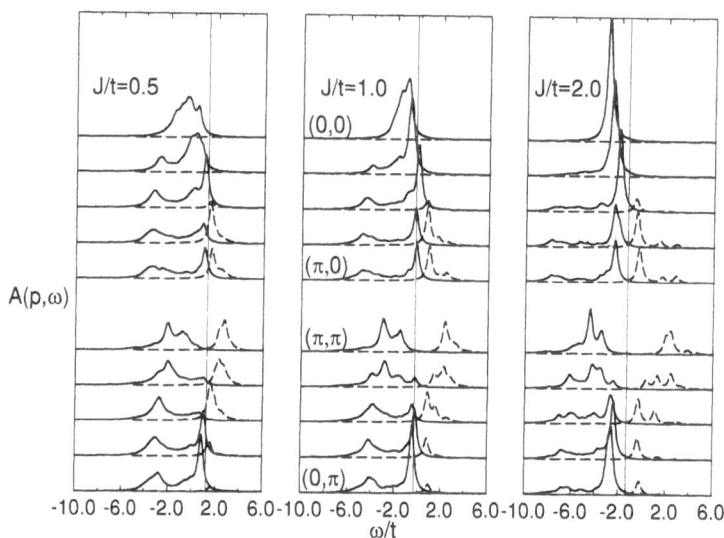

FIG. 14. Photoemission spectra (solid lines) and inverse photoemission spectra (dashed lines) for a 2×8 t-J ladder with 2 holes. The position of the chemical potential is indicated by the thin solid line.

Finally let us remark that shadow bands may also occur in spin-1/2 ladder compounds such as $(VO)_2P_2O_7$ and $SrCu_2O_3$. In Fig. 14 we show the photoemission spectra calculated for an isotropic t-J model on a 2×8 ladder with 2 holes. This system is known to have a spin gap for the parameter values shown in the figure. Furthermore, as the ratio J/t is increased, superconducting correlations are expected to be strong. Indeed, at momentum $(\pi/2, \pi)$ - which is at the Fermi surface for this particular filling level - a sharp low-energy feature is observed to grow with increasing J/t. This peak can likely be associated with the increased $d_{x^2-y^2}$ RVB correlations, corresponding to a Bogoliubov quasiparticle excitation made of an electron and a hole.

Also, in the bonding band ($\mathbf{p} = (0,0)$ to $(\pi, 0)$) PES spectral weight is observed above the (non-interacting) Fermi momentum $(\pi/2, 0)$, indicating that shadow features may also exist in ladders.

VI. CONCLUSIONS

We have attempted to give a brief survey of present (mostly numerical) calculations of valence band photoemission spectra in strongly correlated electronic systems related mostly to the 2D cuprate high-T_c materials. Generic features due to strong electron-electron interactions, such as quenching of the bandwidth, antiferromagnetically induced spectral weight and extended flat regions in the quasiparticle dispersion, have been shown to be present in the regime of small hole doping levels. A quantitative comparison with experiments is possible, and confirms the validity of the low-energy Hamiltonians which are studied.

We wish to thank B. Normand, T.M. Rice, and D. Duffy for useful discussions, and acknowledge the Swiss National Science Foundation for financial support.

Some of the material for this lecture note has been taken from the following references : S. Haas et al., Phys. Rev. Lett. 74, 4281 (1995); A. Moreo et al., Phys. Rev. B 51, 12045 (1995); S. Haas, Phys. Rev. B 51, 11748 (1995); D. Duffy and A. Moreo, Phys. Rev. B 52, 15607 (1995); S. Haas and E. Dagotto, Phys. Rev. B 52, R14396 (1995).

[1] P. Aebi et al., Phys. Rev. Lett. **72**, 2757 (1994).
[2] H. Ding et al., Phys. Rev. Lett. **74**, 2784 (1995); 1425 (E) (1995); H. Ding et al., Phys. Rev. Lett. **76**, 1533 (1996).
[3] Z.-X. Shen et al., Phys. Rev. Lett. **70**, 153 (1993).
[4] E. Dagotto, Rev. Mod. Phys. **66**, 763 (1994).
[5] B. O. Wells et al., Phys. Rev. Lett. **74**, 964 (1995).
[6] A. Kampf and J. R. Schrieffer, Phys. Rev. **B 42**, 7967 (1990).
[7] E. Dagotto and R. Schrieffer, Phys. Rev. B **43**, 8705 (1991).
[8] T. Imai et al., Phys. Rev. b **47**, 9158 (1992).
[9] S. Haas, A. Moreo and E. Dagotto, Phys. Rev. Lett. **74**, 4281 (1995).
[10] B.I. Shraiman and E.D. Siggia, Phys. Rev. Lett. **61**, 467 (1988).
[11] A. Nazarenko et al., Phys. Rev. B **51**, 8676 (1995).
[12] R. Eder and Y. Ohta, Phys Rev. B **50**, 10043 (1994).
[13] W. Stephan and P. Horsch, Phys. Rev. Lett. **66**, 2258 (1991).
[14] R. Eder and Y. Ohta, Phys. Rev. Lett. **72**, 2816 (1994); R. Eder and Y. Ohta, cond-mat SISSA preprint 9407097.
[15] J.R. Schrieffer, X.G. Wen and S.C. Zhang, Phys. Rev. Lett. **60**, 944 (1988); J.R. Schrieffer, X.G. Wen and S.C. Zhang, Phys. Rev. B **39**, 11663 (1989).
[16] S. Trugman, Phys. Rev. Lett. **65**, 500 (1990).
[17] E. Dagotto, A. Nazarenko and M. Boninsegni, Phys. Rev. Lett. **73**, 728 (1994); E. Dagotto, Rev. Mod. Phys. (to be published); E. Dagotto, J. Riera, Y. C. Chen, A. Moreo, A. Nazarenko, F. Alcaraz, and F. Ortolani, Phys. Rev. **B 49**, 3548 (1994).
[18] P.W. Anderson, Physica B **199**, 8 (1994), and references therein.
[19] R.R.P. Singh and R.L. Glenister, Phys. Rev. B **46**, 14313 (1992).
[20] D. Duffy and A. Moreo, Phys. Rev. B **51**, 11882 (1995).

[21] D. Poilblanc and E. Dagotto, Phys. Rev. B **42**, 4861 (1990).

[22] C.G. Olson et al., Science **245**, 731 (1989); C.G. Olson et al., Phys. Rev. B **42**, 381 (19900; D.S. Dessau et al., Phys. Rev. Lett. **71**, 2781 (1993).

[23] The near-degeneracy between $(\pi/2, \pi/2)$, $(\pi, 0)$ and $(0, \pi)$ in the cuprates leads to a quick washing out of pockets from $(0, \pi)$ to $(\pi, 0)$ at small temperatures. In general, the existence of pockets along the $k = k_x = k_y$ direction is a feature very difficult to see. In small lattices, .i.e. 4×4, 8×8, not even at $T = 0$, the pockets cannot be observed at the mean-field level along the $k_x = k_y$ direction.

[24] S. Uchida et al., Phys. Rev. B **43**, 7942 (1991).

[25] D. Duffy and A. Moreo, Phys. Rev. B **52**, 15607 (1995).

[26] A. Moreo et al., Phys. Rev. B **41**, 2313 (1990).

[27] M. Hase, I. Terasaki and K. Uchinokura, Phys. Rev. Lett. **70** , 3651 (1993).

[28] H. Kuroe et al., Phys. Rev. **B 50**, 16468 (1994). See also H. Völlenkle et al., Monatschefte der Chemie **98**, 1352 (1967).

[29] J. E. Lorenzo et al., Phys. Rev. **B 50**, 1278 (1994).

[30] ·J. Riera and A. Dobry, preprint.

[31] G. Castilla, S. Chakravarty and V. J. Emery, preprint.

[32] E. Dagotto et al., Phys. Rev. **B 41**, 9049 (1990).

[33] T. Ami et al., Phys. Rev. **B 51**, 5994 (1995); and references therein.

Van Hove Scenario of High-T_c Superconductivity

J. González[1]

Instituto de Estructura de la Materia, CSIC
Serrano 123, 28006 Madrid, Spain

Abstract

In this lecture we review the properties of interacting electrons near two-dimensional van Hove singularities in the framework of a renormalization group approach. We carry out the discussion making contact with the phenomenology of the copper-oxide superconductors, showing that several properties, like the development of a flat band and the pinning of the Fermi level at the van Hove singularities, can be reproduced unambiguously.

1   Introduction

In recent years we have started to increase significantly our knowledge about the electronic properties of high-T_c superconductors. In particular, angle resolved photoemission spectroscopy (ARPES) has been able to provide accurate enough pictures of the dispersion relation for the different phases of the copper-oxide materials[1]. As is well-known, these have a very rich phase structure. At zero temperature the carrier free materials are antiferromagnetic insulators and it is only upon doping that they reach the superconducting phase, showing correspondingly metallic properties above the critical temperature. The electronic dispersion of a typical carrier free material obtained by ARPES, like that of $Sr_2CuO_2Cl_2$, uses to show sharp isotropic peaks about the points $(\pm\pi/2, \pm\pi/2)$[2], what is a reflection of the antiferromagnetic properties of the material. In contrast to this characteristic shape, the hole-doped compounds have a typical dispersion relation with extended saddle points at the symmetry points $(\pi, 0)$, $(0, \pi)$, very close to the position of the Fermi level[1, 3]. Given that these features appear to be a constant for the hole-doped materials, it becomes difficult to imagine how by the only effect of doping the dispersion relation of a given compound may undergo such a drastic change. In fact, it is the impossibility of making the transition from one typical shape to the other by a rigid shift of the Fermi level what makes so difficult to reconcile the two pictures for the carrier free and the doped materials. This certainly poses a puzzle, at the theoretical level, which is harder to solve as the traditional many-body methods are not well-suited to study the evolution of

---

[1] e-mail: emgonzalez@iem.csic.es

the Fermi surface or the dispersion relation under changes in the coupling constant or the doping level.

In this lecture we are not going to deal with such a big problem as trying to predict the evolution of the dispersion relation from the antiferromagetic regime, but we are going to concentrate on a smaller scale problem that is however related to the former. We are going to study the stability of what seem to be universal features of the metallic phase for the hole-doped compounds. That is, we want to understand why under certain circumstances extended saddle points are likely to develop and, once they are formed, the flat band remains attached close to the Fermi level. This may also help to address the issue of the high-T_c superconductivity of the compounds, as we will see in what follows. With regard to the other end of the problem, that is the description of the antiferromagnetic phase, the features in the dispersion relation can be easily understood on theoretical grounds. The antiferromagnetic properties are explained by strong correlation effects inside the two-dimensional layers, which require the introduction of the Hubbard model (or its strong coupling version, the $t - J$ model) for the appropriate study of the electronic properties. The peaks of the dispersion relation at $(\pm \pi/2, \pm \pi/2)$, for instance, have been reproduced in numerical studies of the dispersion of one hole in the Hubbard model. Although the agreement between theory and experiment is not so good in other regions of the Brillouin Zone, it has been shown that it may be significatively improved by inclusion of next-to-nearest neighbor hopping t' in the Hubbard model[4], that is also expected from a careful consideration of the model parameters for the two-dimensional layers of the cuprates.

2   2D van Hove singularities

Any lattice in two dimensions has always a number of points in momentum space where the density of states diverges logarithmically. These are typical cases of the so-called van Hove singularities, which in two dimensions are realized as saddle points of the dispersion relation. The square lattice has two of such singularities and the hexagonal lattice has three of them. Taking into account the remarks at the end of the introduction, we will focus our attention on a square lattice model including nearest neighbor hopping t and next-to-nearest neighbor hopping t' (with $0 < t' < 0.5t$ in order to avoid the degeneration of the Fermi sea into two straight lines). The generic form of the contour map for the one-particle energy levels (in the non-interacting case) is given in Fig. 1. In that plot one may notice the location of the two inequivalent saddle points of the dispersion relation at the symmetry points $(\pi, 0), (0, \pi)$. We will be interested in studying the stability of the situation in which the Fermi energy is at the level of the two van Hove singularities. We see that in such case the main effect of the next-to-nearest neighbor hopping is that of curving the straight Fermi line shape characteristic of the $t' = 0$ model. As a consequence of that, the antiferromagnetic correlations are much less enhanced as the value of t' is increased. This means that, from the physical point of view, the value of t' plays an important role, together with the strength of the interaction,

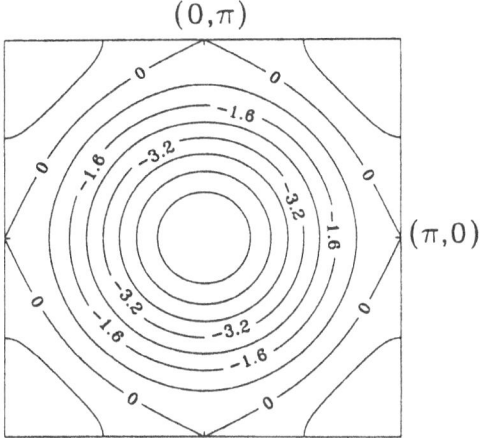

Figure 1: Contour map for the one-particle levels of the noninteracting $t-t'$ model.

in determining the phase in which the system is placed.

The subject of interacting electrons near van Hove singularities has been studied by a number of authors and it goes back to the early days of high-T_c superconductivity[5, 6]. More recently a more refined scenario has been worked out in order to explain some of the abnormal properties of the metallic regime of the hole-doped cuprates[7], as well as to propose a purely electronic mechanism of the superconductivity[8]. The novelty of the study that we present in this lecture is to rely on a renormalization group approach, that is well-suited to deal with the singularities which plague a standard perturbative treatment[9, 10]. Actually, it is easily realized that the traditional many-body approach to interacting fermion systems cannot be applied to the present model, as one of its assumptions is that the dispersion relation admits a linear approximation near the Fermi level. In the usual perturbative approach one starts by approximating the one-particle Green function $G^{(0)}(\omega, \mathbf{p})$ close to the Fermi energy ε_F, which becomes in the case of an isotropic dispersion relation $\varepsilon(\mathbf{p})$

$$
\begin{aligned}
G^{(0)}(\omega, \mathbf{p}) &= \frac{1}{\omega - \varepsilon(\mathbf{p}) + \varepsilon_F + i\epsilon\mathrm{sgn}\omega} \\
&\approx \frac{1}{\omega - v_F(|\mathbf{p}| - p_F) + i\epsilon\mathrm{sgn}\omega}
\end{aligned}
\tag{1}
$$

In the present case, however, most part of the modes near the Fermi level are close to any of the two van Hove points, where the dispersion relation becomes

$$
\varepsilon(\mathbf{p}) \approx p_x^2/(2m_x) - p_y^2/(2m_y)
\tag{2}
$$

As long as $\varepsilon(\mathbf{p})$ does not admit any linear approximation, it seems unclear how to proceed in the usual many-body approach. We will see, on the other hand, that the renormalization group approach provides a rigorous framework to treat the model (analysis of cutoff dependences, scaling of coupling constants).

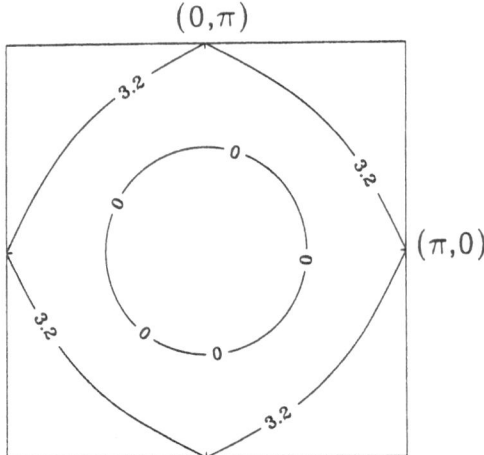

Figure 2: Plot of an ideal isotropic Fermi line (contour marked with zero energy).

3 Renormalization of fermion systems

Before going ahead with the model of the van Hove singularities, we briefly review what is the status of Fermi liquid theory in the context of the renormalization group approach. The renormalization group methods applied to fermion systems have been recently developed by Shankar[11], establishing that there can only be a few ways of making relevant perturbations to the Fermi liquid picture. This necessarily holds under very general conditions, like the isotropy of the dispersion relation at the Fermi surface and the short-range character of the interaction.

We shift therefore the discussion to the case of a circular Fermi line, as represented in Fig. 2. The analysis begins at the classical level, by identifying among all the possible channels given by incoming momenta $\mathbf{p}_1, \mathbf{p}_2$, and outgoing momenta $\mathbf{p}_3, \mathbf{p}_4$, those in which the interaction strength $U(\mathbf{p}_1, \mathbf{p}_2, \mathbf{p}_3, \mathbf{p}_4)$ does not change under reduction of the energy scale measured with respect to the Fermi level. This gives what are called marginal interactions, which are those which we have to keep as long as in the RG approach we are interested in a sensible low-energy theory of states close to the Fermi line. The first important remark made by Shankar, that we state here without proof, is that the condition for a given four-fermion interaction U being marginal is that the four momenta $\mathbf{p}_1, \mathbf{p}_2, \mathbf{p}_3$ and $\mathbf{p}_4$ sit all at the Fermi line. Given that they also have to satisfy momentum conservation, it is realized that one is left with only two possibilities to fulfill both conditions. One of them is that $\mathbf{p}_1 \approx \mathbf{p}_3$ (or equivalently $\mathbf{p}_1 \approx \mathbf{p}_4$), which amounts to having an almost zero momentum transfer in the interaction between the two particles. The other possibility is that $\mathbf{p}_1 \approx -\mathbf{p}_2$, that is, the momenta of the incoming particles add to zero. Quite remarkably, by means of this analysis we recover the interaction which plays the central role in Fermi liquid theory, which is concentrated on the forward scattering channel, as well as the BCS channel as a possible perturbation[11]. All the rest of interactions scale to zero as the energy is scaled towards the Fermi level and therefore they are irrelevant in the low-energy limit.

Of course, the next step in the analysis is to check whether the above picture

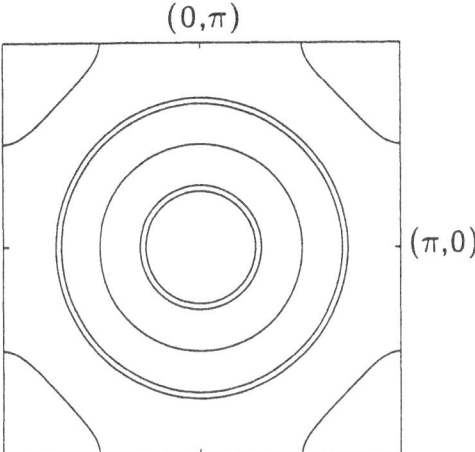

$(0,\pi)$

$(\pi,0)$

Figure 3: Plot of the two thin slices with high-energy states above and below the Fermi line.

is modified or not when taking into account the effect of quantum corrections. For this purpose we adopt the point of view of Wilson, in which the influence of the high-energy states far from the Fermi level on the modes close to the Fermi line is studied by integrating out the states contained in two thin slices between energies $-\Lambda, -\Lambda + d\Lambda$ and $\Lambda - d\Lambda, \Lambda$, as represented in Fig. 3. That is, the cutoff Λ sets the distance to the Fermi level. The operation of integrating out the high-energy states amounts, in practice, to perform a diagrammatics in which only the states belonging to the mentioned slices may appear as intermediate states. Let us consider, for instance, the quantum corrections to the forward scattering channel of the four-fermion interaction vertex. A typical contribution is shown by the diagram in Fig. 4, where it is understood that the states in the particle hole bubble are high-energy states. If we are in a situation in which $\mathbf{p}_1 \approx \mathbf{p}_3$ this means that the momentum transferred to the loop is very small. However, it is clear from Fig. 3 that it takes a large momentum to create a particle-hole excitation by promoting a particle from the filled slice to an empty state in the upper slice. Thus, there is no way of building up the diagram in Fig. 4 with high-energy intermediate states. There are two more diagrams which may correct the forward scattering channel, but it can be checked that they do not give any contribution either[11]. The claim is that the interaction in the forward scattering channel is not renormalized by quantum corrections to all orders in perturbation theory. Regarding the other interaction that was not irrelevant at the classical level, it turns out that the BCS channel does get renormalized by high-energy processes[11]. The scaling at low energies is towards decreasing values of the interaction, which means that for an originally repulsive interaction, $V > 0$, the renormalization group flow is arrested at $V = 0$.

This brief analysis gives an idea of the stability of the Fermi liquid theory in the renormalization group framework. In fact, one of the main conclusions that follows from Shankar's work is that there are only two types of relevant perturbations which may destabilize the Fermi liquid fixed-point, namely the already mentioned

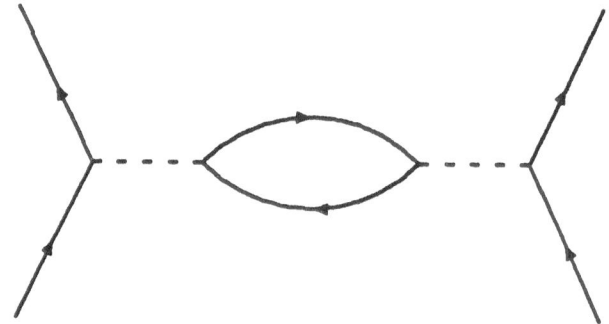

Figure 4: Contribution to the four-fermion interaction vertex.

pairing interaction with $V < 0$ and a charge-density-wave type of perturbation. We will see to what extent the model with the Fermi level close to the van Hove singularities provides a wayout to this well-established picture.

4   Renormalization of the van Hove singularity

We now turn to the model with the dispersion relation shown in Fig. 1, but with the Fermi energy at the level of the two van Hove singularities. We would like to follow the same steps which we have taken in the renormalization group program applied to Fermi liquid theory. The main difference with respect to the previous situation is that now the dispersion relation is not isotropic at the Fermi level, but it rather looks as shown by the contour energy map in Fig. 5. We will not make any special assumption about the interaction apart from requiring it to be short-ranged, as in Fermi liquid theory.

What we are looking for by means of the renormalization group approach is an effective theory for the modes that lie very close to the Fermi level. These may be selected in practice by impossing an energy cutoff above and below the Fermi energy, leading to a region in momentum space around the Fermi line with the shape shown in Fig. 5. We have first to make sure that it is possible to write down such a theory at the classical level, that is, that there is an effective action made of terms that remain scale invariant when the cutoff is reduced towards the Fermi line. This will guarantee that the whole picture is not spoiled by a coupling constant that blows up, as well as that we remain with a nonvanishing interaction as we proceed to describe processes with lower energies.

At very low energies almost all the states are concentrated around any of the two van Hove singularities, and the dispersion relation may be approximated about each of them by $\varepsilon(\mathbf{p}) \approx p_x^2/(2m_x) - p_y^2/(2m_y)$. A quite general expression for the

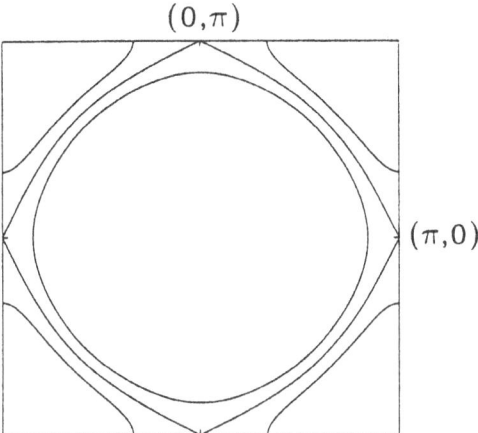

Figure 5: Contour energy map of the dispersion relation close to the level of the van Hove singularities.

effective action of the model is given by the following terms

$$
\begin{aligned}
S \;=\; & \int d\omega d^2 k \sum_\alpha (\omega \, a_\alpha^+(\mathbf{k}, \omega) a_\alpha(\mathbf{k}, \omega) \\
& - (k_x^2/(2m_{x,\alpha}) - k_y^2/(2m_{y,\alpha}))\, a_\alpha^+(\mathbf{k}, \omega) a_\alpha(\mathbf{k}, \omega)) \\
& - \frac{U}{2} \int d\omega d^2 k \; \sum_\alpha \rho_\alpha(\mathbf{k}, \omega)\, V(\mathbf{k}) \sum_\alpha \rho_\alpha(-\mathbf{k}, -\omega)
\end{aligned}
\tag{3}
$$

where $a_\alpha(a_\alpha^+)$ is an electron annihilation (creation) operator (α stands for spin and also labels the singularity), ρ_α is a momentum density operator and $V(\mathbf{k})$ is a short-ranged interaction potential. We stick, for the time being, to the case of a spin-independent interaction. Under a reduction of the cutoff $\Lambda \to s\Lambda$, the question is whether one can find a scaling of the momenta and the fields so that all the terms in (3) remain independent of s. The answer is affirmative, and the appropriate scaling to be applied is[9, 10]

$$
\begin{aligned}
\omega &\;\to\; s\omega \\
\mathbf{k} &\;\to\; s^{1/2}\mathbf{k} \\
a_\sigma(\mathbf{k}, \omega) &\;\to\; s^{-3/2} a_\sigma(\mathbf{k}, \omega)
\end{aligned}
\tag{4}
$$

The transformation rule for the momenta and the fields is obtained by the condition that the kinetic terms in (3) remain finite at low energies. The important result is that with the above scaling the interaction term also becomes independent of the s factor (as long as $\rho_\alpha(\mathbf{k}, \omega)$ is the integral of a bilinear of electron fields). The interaction does not vanish neither grows large at low energies, which provides us with a stable picture at least at the classical level.

Paralleling the discussion carried out for Fermi liquid theory, we study next the behavior of the interaction coupling constant when quantum corrections are

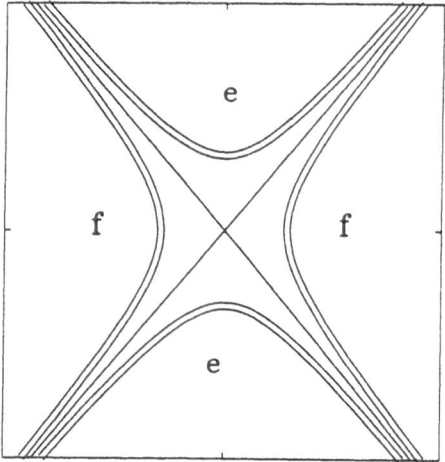

Figure 6: Plot of the thin slices with high-energy states. The quadrants marked with "f" have all the states filled while those marked with "e" are empty.

turned on in the theory. In this section we analyze the effect of the integration of high-energy modes on a given singularity alone. For this purpose, we amplify the region around it and take two thin slices in momentum space at a distance Λ in energy above and below the Fermi level, which now have the form shown in Fig. 6. Opposite to what happened in the renormalization of Fermi liquid theory, the states from the two slices do produce a contribution as intermediate states in the particle-hole polarizability in Fig. 4. The polarizability can be computed exactly at small momentum transfer, displaying the explicit dependence on the cutoff Λ,[9, 10]

$$
\begin{aligned}
\operatorname{Re} \Gamma^{(2)}(\mathbf{k}, \omega) &= \frac{1}{2\pi^2} \log \frac{|\varepsilon(\mathbf{k})|\Lambda}{|\omega^2 - \varepsilon(\mathbf{k})^2|} + \frac{1}{2\pi^2} \frac{\omega}{\varepsilon(\mathbf{k})} \log \left| \frac{\omega - \varepsilon(\mathbf{k})}{\omega + \varepsilon(\mathbf{k})} \right| \\
\operatorname{Im} \Gamma^{(2)}(\mathbf{k}, \omega) &= \frac{1}{4\pi} \left(\frac{|\omega + \varepsilon(\mathbf{k})|}{\varepsilon(\mathbf{k})} - \frac{|\omega - \varepsilon(\mathbf{k})|}{\varepsilon(\mathbf{k})} \right)
\end{aligned}
\tag{5}
$$

Quite remarkably, the imaginary part is consistent with the "marginal Fermi liquid" hypothesis of Ref. [12]. As a consequence of the above dependence on Λ, the interaction in the forward scattering channel is corrected by the influence of the high-energy modes. The reason is that now, no matter how small is the momentum injected into the particle-hole bubble, it is possible to build particle-hole excitations across the assymptotes in Fig. 6.

The basic principle of the RG approach is that the vertex functions have to be independent of the cutoff used in the theory, since they are used in the computation of physical observables. In the case of the four-point vertex function, by demanding the Λ-independence of the whole expression we get

$$
\Lambda \frac{d}{d\Lambda} \left\{ U - \frac{1}{\pi^2} mU^2 \log \frac{|\varepsilon(\mathbf{k})|\Lambda}{|\omega^2 - \varepsilon(\mathbf{k})^2|} + \ldots \right\} = 0
\tag{6}
$$

This gives the dependence of the effective coupling constant, up to order $(2mU)^2$,

$$\Lambda\frac{d}{d\Lambda}U(\Lambda) = \frac{1}{\pi^2}\, mU^2 \qquad (7)$$

That is, as the high-energy cutoff is reduced towards the Fermi line the renormalized coupling constant tends to vanish. From the physical point of view, this is a consequence of the strong screening effects present in the model. This nontrivial scaling of the interaction is incompatible with Fermi-liquid behavior, as in a normal Fermi liquid the effective interaction near the Fermi surface is insensitive to the high-energy cutoff of the theory.

Given that the flow of the renormalized coupling constant is bound at low energies, it may seem that we have reached our goal of finding a sensible effective theory in this limit of the RG approach. There are, however, higher order processes that may affect the behavior of other parameters. The fact that the hopping parameter t ($\equiv 1/(2m)$) is not renormalized at the one-loop level, for instance, is just an accident. In order to study the corrections that the dispersion relation may suffer at the quantum level one has to compute the electron self-energy Σ, which is given by the usual definition

$$\frac{1}{G} = \frac{1}{G^{(0)}} - \Sigma \qquad (8)$$

Thus, the terms in Σ proportional to the frequency ω modify the scale of the electron wavefunction, while those proportional to $\varepsilon(\mathbf{k})$ that do not match with the former provide a correction to the dispersion relation. The first nonconstant contributions to Σ are given by the diagrams in Fig. 7. The important result is that these diagrams show a dependence from the high-energy modes that are integrated as intermediate states. The explicit expression for $1/G$ to this order reads[9, 10]

$$\begin{aligned}
\frac{1}{G} &= \frac{1}{G^{(0)}} - \Sigma \\
&\approx Z_\Psi^{-1}(\Lambda)\left(\omega - t(\Lambda)(p_x^2 - p_y^2) + \varepsilon_F\right) \\
&\quad + Z_\Psi^{-1}(\Lambda)\left(\omega - \varepsilon(\mathbf{p})\right)\frac{1}{8\pi^4}\frac{U^2}{t^2}\left(\log 2 \log^2(\Lambda/\varepsilon_F) - c_1\, \log \Lambda\right) \\
&\quad + Z_\Psi^{-1}(\Lambda)\varepsilon(\mathbf{p})\frac{1}{8\pi^4}\frac{U^2}{t^2}\left(\frac{3}{4}\log^2(\Lambda/\varepsilon_F) - (c_1 + c_2)\, \log \Lambda\right) + O\left((U/t)^3\mathbf{p}\right)
\end{aligned}$$

where we have introduced the wavefunction renormalization constant $Z_\Psi(\Lambda)$, defined by the scale of the original electron field compared to that of the cutoff-independent electron field $\Psi_{bare}(\Lambda) = Z_\Psi^{1/2}(\Lambda)\Psi$. Requiring the cutoff-independence of the electron Green function G, we obtain the flow equations

$$\Lambda\frac{d}{d\Lambda}\log Z_\Psi(\Lambda) = \frac{1}{8\pi^4}\frac{U^2}{t^2}\left(2\log 2\, \log(\Lambda/\varepsilon_F) - c_1\right) \qquad (10)$$

$$\Lambda\frac{d}{d\Lambda}t(\Lambda) = \frac{1}{8\pi^4}\frac{U^2}{t}\left(\frac{3}{2}\log(\Lambda/\varepsilon_F) - c_1 - c_2\right) \qquad (11)$$

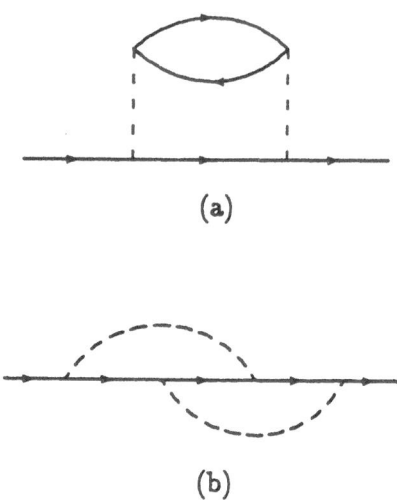

Figure 7: Two-loop corrections to the electron self-energy.

In the above equations $c_1 \approx -5.896$ and $c_2 \approx 3.839$, and we have not taken the Fermi energy strictly equal to zero though it is supposed to be smaller than any other energy scale in the problem. Therefore, the conclusion that we reach by inspection of (10) and (11) is clear. In the first place, the effective scale of the electron field gets reduced as the cutoff is sent towards the Fermi line. This effect goes in the same direction as that predicted on phenomenological grounds by the "marginal Fermi liquid" hypothesis. Regarding the hopping parameter t, we see that it is renormalized towards decreasing values at low energies. From the physical point of view, this is interpreted as a tendency of the dispersion relation to become flatter near the van Hove singularity, which is therefore reinforced at the quantum level. An important technical detail is the presence of the cutoff Λ at the right-hand-side of the flow equations (10) and (11). This is unusual and it means, in general, that there is an auxiliary energy scale that has to be determined to ascertain a predictable flow of the couplings. In the present case, though, there is no freedom left in that choice since the scale that plays that role, ε_F, is a dynamical energy scale with its own flow equation, which has to be solved in conjunction with (10) and (11)[9, 10].

Though the renormalized hopping parameter t decreases when measured at low energies, this is a higher-order effect that does not modify the leading behavior of the effective interaction strength U/t, given by (7). This means that there is a well-defined weak coupling regime in which the properties of the model can be studied within the perturbative renormalization group framework we have proposed. In this weak coupling regime of the model with spin-independent interactions, the electronic properties do not differ much, apart from slight deviations, from that of Fermi liquid theory.

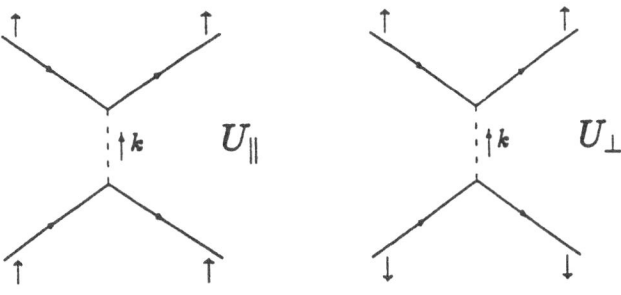

Figure 8: Interactions between currents of like spin $(U_\parallel)$ and currents of different spin $(U_\perp)$.

5 Spin-dependent interactions

At this point it becomes clear that if there may appear any kind of instability in the model it has to be due to the introduction of the spin or to the interaction between the states at the two different singularities. In order to investigate the first possibility we allow for two different interactions between currents of like spin and currents with opposite spin, respectively. These are shown diagrammatically in Fig. 8. The convention of using the two coupling constants $U_\parallel$ and $U_\perp$ does not lead to a manifestly rotation invariant scheme, but it is convenient for computational purposes. One may parallel the above computation of the quantum corrections taking now into account the two different coupling constants, which are mixed in the renormalization group procedure. It is not difficult to see that the coupled flow equations have the form

$$\Lambda \frac{d}{d\Lambda} U_\parallel = \frac{1}{4\pi^2} \frac{1}{t} \left(U_\parallel^2 + U_\perp^2 \right) \tag{12}$$

$$\Lambda \frac{d}{d\Lambda} U_\perp = \frac{1}{2\pi^2} \frac{1}{t} U_\parallel U_\perp \tag{13}$$

The integral of equations (12) and (13) gives the RG orbits in the $(U_\parallel, U_\perp)$ space, which have a nontrivial structure. They are plotted schematically in Fig. 9. The line $U_\parallel = U_\perp$ reproduces the instance studied in the previous section, namely the case of spin-independent interactions, and describes a flow that is arrested at the origin in the low-energy limit $\Lambda \to 0$ (starting from purely repulsive interactions). The case in which the initial interactions are $U_\parallel > U_\perp$ does not differ much, as the renormalized couplings have the origin as attractive fixed-point. However, the situation in which $U_\parallel < U_\perp$ gives rise to an unstable flow, since reducing the cutoff to the Fermi line leads to an effective theory in which $U_\perp$ grows large remaining positive, while $U_\parallel$ grows large in absolute value becoming negative. We are facing

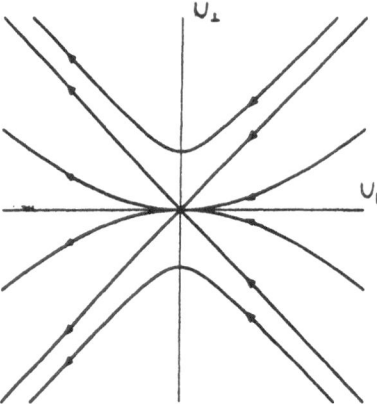

Figure 9: Renormalization group flow in the $(U_\parallel, U_\perp)$ space.

therefore a situation in which our perturbative renormalization group approach may not be reliable. It is pertinent, though, to ask for the physical interpretation of the divergent flow, which points at a different phase than that studied before.

The way to identify this new phase requires the knowledge of the properties of the ground state. These in turn are characterized by looking at appropriate correlations which may signal the appearance of superconductivity, antiferromagnetism, ferromagnetism, etc ... One may proceed in the same fashion as for one-dimensional electron systems, where renormalization group methods are also used and the strongest correlation at low energies determines the nature of the ground state[13]. The main difference with the one-dimensional analysis is that in two dimensions the renormalization only takes place for very definite kinematics. In this respect, we have to remind that the above divergent flow only affects to the forward scattering channel.

By inspection of all the possible divergent correlations in the model, one may check that the above unstable flow is closely connected to a divergence of the ferromagnetic response function. This is defined by the expectation value

$$R(\mathbf{q},\omega) \sim \langle (\rho_\uparrow - \rho_\downarrow)(\rho_\uparrow - \rho_\downarrow) \rangle \tag{14}$$

at very small momentum $\mathbf{q}$. The zeroth-order approximation to this quantity, for instance, is proportional to the expression for the particle-hole polarizability that we have studied in the previous section, showing a characteristic logarithmic dependence. This means that, if we were to study the response function by perturbation theory, we would find increasingly divergent contributions to higher orders in the coupling constant. To handle this problem one may ressort again to renormalization group methods, with which one seeks the functional form of the response function by studying its cutoff dependence. By collecting the first diagrams up to order $U_\parallel$ and $U_\perp$, we obtain the equation

$$\frac{\partial R}{\partial \Lambda} = -\frac{1}{\pi^2 t}\frac{1}{\Lambda} + \frac{2U_\perp(\Lambda)}{\pi^2 t^2}\frac{1}{\Lambda}R \tag{15}$$

This equation looks similar to those obtained in the context of one-dimensional electron systems. The main difference is that we are interested now in a regime in which the renormalized coupling constant $U_\perp(\Lambda)$ blows up. Had the renormalized coupling flowed to a finite fixed-point, the divergence of R would have adopted a power-law form at small frequency as in one dimension. One may easily convince that a divergent $U_\perp(\Lambda)$ in equation (15) leads to a finite value of the frequency at which $R(\mathbf{q}, \omega)$ blows up. This energy scale sets the appearance of ferromagnetism in the system.

6 Umklapp processes

We discuss now the other source of instability, which is due to the possibility of transferring a pair of electrons with opposite spin from one van Hove singularity to the other. These processes are allowed since the momentum which is exchanged in the interaction is of the order of one of the reciprocal vectors of the lattice. Thus, there are two different BCS channels as depicted in Fig. 10. One of them corresponds to the scattering of pairs within the same singularity, while the other corresponds to inter-singularity scattering. Again the important question is the way in which the quantum corrections affect the bare couplings and, in particular, whether the interaction for the low-energy modes is sensitive to the integration of high-energy modes that lie at a distance of the order of the cutoff from the Fermi line. The distinctive feature of the model for the van Hove singularities is that the one-loop corrections to the couplings show a dependence $log^2\Lambda$ on the cutoff, that differs from the conventional $log\,\Lambda$ dependence of the BCS channel. This poses a problem from the theoretical point of view, since there is no straightforward way in which the renormalization group procedure can be followed with such unconventional scaling. The situation is similar to that already faced in the renormalization of the electron wave-function and the hopping parameter t, in that the cutoff Λ appeared at the right-hand-side of the flow equations. A proper solution to this unconventional scaling has to be achieved again by incorporating the renormalization of the Fermi energy ε_F, since the additional energy scale which enters in the flow equations is actually ε_F/Λ[9, 10].

As long as we are only interested in studying the qualitative features of the RG flow, we may resort though to a less refined alternative that leads to the correct physical conclusions. What we do here is to scale the quantum corrections with $log^2\Lambda$, what in general may not provide a proper low-energy effective theory but in the present context it gives a good approximation to the flow, since the renormalized Fermi energy scales in step with the cutoff. The important point is that the intra-singularity V and inter-singularity V' couplings mix up in the renormalization process. The coupled flow equations have the form

$$\frac{\partial V}{\partial l^2} \;\propto\; V^2 + V'^2 \tag{16}$$

$$\frac{\partial V'}{\partial l^2} \;\propto\; 2VV' \tag{17}$$

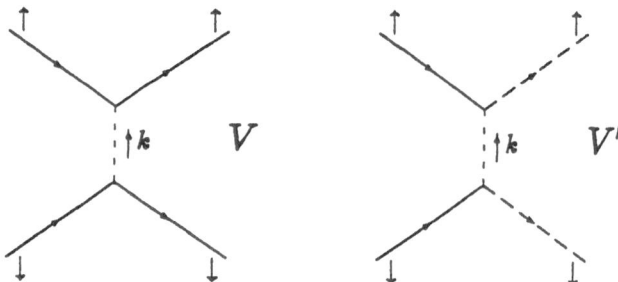

Figure 10: Different BCS channels. The full lines and the long-dashed lines represent the propagation of electrons near the two different singularities.

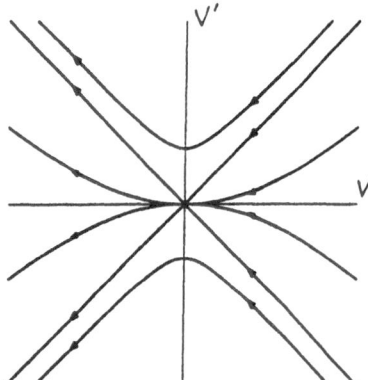

Figure 11: Renormalization group flow in the (V, V') space.

with $l \equiv log\, \Lambda$. This system has the same form than that given by equations (12) and (13), and the integral in (V, V') space is also quite similar to that found in the previous section. The orbits, which are parametrized with respect to l^2, are represented in Fig. 11.

We face the situation in which the bare couplings V and V' are repulsive, that is, $V > 0$ and $V' > 0$. Then we find two different scenarios, depending on whether originally $V > V'$ or $V < V'$. The first case corresponds to the conventional picture in which a purely repulsive interaction is renormalized to zero as more and more high-energy states are integrated out. The case in which $V' > V$ presents however a novel feature, since at sufficiently low energies the V' coupling is renormalized to large values while V becomes negative and large in absolute value[9]. Thus, in such regime the process of integrating out high-energy modes shows an attractive pairing channel at low energies, though the original bare interaction has been chosen purely repulsive. This is nothing but a new version of the Kohn-Luttinger mechanism, which was proposed many years ago when studying the possibility of superconductivity from screened repulsive interactions[14]. The physics is very much the same, except that, in the present case, the anisotropy of the screening matches well the phase space available for the formation of Cooper pairs[9].

7 Fermi level renormalization

We now address the question of the renormalization of the Fermi level in our model. In general, the computation of the electron self-energy produces some frequency and momentum-independent terms, which are to be understood as a correction to the chemical potential of the noninteracting theory. In the renormalization group framework one has to study the dependence that this constant terms bear on the cutoff at which the high-energy modes are integrated out. In this process the Fermi level is corrected at each step of the integration, so that it makes sense to talk in terms of the renormalization group flow of the Fermi energy. In principle one may think that this phenomenon of renormalization is much more involved, as it should envisage the deformation of the whole Fermi line by the interaction with the high-energy degrees of freedom. In the more conventional systems, though, there uses to be some symmetry that keeps invariant the form of the Fermi line, making pertinent the consideration of the corrections to its energy level. This is the case, for instance, of the standard Fermi liquid theory with an isotropic dispersion relation. In the model of the van Hove singularity and focusing the renormalization process near the saddle points, there is also an $SO(1,1)$ invariance of the lines of constant energy which protects the form of the dispersion relation at the quantum level.

It has to be stressed, however, that the issue of renormalization of the Fermi energy can only be relevant for systems in which to some extent there is a variable number of electrons. This is clear from the interpretation that the constant terms of the electron self-energy directly modify the value of the noninteracting chemical potential. The distinction between bare chemical potential and renormalized Fermi energy makes sense only if the number of electrons is not fixed, as it is the case when the system is placed near a particle reservoir at constant chemical potential. The high-T_c superconductors seem to adhere to this picture, since the conduction properties refer to copper-oxide layers in these materials while it is the three-dimensional structure what furnishes extra-charge into the two-dimensional planes.

We begin by reminding what is the situation of the renormalization of the Fermi level in Fermi liquid theory, that is, in the case of an isotropic dispersion relation. The first order corrections to the chemical potential come from the Hartree diagram, which produces just a constant shift at each renormalization step. We recall that the Fermi energy ϵ_F is actually a relevant parameter of the theory, since by dimensional arguments it scales like ϵ_F/Λ under a reduction of the energy scale ($\Lambda \to 0$). The relative shift of ϵ_F due to the interaction with the high-energy part of the Fermi sea goes in the opposite direction. As stated by Shankar[11], the evolution of the renormalized Fermi energy $\tilde{\epsilon}_F = \epsilon_F/\Lambda$ as the cutoff is sent towards the Fermi line is given by the flow equation

$$\frac{\partial \tilde{\epsilon}_F}{\partial l} = -\tilde{\epsilon}_F + \frac{U}{2\pi t} \tag{18}$$

At weak coupling, the flow of $\tilde{\epsilon}_F$ may only have an unstable fixed-point which is determined by the interaction and the density of states. This is rather puzzling, since it seems that under a small perturbation the system should flow to a regime with a vanishing density of electrons or to infinite density. Again one has to bear in mind that these possibilities could be realized only in the ideal case of an unconstrained number of electrons. In practice, however, one deals with the opposite situation, in which the system is closed and the density of electrons is fixed from the start so that there is no room left for variations in the Fermi energy.

In contrast to the previous example, the renormalization of the Fermi level close to a van Hove singularity produces a nontrivial stable flow. Now the density of the states which are integrated out is very sensitive to the proximity of the van Hove singularity. Taking into account the dependence on Λ of the Hartree diagram we find the scaling at the one-loop level[9, 10]

$$\frac{\partial \tilde{\epsilon}_F}{\partial l} = -\tilde{\epsilon}_F + \frac{U}{8\pi^2 t} \log \left(\frac{1}{1 - \tilde{\epsilon}_F} \right) \tag{19}$$

where ϵ_F is measured with respect to the van Hove singularity. The high-energy states which are removed lie at energy Λ below ϵ_F, and the second term in the r.h.s. of (19) accounts for the divergent density of states near the singularity. Unlike in normal metals, the second term does not scale in step with the first. Thus, for repulsive interactions and $\epsilon_F > 0$, there may be a nontrivial, *stable*, fixed point, at which the flow in (19) is arrested. As Λ is further reduced, $\tilde{\epsilon}_F = \tilde{\epsilon}_F^*$ remains unchanged. In dimensionful units, it means that the Fermi energy gets pinned at the van Hove singularity[9, 10].

A simpler but more physical explanation of the pinning mechanism can also be given. The argument is suited for a system that may receive extra-charge from a reservoir (as is the case of the copper-oxide compounds). The number of particles in the reservoir is much larger than that in the system, so that the flow of particles into the latter is controlled by the chemical potential μ of the reservoir. This quantity acts as a kind of external "pressure" of particles. However this external pressure is not seen in all its magnitude in the two-dimensional system, since it is partially counterbalanced there by the repulsion between the electrons. The competition between the two effects is what sets the value of the Fermi level ε_F in the system. By using a simple one-loop approximation we may express the correction to the chemical potential as the frequency and momentum-independent contribution of the Hartree diagram

$$\varepsilon_F \approx \mu + i \frac{U}{8\pi^3} \int d\omega \, d^2 k \, G^{(0)}(\mathbf{k}, \omega) \tag{20}$$

In terms of the density of states $n(\omega)$ we may write this expression in the form

$$\varepsilon_F = \mu - \frac{U}{4\pi} \int^{\varepsilon_F} d\omega \, n(\omega) \tag{21}$$

We may obviate the influence of the particular shape of the dispersion relation by differenciating (21) with respect to μ, so that

$$\frac{d\varepsilon_F}{d\mu} = 1 - \frac{U}{4\pi}n(\varepsilon_F)\frac{d\varepsilon_F}{d\mu} \tag{22}$$

Then, we arrive at the differential equation

$$\frac{d\varepsilon_F}{d\mu} = \frac{1}{1 + \frac{U}{4\pi}n(\varepsilon_F)} \tag{23}$$

which expresses how the filling level changes in the system under variations of the external chemical potential. Equation (23) shows that when the Fermi energy ε_F is very close to a level with a divergent density of states, as is the case of the van Hove singularity, the Fermi level is very weakly influenced by changes in the chemical potential of the reservoir. Of course, this does not prove that the level of filling has necessarily to fall close to the van Hove singularity, but the simple argument outlined gives a notion of the stability of the model with the particular filling considered in this paper. On the other hand, with the methods employed above it is possible to show that the renormalization of the Fermi energy ε_F keeps up with the integration of the high-energy degrees of freedom. The result leads to the dynamical effect of pinning the Fermi level to the van Hove singularity as $\Lambda \to 0$. Quite remarkably, this feature is in correspondence with the experimental observation of many hole-doped copper-oxide compounds, for which the Fermi level is found very close to a pronounced peak of the photoemission spectra.

8   Conclusion

To summarize, we have seen that the renormalization group approach is well-suited to handle the divergences that appear in the model with van Hove singularities. In particular, it is able to reproduce a number of features that are observed experimentally in the copper-oxide superconductors:

- The renormalization of the hopping parameter t shows the tendency to the formation of flat bands.

- The Fermi level is renormalized at low energies towards the level of the van Hove singularity. This flow of the Fermi energy, unusual in condensed matter systems, may explain the existence of a whole family of compounds where experiments show that the Fermi level is close to a van Hove singularity[1].

- The attenuation of the electron quasiparticles that we have found in our model is in agreement with a phenomenological approach to the copper-oxide superconductors[12]. The logarithmic corrections to the scale of the electron field are also consistent with another known property, that is, the abnormal behavior of the quasiparticle lifetime in the van Hove scenario[7].

Further analysis is needed to elucidate the competition between the different phases that may appear in the model, that is susceptible of showing ferromagnetism and superconductivity. Quite remarkably, a signal of ferromagnetism has been found recently in quantum Monte Carlo studies of the $t - t'$ Hubbard model, with the filling level at the van Hove singularity and values of t' close to 0.5 [15]. On the other hand, for moderate values of this parameter, quantum Monte Carlo results[16] as well as exact diagonalization computations[17] show that the dominant correlation in that model points at superconductivity with a d-wave order parameter. It would be worthwhile to map these phases into the continuum limit by means of the renormalization group approach we have been discussing, to clarify under which conditions a purely electronic mechanism for superconductivity may appear in the model of a van Hove singularity.

References

[1] See Z.-X. Shen, W. E. Spicer, D. M. King, D. S. Dessau and B. O. Wells, *Science* **267** (1995) 343, and references therein.

[2] B. O. Wells et al., Phys. Rev. Lett. **74**, 964 (1995).

[3] K. Gofron et al., Phys. Rev. Lett. **73**, 3302 (1994).

[4] A. Nazarenko, K. J. E. Vos, S. Haas, E. Dagotto and R. J. Gooding, Phys. Rev. B **51**, 8676 (1995).

[5] J. Labbé and J. Bok, *Europhys. Lett.* **3** (1987) 1225. H. J. Schulz, *Europhys. Lett.* **4** (1987) 609. J. Friedel, *J. Phys.* (Paris) **48** (1987) 1787; **49** (1988) 1435. J. E. Dzyaloshinskii, *Pis'ma Zh. Eksp. Teor. Fiz.* **46** (1987) 97, [*JETP Lett.* **46** (1987) 118]. P. A. Lee and N. Read, *Phys. Rev. Lett.* **58** (1988) 2691.

[6] R. S. Markiewicz, *J. Phys. Condens. Matter* **2** (1990) 665. R. S. Markiewicz and B. G. Giessen, *Physica* (Amsterdam) **160C** (1989) 497.

[7] P. C. Pattnaik, C. L. Kane, D. M. Newns and C. C. Tsuei, *Phys. Rev. B* **45** (1992) 5714.

[8] D. M. Newns, H. R. Krishnamurty, P. C. Pattnaik, C. C. Tsuei and C. L. Kane, *Phys. Rev. Lett.* **69** (1992) 1264, and references therein.

[9] J. González, F. Guinea and M. A. H. Vozmediano, Europhys. Lett. **34**, 711 (1996).

[10] J. González, F. Guinea and M. A. H. Vozmediano, Nucl. Phys. B, to be published.

[11] R. Shankar, *Rev. Mod. Phys.* **66** (1994) 129. See also J. Polchinski, in *Proceedings of the 1992 TASI in Elementary Particle Physics*, J. Harvey and J. Polchinski eds. (World Scientific, Singapore, 1992).

[12] C. M. Varma, P. B. Littlewood, S. Schmitt-Rink, E. Abrahams and A. E. Ruckenstein, *Phys. Rev. Lett.* **63** (1989) 1996.

[13] H. J. Schulz, *Interacting Fermions in One Dimension: From Weak to Strong Correlation*, in *Correlated Electron Systems*, Vol. 9, ed. V. J. Emery (World Scientific, Singapore, 1993).

[14] W. Kohn and J. M. Luttinger, Phys. Rev. Lett. **15**, 524 (1965).

[15] R. Hlubina, S. Sorella and F. Guinea, report May 1996.

[16] T. Husslein, I. Morgenstern, D. M. Newns, P. C. Pattnaik, J. M. Singer and H. G. Matuttis, report August 1996.

[17] J. González and J. V. Alvarez, report September 1996.

Springer
and the
environment

At Springer we firmly believe that an international science publisher has a special obligation to the environment, and our corporate policies consistently reflect this conviction.
We also expect our business partners – paper mills, printers, packaging manufacturers, etc. – to commit themselves to using materials and production processes that do not harm the environment. The paper in this book is made from low- or no-chlorine pulp and is acid free, in conformance with international standards for paper permanency.

Springer

Lecture Notes in Physics

For information about Vols. 1–461
please contact your bookseller or Springer-Verlag

New Series m: Monographs